The
GAFF RIG
Handbook

Glory of the gaff rig.
The 158ft racing schooner *Rainbow* of 1898 sweeping
through Cowes Roads setting 14,000 square feet.
Photo: Beken of Cowes

The GAFF RIG

Second Edition

Handbook

History · Design · Techniques · Developments

John Leather

WITH DRAWINGS BY THE AUTHOR

ADLARD COLES NAUTICAL
LONDON

To the mariners and shipbuilders of Rowhedge

Published by Adlard Coles Nautical
an imprint of Bloomsbury Publishing Plc
50 Bedford Square, London WC1B 3DP
www.adlardcoles.com

First published as *Gaff Rig* in 1970 by Adlard Coles Ltd
Reissued 1989
Extended edition published as *The Gaff Rig Handbook*
by Waterside Productions Ltd in 1994
Second editon published 2002
Reissued 2009
Reprinted 2012 and 2014

ISBN 978-1-4081-1440-7

A CIP catalogue record record for this book is available from the British Library.

This book is produced using paper that is made from wood grown in managed,
sustainable forests. It is natural, renewable and recyclable. The logging
and manufacturing processes conform to the environmental regulations
of the country of origin.

Printed and bound in Great Britain by CPI Group (UK) Ltd, Croydon CR0 4YY

Note: while all reasonable care has been taken in the publication of this book,
the publisher takes no responsibility for the use of the methods
or products described in the book.

CONTENTS

About the Author

JOHN LEATHER is descended from Lancashire and Essex maritime families whose background was shipbuilding, marine engineering, deep sea seafaring, and professional yachting. He is a naval architect by profession and has been involved with the design, construction and surveying of ships of many types, sizes and complexity throughout his career. He is a Fellow of the Royal Institution of Naval Architects and a Chartered Engineer. For his own pleasure and use he has, over the years, designed numerous yachts and boats, many of them gaff rigged, and has sailed in many others.

Gaff Rig, his first book, won the *Daily Express* Best Book of the Sea Award in 1970 and his second, *The Northseamen*, won second prize in 1971. He is author of 13 other maritime books and a number of technical papers.

In retirement John Leather continues to sail, to write, and to enjoy the coast and countryside of East Anglia.

INTRODUCTION

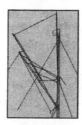

CONTEMPORARY SEASCAPES are more likely to reveal yachts under sail or power than the now much less numerous merchant vessels or warships. Although the present sailing yacht is often a wonder of bermudian rig sailing ability, seaworthiness and accomodation comforts, and is built with a fibre reinforced plastic hull and aluminium-alloy masts and spars for reduced maintenance, a large and continually growing segment of leisure sailors prefer the much older gaff rig along with the more conservative hull forms and often wood construction which frequently go with it. This is not simply reaction to the technical advances and many undoubted advantages of contemporary yacht design and use. It usually stems from a strong desire to own, experience and savour the advantages of the older rig and its traditions.

This book was first published over 30 years ago when the revival of interest in traditional craft and their rigs and in yachts and pleasure boats of the past was getting into its stride. Since then interest and enthusiasm have developed beyond all expectation in countries all over the world, sometimes in the most unexpected places. This resurgence has led to a continuing need for knowledge of gaff rig and its application to various types of craft and this book is written for all interested in its past, present and future. It outlines the practical aspects of masts, spars, running and standing rigging and sails, and contrasts many types of British, American, Danish and French gaff rigged commercial sailing craft, fishing vessels and yachts, besides examining some economic and technical factors influencing their development. It also provides some historical background to the various craft setting the rig for work or pleasure and to those who designed, built and sailed them. During the past 31 years

GAFF RIG has become internationally regarded as a handbook for those designing, building, rigging or just dreaming about gaff rigged craft, which is pleasing to the author and has encouraged its re-publication in expanded form.

About 90 years ago the use of gaff rig went into a decline which lasted for three decades, followed by a further 20 years of stagnation. But, since the early 1960s the rig has experienced a dramatic re-birth, initially, perhaps, as a reaction against the widespread use of bermudian rig in yachts. The rig's decline was affected by the rapid rise in use of the marine motor by fishing and other small commercial vessels, leading to the virtual extinction of craft working under sail and also by the widespread adoption of the bermudian rig for yachts. The unquestionably superior windward performance of bermudian rig applied to a suitable hull form resulted in almost blind acceptance of that rig as being best for yachts of almost all types, and for all purposes and conditions. There are many bermudian rigged yachts of great beauty, but some cruising yachtsmen prefer gaff rig for practical conditions of sail handling and lying more easily at anchor in a breeze. To others a well-built wooden craft of superior traditional design, with gaff rig, is also a thing of beauty, apart from its functional utility. Some are attracted by the rig's connection with working craft of the past, enjoying its appeal without the grinding toil endured by their crews.

Gaff rigged working vessels were built and sailed with remarkable skill, often by humble men. Although long since superseded by powered craft their memory, and that of their crews, still commands respect for products of the endeavours of usually small communities, earning a hard living from the sea and breeding the best qualities of seafaring. To them the gaff rig was a tool of trade whose handling was often drudgery, but could also be an art of pride, excelled from competence to perfection by some seamen, often from unlikely places.

Gaff rig propelled types as widely contrasting as great racing cutters setting 14,000 square feet of sail on one mast, to the humble 18 foot waterman's boat, beating out on an errand to some ship in an estuary. Attempts to catalogue craft by the present name of rig are difficult and often unsatisfactory, but this is the only convenient method of setting down in one book comparisons of a few of the thousands of types of gaff rigged craft which have been built. Types changed, sometimes merged, and new ideas and materials evolved new types which often defy classification by researchers who tend to ignore that a vessel's use was more important to her owners than her name. Names then were as accidental as now, originating from use, shape or arrangement of hull, or differences in rig. New types emerged from old when builders succeeded in blending the best qualities in craft which set a standard and become fashionable. Specialisation of type increased tremendously during the the 19th century when expanding trade and populations brought into being very large numbers of fore and aft craft for commerce, fishing and pleasure. Previously small craft were often built for general service and were used indiscriminately for fishing, cargo carrying, or ocean voyaging. The history of development is further complicated as small fore and aft vessels were generally considered unimportant and therefore poorly recorded before photography.

The traditional rigs were evolved by sailors usually guided by precedent and always with an awareness of the dangerous power of wind and waves, and of the seas' contempt of radical rigs and extreme gear; so strength and safety usually came first. Changes in rig were usually the work of unrecorded individuals seeking to increase sail area, obtain greater windward ability or ease of handling, and often influenced by rigs and gear seen in craft and ports of other countries. Each country, every port, and often individual skippers and owners had distinct ways of fitting rigging and setting sails. Rig development was at first often temporary; old pieces of sailcloth would be used as makeshift sails supported by temporary rigging and, if successful, would be re-cut until gradual evolution

brought purposely made sails and rigging; provided it was a logical working arrangement.

The single squaresail was the earliest northern sail. At first this could not be carried to windward but its use was later improved in various ways, enabling it to stand fairly close to the wind. When set in small craft, without topsails, the squaresail was taller and comparatively narrow to obtain a good leading edge, which was further tautened on the wind by a wood prop termed a "vargood" or "fargood", later a bowline of rope was used to stretch the luff taut on the wind; hence the terms "sailing on a bowline" and "keeping a taut luff". When more sail was needed in small craft a second smaller squaresail was added forward. Squaresails were too clumsy for frequent tacking in confined waters and ingenuity devised the fore and aft rig for improved manoeuvrability.

The spritsail is presumed to have evolved from the squaresail, at first retaining its low aspect ratio, spread by a sprit instead of a yard, and with the luff laced to the mast. The sail could by furled, or reduced in effective area, by brails which drew it to the mast, throat and headrope. Tacking became much easier and sailing to windward in narrow waters was practical. The spritsail was much used by the Dutch, who may have evolved it and certainly developed it. As a nation they had tremendous influence on the early gaff rig and many other aspects of seafaring.

Gradual lengthening of the luff, peaking-up of the head and broadening of the foot improved the spritsail's windward ability, and a forestaysail was added, filling the space between forestay and mast to make an elementary form of sloop. Later, another headsail that came to be known as the jib was set forward of the bow, probably at first on a temporary unstayed spar forming a bowsprit; rather as some Dutch botters used to do. With efficiency in windward work and manoeuvring proved, the jib and bowsprit became permanent features of many rigs.

Further change evolved the gaff sail, which appeared in two basic forms which eventually merged to form the "modern" gaff rig. One was the standing gaff which evolved from the spritsail and was a sail, long in the head with an almost vertical leech and not fitted with a boom. It retained brails for furling. The change from spritsail to standing gaff came from the spar, not the shape of the sail, as the sprit was set progressively higher up the mast to shorten its length and reduce weight until it assumed the length and angle of the head of the sail, and was held aloft by throat and peak halliards, but was still attached to the mast by a lashing. This probably changed to a half-jaw having a shoulder to take thrust with a lashing to pull it to the mast. Eventually the basic form of wooden gaff jaws was devised and have remained in use since. The standing gaff sails had various halliard arrangements, but were not intended for frequent lowering, relying on the brails for furling and a row of reef points for reefing. The brail blocks were seized to the halliards, above the gaff, after it was hoisted, emphasising the rig's permanency.

The short-gaffed sail was a parallel development of uncertain origin but also much used in Holland. It had a gaff less than half the average length of the standing gaff type, supporting a tall, narrow loose-footed sail fitted with a boom and hoisted by a single halliard set well out along the gaff to draw it towards the mast when set. The luff was laced, and later hooped, to the mast. Probably this gaff was originally a short yard or headstick at the head of a triangular sail, possibly the "shoulder of mutton" sail once extensively used in open boats. It was certainly not a modification of the sprit or the long gaff rig. Later the halliard was actually arranged to give a peak and throat purchase, although still in one length, but sometimes separate halyards were fitted.

The use of a boom on a gaff sail doubtless developed from the use of boat-hooks and other temporary bearing out spars to extend the foot of a sprit or gaff sail when running. Eventually it would be found that a boom could spread the mainsail clear beyond the stern to gain valuable sail area. At first the booms would be lashed to the mast, the mainsail being secured only to the after end and tensioning of the sail foot probably being achieved by setting back on a mast rope. Eventually horns were fitted to the forward end of the boom and a clew outhaul was devised to tension the sail's foot, creating a flatter setting sail, more efficient for windward work, but which could not be brailed and therefore needing rows of reef points for reduction of area, and easy running and powerful halliards to quickly adjust the gaff, which was now able to be stowed with the sail.

By the 1600s spritsails, standing gaffs and short-gaffed sails were in common use by seamen of the North Sea and Baltic. The long gaff sail evolved from the standing gaff to be more readily hoisted and lowered and

to obtain a flatter setting and more powerful sail, besides relieving chafe of halyards, at anchor. The fitting of a boom further assisted control of sail shape by the outhaul, and increased area and driving power. This was the form of gaff most used in Britain.

Early developments and application to larger vessels brought complexity of rig, but the limitations of strength and inefficiency of cordage, canvas and fittings restricted the height and area of fore and aft sail plans into the 19th century. To increase area without increasing mast and spar dimensions required the use of squaresails, principally topsails and topgallants which were regularly carried in fore and aft craft as small as 50 feet length; the inevitable increase in top weight being accepted as penalty for speed off and reasonably close to the wind.

Experiment, change and improvement were constant, but slow in effectiveness until the end of the 18th century but the gaff rig developed rapidly throughout the 19th, particularly when large-scale manufacturing of sail cloth, cordage, blocks and improved ironwork allowed larger and taller sail plans, accelerated by the invention of iron wire rope and its use in British ships' rigging by 1844. This was followed by the invention of plough steel wire rope in England, which was applied to ships' rigging in the 1860s and, becoming cheaper quickly spread to the smaller craft, replacing the ever-stretching hemp shrouds and improving the strength, reliability and wind resistance of rigs.

In North America the basic fore and aft sails were also used in colonial times but for unknown reasons the tradition of the short-gaffed sail was strongest and persisted in many American small craft until the late 19th century, usually with the sail foot laced to the boom throughout its length; a feature rarely favoured in British working craft, but finally firmly adopted in British yachts during the 1880s.

The growth of yachting in England after 1800 and later in North America and elsewhere, particularly the search for speed and efficiency in racing, brought improvements in rig and hull form which were at first imitated to advantage in some commercial gaff rigged craft. However since about 1850 the rigs of yachts have been increasingly devised by designers rather than sailors; although both are sometimes combined in one person. The designer has usually always tended to be more extreme than the sailor and is more influenced by yachting fashion and latterly bound by rating rules. These factors have gradually erased local and even national characteristics from yacht rigs and today they are generally alike. However, yachting's claims to have "invented" sails or gear must be cautiously regarded; bermudian rig, cut-away forefoots and deadwoods, and overlapping headsails have all been revived or rediscovered by designers and builders of yachts during the past 150 years and hailed as revolutionary principles. For instance, the genoa headsail supposed to have been first used by a yachtsman racing at Genoa in the 1920s was antedated by at least a century in working craft. E.W. Cooke's engraving of "Mackerel Boats Coming In" at Brighton, 1830, depicts tubby, clench-built, leeboard craft about 30 feet long, carrying large staysails which well overlapped the after shrouds and sheeted on the quarter, in many cases replacing all other headsails. It is interesting to note that these Brighton "hogboats" set a sprit mainsail and mizzen, with the heel of the main-sprit three-quarters up the sail luff; this was a survival of the earlier transition of the long sprit to the gaff.

The manufacture of reasonably priced marine motors plus economic depressions caused rapid decline in sailing commercial and fishing craft after 1920, and ten years later the gaff sail, which had dominated rigs of fore and aft craft for three centuries, was displaced from yachtsmen's favour by the triangular jib headed or bermudian sail which had been extensively used in the Caribbean sloops and schooners since at least the end of the 18th century, often with "modern" proportions of luff to foot length. The rig was introduced to England about 1800, probably by small naval schooners built in the West Indies, and was later used in Exmouth pleasure boats and small yachts such as R.C. Leslie's yawl *Rip Van Winkle* which he built at Sidmouth, Devon in 1862. Later, at infrequent intervals, small sloop and cutter yachts were rigged with Bermudian mainsails laced or hooped at the mast below the hounds and working on a wire jackstay above.

Racing-yacht designers rediscovered the rig's efficiency for windward sailing about 1910 and developed spars, rigging and fitting, for lightness with strength.

Bermudian rig rapidly proved superior to the gaff for racing, besides being easier to handle with a smaller crew; features which soon commended it to cruising yachtsmen.

During the past 40 years there has been a vigorous revival of gaff rig, which had lagged in development because it had largely been out of fashion since the 1920s. Its future is best served by designing gaff rigs to ease handling for crews and increasing windward efficiency by using new materials and methods. Instead of regarding the gaff sail with antiquarian wistfulness owners should be encouraged to experiment and fit aluminium or plastic spars, lightweight plastic blocks and sails and ropes of synthetic materials. A gaff mainsail came to be a miserable baggy thing, cut without much thought and set badly, whereas it should be made and set with similar care to its windward efficiency as the bermudian racing sail. Much may be gained by applying modern sailmaking techniques to gaff rig.

Minimum weight aloft is as important to the gaff as any other rig and here modern lightweight blocks, ropes and fittings combine with aluminium alloy or plastic gaffs and topmasts can contribute to improved performance in old craft, and new ones can be designed to take full advantage of them. Much development could be carried out in lighter masts for use with gaff rig. There is difficulty in achieving sufficient strength in a mast groove to take the thrust of the sliding gaff swivel fitting that replaces jaws. With the mast extended to form a topmast the topsail luff could be set on slides.

Greater attention should be paid to hull balance and compatibility with the sail plan. Mainsails should have longer luffs, with moderate length or short gaffs, booms and bowsprits, where these are still required. Consequently, gaff rig headsails would be longer in the luff and steeper in rake than formerly.

A modern cutter rig with or without topsail is best for small and moderate size cruising yachts. For larger craft the ketch rig with gaff main and bermudian mizzen is simplest and easily handled. The wishbone ketch or schooner is next choice as the wishbone gaff sails, of divided area, enables a large sail plan to be handled by a small crew. The rig also has a smaller heeling moment than traditional gaff sails.

Gaff rig offers many opportunities for experiment which seldom seem to be taken advantage of by those designing, owning or sailing the craft which set it. However, as most of these are owned and maintained out of sentiment, convenience of working and speed are usually balanced against budgets, but the rig's handiness, traditions and usually attractive appearance will ensure its survival in various forms for yachts and boats far into the future.

CHAPTER ONE

MASTS AND SPARS

THE DIAMETER OF A MAST is dependent on its height and the type of rig. Lower masts for gaff rig should be of parallel diameter from deck to hounds. Above the hounds the masthead may taper according to the pull imposed upon it by the angle of the gaff, arrangement of halliards, and whether a topsail or topmast is carried. The average proportion of lower mast diameter to height (h) between deck to hounds for English cruising yachts up to 50 feet length is ·024h. An old general rule for lower mast diameter was $^7/_8$in for every foot of beam; another was $^7/_{32}$in for every foot of length from deck to head of mast. Masts are usually stepped with a rake varying with rig and individual craft, to obtain optimum performance. For craft of 50 feet length and above the usual rake per foot of height for lower masts is $^1/_4$in for cutters and sloops, $^3/_8$in for ketches and $^1/_2$in for schooners. Two-masted craft usually have the forward mast at a greater rake than the after. This is most pronounced in schooners. Cutters and sloops may be greatly affected in balance by mast rake and their masts are often fitted at right angles to the waterline.

Gaff rig masts may be solid or hollow timber, or aluminium. Steel masts and booms were used in the large yachts of the past but this need not now be considered. Solid masts may be shaped from a tree in the round, cut from a long baulk of timber, or built up from pieces of timber. A mast shaped from a tree is usually the cheapest for small craft. Typically, a straight black spruce or fir tree is selected, Norway spruce being a favourite.

Both ends of the tree should be examined to locate the heart which should be as nearly as possible in the centre of the tree at each end. Generally trees of this type have no internal faults, but this can be verified by listening with an ear close to one end of the tree while the other is tapped. The sound should carry with clarity in a spar of considerable length. The tree should be of slightly larger diameter than the finished mast size and when the bark and sapwood are removed it should approximate to the finished diameter.

When planed and sandpapered to size it should be treated immediately with a liberal mixture of one part boiled linseed oil with two parts of paraffin (kerosene), and afterwards stood on end to drain moisture to its heel. Due to greater shrinkage of the outer wood fibres, longitudinal shakes and splits will appear in the surface but provided the linseed oil/paraffin mixture is applied for some time, these should not become serious and are not a weakness unless of great depth. However, there should be no shakes or flaws across the grain and any knots should be small and firm. Longitudinal shakes should afterwards be filled with a soft mastic stopping which will readily squeeze out when the timber swells. Large knots or flaws should be particularly avoided around the mast wedging position at deck, at the hounds, and at the halliard attachments. It is customary for the butt of the tree to be used for the head of a grown lower mast. Grown masts were widely used in small craft where shrouds were not fitted, particularly in Holland and North America. These were fitted with the butt at the keel to retain the natural resilience of the tree as a mast, helping it to stand without shrouds.

Oregon pine is the most suitable timber for solid spars cut from baulks, being light and extremely tough. Pitch pine is also strong but much heavier and does not have such straight grain as Oregon. Spruce is a stronger timber than either but is difficult to obtain in length and can be badly chafed by gaff jaws. Red, yellow or white pine are also used, if obtainable.

Many American catboats now have glass reinforced plastic hulls, aluminium alloy masts and spars and a Dacron sail. Fleet racing is keen in some parts of New England.

If the mast is large and long it will nowadays almost certainly have to be built from two or more pieces, scarphed and glued together with the scarphs carefully placed clear of deck, hounds, gaff jaws, etc; and the grain selected to avoid distortion of the completed mast.

Hollow wooden masts have been built since the mid-19th century but are now rarely used in gaff rigged yachts, though they saved much weight aloft in large yachts and are more rigid than solid masts. They are expensive and have to be made solid in way of the step, the deck, hounds, gaff jaws and mastbands or halliard bolts. Hollow wooden top-masts were sometimes fitted on the top of solid wooden lower masts, socketed into a special steel head fitting.

Aluminium alloy masts have been little used, as yet, with gaff rig. Some American glass reinforced plastic catboats have them and should provide a good test from

the lack of staying and great twist imposed on the mast by cat rig. The principal disadvantage is that the gaff has to work up the mast in the sail track groove on a small swivel fitting which must withstand great strain and is liable to fatigue and break. Gaff jaws should not be used as they would quickly chafe the alloy and eventually cause mast failure.

Masts may be stepped through the deck onto a chock on the keel or floors, or in a tabernacle on deck. Stepping on the keel or floors is traditional from the days when few craft, except racing yachts, had their masts removed for long periods. This method can place great strain on the garboards and adjacent structure when the craft is driven hard, and may cause leaks. Masts stepped thus are stiffer than those stepped on deck. The mast heel is shouldered down to a tongue which fits into a similarly sized slot in the chock, holding the mast from turning. Traditionally, a silver coin should be placed in the step, under the mast, for good luck. Masts cut from the baulk are usually left octagonal in shape in way of the deck to facilitate the fitting of hardwood mast wedges at the deck chock, known as partners. If the mast is round throughout, shaped wedges have to be used, which are more awkward to fit and keep tight. When the mast is stepped and set up a canvas or Terylene mast coat is fitted snugly around it to cover the wedging and prevent water getting below. This is shaped as a flared skirt, is tightly laced around the mast with small cord, and its lower edges are turned over and tacked under a circular wood or metal ring which is screwed to the deck, making a watertight seal. In large craft the mast wedges are usually carefully caulked and payed or stopped over to make them tight. A mast coat is not then required.

Masts stepped on deck in a correctly designed tabernacle supported by a compression pillar do not cause bottom or deck leaks, provided the deck is suitably reinforced. They are more easily stepped and unstepped and may be lowered, in sheltered waters, for attention to rigging or passing bridges. However, deck stepped masts must be of greater diameter than those stepped through the deck, approximating to a 25 per cent increase in weight, and must be more completely stayed if acceptable lightness is desired. Wooden tabernacles are often fitted in traditional craft, but galvanised steel is superior in strength, lightness, ease of construction and neatness. The boom gooseneck and pinrails set well out from the mast may be incorporated in steel tabernacles.

MAST IRONWORK The boom gooseneck connects the boom to the mast. The conventional pattern has a single band, in two pieces, cramped together around the mast by two bolts and has a swivelling gooseneck on the after side and metal belaying pins on each side. This is an inefficient fitting as the band has to be cramped so tightly to hold it in place on the mast that the wood fibres are crushed, usually admitting water and causing rot. Also, the gooseneck often has a spike to drive into the boom-end instead of a ferrule or cup end, with long arms which can be fastened through the boom. The belaying pins are too close to the mast causing halliards to bear on the mast hoops and be too close together when belayed.

An improved fitting has two mastbands cramped as described, with the gooseneck sliding vertically on a rod between them. This distributes the strain evenly on the mast, increases the frictional area of the mastbands and reduces the crushing. Screw or bolt fastenings are

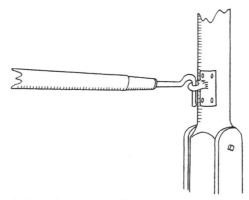

Figure 2. Simple iron gooseneck

sometimes fitted in mastbands to keep them in place but this weakens the mast and all mastbands should be a driving fit or rely on an adequate area of friction, if cramped. The sliding gooseneck enables a tack tackle to set the luff taut and relieves wringing strains on the mastbands.

Traditionally rigged craft have a pair of wooden horns fitted to the forward end of the boom (figure 1), lodging on a semi-circular wood mast collar supported by wood thumbs. The horns are covered with leather or hide on the inner faces, against chafe, and the mast should be lightly greased. In large craft a parrel line with lignum vitae balls secures the ends of the horns around the mast. This arrangement can be very noisy at night or in a seaway.

When a tabernacle is fitted it is usual to fit the gooseneck on its after side, particularly with loose-footed sails. Traditional Dutch craft often have goosenecks formed by a single hook and eye (figure 2), which is practical and cheap. Catboats sometimes have the gooseneck mounted on a metal pedestal called a crab, just abaft the mast (figure 3), but this is a rather elaborate detail intended to remove boom thrust and twist from an already heavily loaded mast.

THE HOUNDS OR CHEEKS In traditionally rigged craft the shroud eyes are looped over the masthead and rest on wood bolsters set on wooden hounds cheeks transversely bolted through the mast (figure 4). These must be of hardwood, well fitted and very strong. Alternatively a mastband or steel hounds fitting can be made with eyes or lugs to which the shrouds are shackled. It also often incorporates lugs for the spreaders and trestletrees, if a

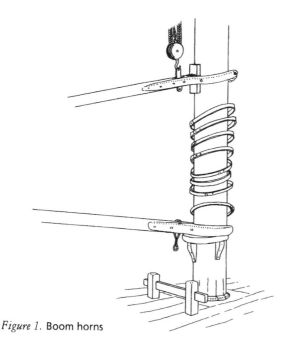

Figure 1. Boom horns

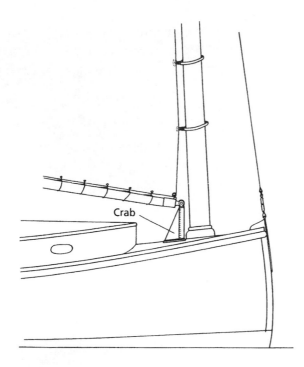

Figure 3. Catboat gooseneck 'crab'

topmast is carried. Although the steel fitting may look neater the mast has to be shouldered down to give a bearing and it is a likely spot for rot to develop.

THROAT HALLIARD ATTACHMENTS The throat halliard upper block should be supported clear of the after side of the mast and cheeks. A crane or throat halliard bolt is usual (figure 5), and takes considerable strain.

Alternatively the block is slung on a wire strop around the masthead, extended out by a wood chock (figure 6).

The peak halliards and topping lift blocks may be shackled to mast eyebands fitted against shoulders on the masthead (figure 7). Mastbands should be as deep as possible to present a large surface and relieve strain on the mast. In large craft a soft rope grommet was fitted under them, against the shoulder, to relieve the compression. Alternatively the blocks are shackled to eyebolts through the masthead. These should have square necks under the eye to prevent turning, and should be slightly staggered in the after side of the masthead to allow the halliard and lift falls to run clear of each other and, incidentally, clear the grain of the mast. Eyebolts have a tendency to cause splitting in a masthead if undue strain is applied. The forestay eye traditionally rests on one of these eyebolts in which a specially forged cup or hook is usually worked (figure 8). Large craft usually carried the two topping-lift blocks shackled to eyeplates bolted through the hounds cheeks (figure 9).

CROSSTREES OR SPREADERS Crosstrees or spreaders are fitted to support topmast shrouds if a topmast or extended pole masthead is arranged. In large craft they are often made of Canadian rock elm and in small of spruce. The span of spreaders should almost equal the beam and they are now usually angled upwards at the ends to bisect the angle between the shroud and topmast

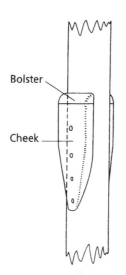

Figure 4. Hounds cheeks and bolsters

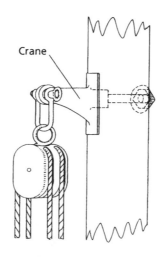

Figure 5. Throat halliard bolt or crane

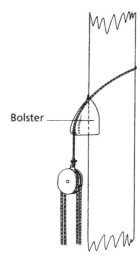

Figure 6. Throat halliard block strop and bolster

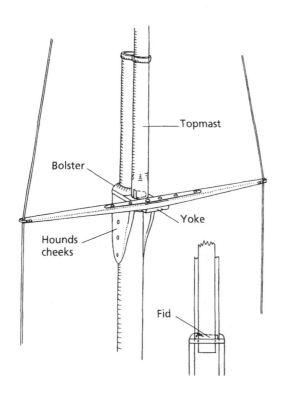

Figure 10. Topmast shroud spreaders and topmast fid

head. If hounds cheeks are fitted in small craft the spreaders are connected across the mast diameter by two flat metal bars through-bolted to the spreaders. Alternatively, the inner end of the spreaders may be socketed into a steel hounds fitting and secured by a

vertical bolt. Large craft with topmasts and traditional masthead arrangement carry the spreaders in one length bolted across the top of the wooden yoke or trestletrees outside the forestay (figure 10).

THE LOWER MASTHEAD The lower masthead is the part of the mast above the hounds in a craft carrying a topmast (figure 11). Its length is also the topmast housing or distance between the masthead capband, which has an eye forged on it to support the topmast, and the trestletrees or the yoke band which supports the topmast heel and the fid retaining it. Traditional English mastheads are of moderate length, often $1/4 - 1/3$ of the mainsail luff in a cutter. Long mastheads are liable to be sprung and shorter ones are preferred for this reason, although they give inferior support to the topmast. In English craft the considerable strain of the peak halliards is evenly transferred to the masthead by three blocks, and the jib halliard block is usually fitted just above the middle peak halliard block and the balloon jib halliard block adjacent to the upper peak block, to further balance the loads.

American sloops preferred short mastheads to save weight and their general use of a single multiple-sheave peak-halliard mast block aggravated the strain.

TOPMASTS Fidded topmasts are usually made of spruce or fir for lightness. The diameter should taper evenly

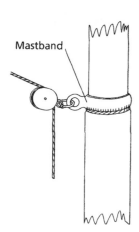

Figure 7. Mast eyeband with grommet under

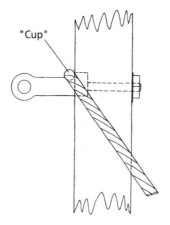

Figure 8. Masthead eyebolt with cup for forestay

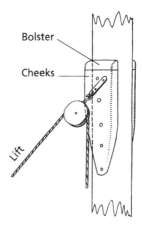

Figure 9. Topping lift blocks on hounds cheeks

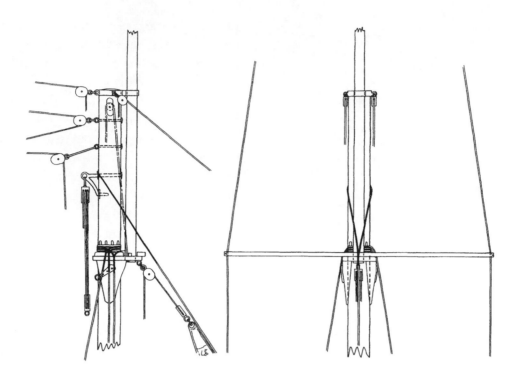

Figure 11. Typical masthead arrangement with topmast

from head to the cap iron at the lower masthead, and slightly from the cap iron to the topmast heel. The average proportion of topmast diameter to height (h) between heel and shoulder for English cruising yachts up to 50 feet length is (·022h) at heel. The topmast head has a small truck with sheaves for flag halliards. A shoulder is formed below the truck and a band is fitted with eyes for topmast stays, backstays and jib topsail and spinnaker halliards. Immediately below this a sheave is usually slotted through the topmast for the topsail halliard. The sheave pin should be well bushed and supported to prevent it biting into the wood under compression. The topmast is held in place by a fid or pin across the trestletrees or yoke, formed by two pieces of wood joined on the forward ends to form a flat-sided hole into which the topmast heel fits, but will just pass through. Topmast weight and rigging compression are thus transmitted to the fid which must be strong. In some craft a steel yoke band replaces or supplements the trestletrees, but arrangements are otherwise similar. The topmast is hoisted and lowered by a heelrope or top-rope. One end of this is made fast to the masthead cap, then rove down outside the lower cap or yoke and through a score or

dumb sheave across the heel of the top-mast, then up through a sheave which is usually slotted obliquely through the masthead just below the capband, and down to the deck. A tripping line is bent to the topmast heel to guide it aloft. A hand must go aloft to trip or release the fid, the topmast is then lowered on the heelrope and the topmast shrouds shortened as described under standing rigging. The heel is lashed to the mast or to one side. Topmasts should be housed in very bad weather or in strong winds when a vessel has a long beat to windward (figure 12). This lowering of a considerable weight and reduction of wind resistance considerably improves behaviour and speed. However, many gaff rigged yachts still setting topsails are rigged with an extended pole masthead forming a topmast integral with the lowermast. Mastbands may be fitted to take rigging intermediate between the masthead and the hounds.

GAFF A gaff should be as light as possible consistent with rigidity required to set the sail properly (figure 13). The sail's angle of peak affects the diameter of a gaff as much as its length. Generally, the greater the angle the less the diameter. The diameter of solid wood gaffs for has been used. Hollow wooden gaffs can be used and were once popular in large yachts, but unless the gaff is large this is a doubtful advantage as solid parts must be left in way of spans and other attachments, giving little weight reduction for small diameters. Aluminium alloy gaffs of mast section, with the luff groove used for the headrope are practical and great weight-savers. Some difficulty will be experienced with bending in a gaff over, say, 14 feet length and it is also difficult to attach the jaws or a sliding saddle.

GAFF JAWS Traditionally made of oak, grown to shape, gaff jaws are now usually laminated from oak or Canadian rock elm. They must be carefully designed to give adequate clearance around the mast or twisting of the gaff by the sail when set. In small craft with mainsail areas up to about 200 square feet the gaff jaws need only be well rounded and covered with hide or leather to protect them, and the mast, from chafe. In larger vessels they should also be leathered and with a tumbler or tongue shaped to the mast radius and fitted on a pivot bolt to take the thrust of the gaff on the mast (figure 14). The gaff-end is rounded to allow the tongue to lie parallel to the mast at all angles of peak. The length and shape of gaff jaws should be carefully considered so that the ends protrude before the mast diameter when the gaff is dropped peak-first, in a hurry. The jaw-ends are connected across the forward side of the mast by a parrel line threaded through a transverse hole in each jaw end and having a number of lignum vitae balls on it to reduce friction when hoisting or lowering. The ends of the parrel line are finished with stopper knots. In larger craft the parrel is sometimes of metal rod, leather covered and bolted to the jaw ends.

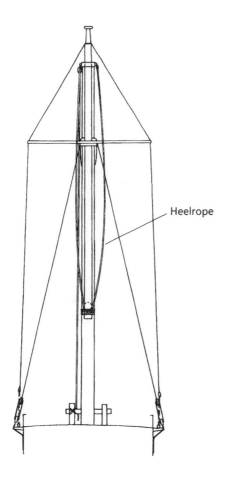

Figure 12. Topmast housed

small English cruising yachts should be about ·017 of the length and for yachts up to 50 feet about ·019 of the length. Where gaffs are subjected to exaggerated strain, such as long and very high-peaked gaffs held close to the mast by multiple spans, a longitudinal reinforcing piece will be needed, glued to the upper surface. Spruce is commonly used for small gaffs, but Oregon pine has greater rigidity for larger craft and even heavy pitch pine

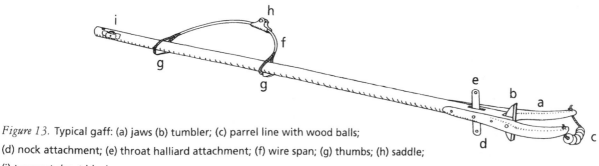

Figure 13. Typical gaff: (a) jaws (b) tumbler; (c) parrel line with wood balls; (d) nock attachment; (e) throat halliard attachment; (f) wire span; (g) thumbs; (h) saddle; (i) topmast sheet block

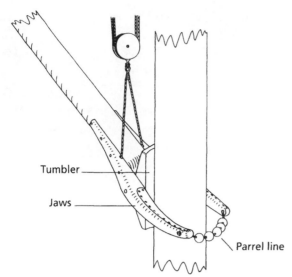

Figure 14. Wood gaff jaws

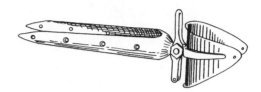

Figure 15. Metal gaff saddle

Some gaff craft have a sheet of copper tacked around the after side of the mast at the position of the jaws with the sail fully hoisted. This is bad practice and can lead to rot and weakness in the mast fibres at the lines of the tacks. If a protective sleeve is considered necessary it can be made with a few layers of glass fibre reinforcement and plastic resin which cures to form a hard coating capable of renewal when required. Iron or steel gaff jaws, covered with hide, were commonly used aboard yachts from about 1870-1914 but are now seldom used.

Gaff saddles have been successfully used in place of jaws for many years. This steel or bronze fitting allows the gaff to pivot independently of the mast and provided the saddle is leathered and fitted with a strong parrel line it is efficient on craft of all sizes (figure 15). It is important that the lugs attaching the throat halliard lower block and those to the throat or knock cringle of the sail are exactly in line to ensure smooth hoisting.

Peak halliard blocks should not be fitted to the gaff on eyebands as they cause crushing of the wood fibres if driven on, leading to weakness which is aggravated if they are also screw-fastened and could lead to the gaff breaking in hard winds. The blocks should be attached by wire spans or strops, see chapter 4. Cheek blocks for the gaff topsail sheet are fitted on the sides of the gaff end and must be strong and well fitted to avoid breaking and jamming the sheet. Traditionally a small metal block on a spike was driven into the gaff end for reeving the ensign halliards. This also served as a peak downhaul.

BOOM A light boom is not thrown about so violently in a seaway and is easier on gear than a heavy one, but lifts quickly when the sheet is started in a breeze, unless a dinghy-style mainsheet is fitted.

Booms may be made solid or hollow, from the timbers described earlier, or from aluminium alloy or steel. Traditional solid booms are round in section and the diameter for use with a laced-footed sail of an English cruising yacht should be about ·02 of the boom length. For a loose-footed sail the diameter should be about ·022 maximum, tapered to about ·02 at the ends. This is necessary to allow for the greater compressive bending forces thrown on them by the clew earring, compared with the laced-footed sail's evenly distributed strain. Reef combs, cleats and outhaul sheave will be through-fastened to the boom and these items must be well fitted and have the minimum number of fastenings required, to avoid weakening the spar. Traditionally, the mainsheet is

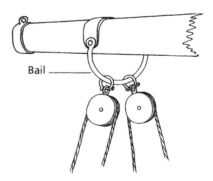

Figure 16. Mainsheet boom bail

attached to the boom by a strop, or in large craft by spans, which distribute the strain. Many American yachts use a bail (figure 16), which concentrates the load and requires fastening, which could be a weakness.

Booms for roller reefing are of circular section having greater diameter at the after end than forward to assist in

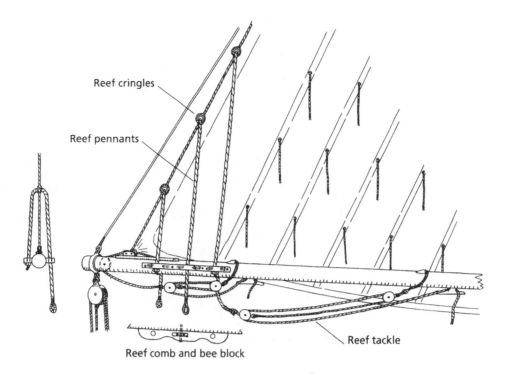

Reef cringles

Reef pennants

Reef comb and bee block

Reef tackle

Figure 17. Traditional reefing arrangements

gathering more of the leech of the sail and preventing the boom from drooping when reefed.

Hollow booms are an advantage in large craft and may be of rectangular section if point reefing is used. This requires the leech pendants to be rove through bee blocks at the after end of the boom. These are small sheaves, preferably of Tufnol, fitted on one or both sides of the boom and positioned to maintain a slight tension on the pendant when it is pulled down to the reefed position. The bee blocks are fitted in wooden 'reef combs' (figure 17), usually of oak or Canadian rock elm shaped to the boom and strongly glued and fastened. The bee block pins should be carried through the boom if possible to give added strength to this very important piece of gear. Cleats for reef pendants, clew outhaul and reef tackles should be well proportioned for quick use in bad weather and darkness, and glued and screwed to the boom, positioned on its sides to minimise risk of striking the crew when gybing.

BOWSPRITS Bowsprits may be of the traditional long, round-sectioned, reeving type, the stout fixed type once favoured in American working craft and yachts, or the modern, short rectangular-sectioned form. In all cases

they are solid spars, though tubular steel bowsprits have been used in large yachts and commercial craft. Long bowsprits need to be stout to resist bending by the pull of headsail and the topmast forestay and Oregon pine is a favourite timber but the heavy pitch pine is sometimes used. Running or reeving bowsprits of the traditionally British type are housed between stout bitts, one of which should be a post carried down and let into the centreline structure and the other a large knee bolted to the beams. The inboard end of the bowsprit is squared to fit between the inner faces of the bitts, where it is secured by a square iron fid which passes through bitts and bowsprit (figure 18). Round fids are used, but are inferior in resistance to bending caused by the bowsprit tending to thrust aft when under sail. In large craft the squared section is carried forward towards the stem and about three fid holes are provided in it at intervals to suit the reefed or withdrawn positions of the bowsprit in hard winds or bad weather. In very large craft the bowsprit may be run out with a heelrope over a sheave in its aft end, usually worked from the capstan in fishing vessels. At one side of the stemhead, usually to starboard, the running bowsprit passes through an iron eye bolted to the stem and known as the gammon iron or span shackle. This takes much

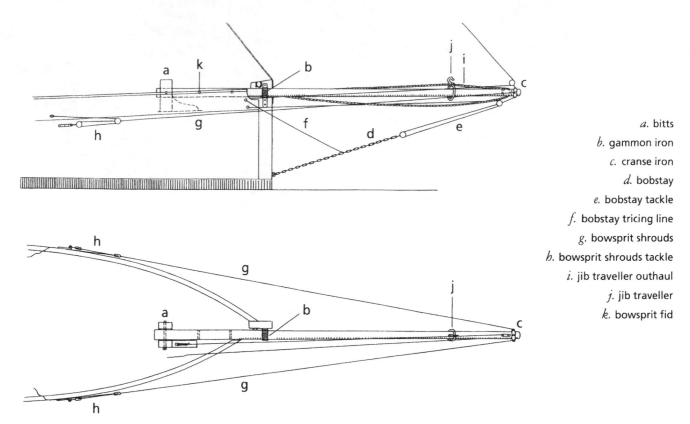

a. bitts
b. gammon iron
c. cranse iron
d. bobstay
e. bobstay tackle
f. bobstay tricing line
g. bowsprit shrouds
h. bowsprit shrouds tackle
i. jib traveller outhaul
j. jib traveller
k. bowsprit fid

Figure 18. Running or reeving bowsprit

strain in a seaway and must be strong. It should be 'woulded' with soft rope passed round it closely to protect the bowsprit from chafe. The outboard end of the bowsprit has a sheave let into it to take the jib outhaul which pulls an iron traveller ring with a curved hook to take the jib tack pendant to the required distance along the bowsprit. The traveller should be of ample dimensions to slide along the spar and be covered with hide or plastic, against chafe. The bowsprit end is shouldered-down for a galvanised iron eyeband with two eyes, the bobstay being shackled to the lower and the topmast forestay to the upper. In some craft the topmast forestay ended in a served eye slipped over the bowsprit end to rest against the shoulder. Larger vessels, particularly yachts with long bowsprits, carried a four-eye band to which the bowsprit shrouds were also attached.

These were often led aft over spreaders known as the whiskers, made of wood or steel rod or tube, usually fitted to the stem at deck and pivoted on pinned fittings if of metal, to adjust to the shroud angle, or bolted rigidly if of wood. These are now obsolete.

Flat, rectangular-sectioned bowsprits are often used with the shorter based modern gaff sail plans and these are described in chapter 2.

Yards for topsails should be as light as possible consistent with strength. Spruce is a good timber and hollow wooden yards were once common, but aluminium alloy tube has obvious advantages. Topsails vary greatly in size, shape and arrangement and the topsail and jackyards should be carefully designed for individual sails. These remarks also apply to spinnaker booms.

CHAPTER TWO

STANDING RIGGING

LIKE ALL THINGS NAUTICAL the design of gaff rigging must be largely based on experience and a clear understanding of a vessel's purpose. There has been a tendency in recent years to reduce the sizes of rigging of bermudian rigged cruisers to racing yacht standards, but this trend should be treated with caution when rigging a gaff boat, which usually needs heavier rigging to absorb strains differing from bermudian rig. Adequate strength of rigging must be equated against weight and these are two conflicting requirements in rigging any sailing craft, where every pound saved aloft increases sail carrying power and stability. Most cruising yachts' rigging is designed or fitted with ample strength, but often with little regard for weight. However, the greater the diameter of a wire rope the more resistant it seems to be to deterioration, and this is often a contributory factor to the apparently excessive size of standing rigging in traditional craft.

SHROUDS The majority of present-day gaff rigged yachts require only two lower shrouds on each side, fitted to the mast close above the position of the gaff jaws when the full mainsail is set. Their lower ends are usually attached to chainplates located just forward and aft of the masts' centreline at deck. Craft not fitted with backstays, and most ex-working craft, have the forward chainplate abreast the mast, and the others spaced well aft to give the best support possible without restricting the boom too much when off the wind. When designing a rig the combined strength of the lower shrouds fitted on one side of the yacht should be made equivalent to the displacement of the yacht, plus a factor of, say, 25 per cent of the displacement.

There are three principal methods of attaching shrouds to a mast

1. By shackling to a mastband or tang.
2. By passing a single eye over the masthead and settling it on a wood bolster on the hounds cheeks.
3. By fitting the shrouds in pairs formed by a wire passing up from the lower eye, round the masthead on the bolster, and down to the next lower eye on the same side. The upper eye is then formed by seizing the parts of the wire together with wire, close under the bolster.

The first method is obvious, but has always been rather unusual in gaff rigged craft, although the second and third are still common.

The single eye is the surest traditional method of attachment. The eye spliced in the shroud should be one and a quarter times the circumference of the mast at the hounds, and should be well served. In large craft it was often further protected by leather sewn around it.

There is a recognised order in which single-eye shrouds are placed over the masthead: the eye of the starboard fore shroud is positioned first; then the port fore shroud (which is made longer by twice the diameter of the shroud wire); followed by the second starboard shroud (the allowance being four times the wire diameter), and last the second port shroud with an addition of six times the diameter of the wire. The forestay is fitted last, over the shrouds, and its arrangement will be discussed later.

The use of paired shrouds was common in working craft and many yachts until seventy years ago, but it is now rarely used. Its advantage lies in reducing the

Figure 19. Deadeyes and lanyards

number of eyes and weight aloft. However, if the wire seizing parts, the shrouds are slackened, or if one of the lanyards or rigging screws breaks, the additional strain might part the seizing and leave the mast unsupported on that side. A common use for this method was to form the second pair of shrouds when three were fitted on each side, the forward shrouds having a single eye. If four shrouds were fitted on each side in large craft, they were usually two pairs, as it was very unlikely that both would be disabled at the same time.

The length of shrouds is measured from the top of the bolster to the eye or deadeye, or from eye to eye if shackled at each end. When measuring from a plan for new shrouds the true length must always be obtained, and any shrouds attaching to chainplates positioned forward or aft of the centreline of the mast must not be measured from the elevation of the rigging plan. The actual offset from the centre of the mast in plan view, to the chainplate, must be set out in conjunction with the height to the point of the mast attachment and the true length obtained, otherwise the shrouds will be short.

If deadeyes are fitted the eye at the lower end of the shroud should be one and an eighth times the

circumference of the deadeye to allow its replacement, if required. Traditionally, the eyes of rigging were covered by parcelling and serving with spun-yarn, covering with canvas and painting, or covering with leather. The first is still common, but the others have largely gone out of use as they were principally necessary in large craft.

If a yacht's shrouds are fitted with lanyards, or deadeyes and lanyards, they should be set up in the order of fitting the shrouds over the masthead; always commencing with the starboard fore shroud.

LANYARDS The use of simple lanyards to set up standing rigging should be confined to craft having sail areas not exceeding approximately 300 square feet and with masts stepped through the deck. In larger sizes, or in any craft with the mast stepped on deck, rigging screws are superior, alternatively, deadeyes and lanyards may be used, although they need frequent attention. Simple lanyards are best made from pre-stretched synthetic rope of sufficient length to give a total of say eight standing parts between the thimble at the lower end of the shroud and the chainplate eye, which should be wide and well rounded to avoid damaging the parts and to allow them to lie evenly when set up. A soft eye is made in one end of the lanyard and a turn is taken with this round the thimble. The lanyard is then passed down and through the chainplate eye, back through the thimble, and so on until sufficient turns are made, when the fall is taken down and made fast round the standing parts.

DEADEYES AND LANYARDS The traditional method of supporting the standing rigging of sizeable working craft and yachts is by tarred hemp rope lanyards set up between circular lignum vitae deadeyes fitted at the lower end of the shrouds, and similar deadeyes attached to the chainplates (see figure 19). Deadeyes and lanyards look natural in working craft, but in all except very large yachts they have a slightly theatrical appearance and a rigging screw will do the work better. However, a slight advantage of the use of rope lanyards instead of rigging screws in larger sailing craft is that they distribute strain evenly between two or more shrouds. If one is set up too tightly there is usually sufficient stretch in the lanyard to transfer some of the strain to the slacker one.

The rim of the deadeye is grooved all round,

allowing the shroud wire to be placed tightly round it, before being seized back on itself above the deadeye. The lower deadeyes are attached to the chainplates by an iron hoop of circular section which is either bolted to the chainplate or, sometimes, shackled to the end of the chain lanyard which is fitted to allow the deadeye to be positioned above the rail. Deadeyes for small craft such as yachts and smacks usually had three holes for lanyards. Below the holes the wood is faired, allowing the lanyard to lay fair around it.

Only the best tarred hemp was used for lanyards or, in later years, in some commercial craft and racing yachts, flexible steel wire rope. Hemp lanyards should be well stretched before use and given a slippery coating of tallow where they are to render through the deadeyes. A Matthew Walker or wall knot is made in one end and the other rove *out* through the forward hole in the upper deadeye, *in* through the forward hole of the lower deadeye, then *out* through the centre hole of the upper deadeye; following on until the hauling part is passed through the aftermost hole in the lower deadeye, when it is set up with a purchase to tauten all parts. Sometimes it is necessary to set up the lanyard as work progresses and this is usually done by taking the parts leading up from the lower deadeye and tailing them on to the throat halliard to obtain a strong purchase. When the parts have settled, and this takes time with new rope, the hauling

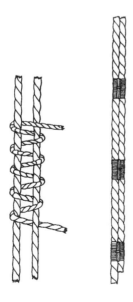

Figure 20. Racking seizing

part is secured to the next part by a racking seizing, after which it is taken up between the eye of the shroud wire and the top of the deadeye; passed round behind the shroud eye; and brought in again between the shroud wire and the top of the deadeye. A purchase is again applied to the fall and, when set up, a racking seizing secures it to the adjacent part. Sometimes this fall is not passed through the shroud eye, but instead, two half-hitches are taken round the shroud, and the fall seized to another part, as described above. It is possible for the fall of the lanyard to be made fast by two turns and seized round the eye of the shroud.

The Matthew Walker knot has been known to part under the strain in large gaff-rigged craft and by the end of the 19th century many yachts were adopting the system of splicing a thimble eye in one end of the lanyard in place of the knot, and this shackled to an eyebolt in the channel, just ahead of the chainplate. The other end of the lanyard is rove *out* through the forward hole in the upper deadeye and continuing as previously described. As this method introduces another part, additional care is required to set it up and it may be necessary to set up say three parts with a tackle and seize them to each other, before continuing to reeve off the remaining parts.

The racking seizing used is made with light rope, varying in length from 2-3 feet in a large yacht to 1ft 6in to 2ft in small, making a running eye at the top of the seizing and passing the other end in and out between the two parts to be seized, as shown in figure 20. The seizing is then pushed together and the end secured.

If flexible steel wire is used for lanyards it is usual to splice one end to the shroud above the deadeye, in place of the Matthew Walker knot. The other end is passed through the deadeye and the fall finished off, as for rope.

The supple pre-stretched synthetic rope appears to be ideal for lanyards, except that it tends to be very slippery, even in the stapled varieties, and seizings to the shrouds or other parts tend to draw. However, with care it can be satisfactory. It is desirable to heat the stopper knot to secure it from pulling apart.

Sheerpoles, also in the past called sheerbattens or sheer stretchers, are often fitted across the shrouds just above the deadeyes or rigging screws. They are seized to the shrouds and prevent them untwisting under strain. Where deadeyes and lanyards are used the sheerpole should be fitted before these are rove and set up as the finishing hitches of the fall must be made above them. Sheerpoles may be of hardwood or metal. Wooden ones are often favoured on traditional types, sometimes with a vertical belaying pin between each shroud, to which halliards and lifts can be made fast.

When rigging was universally set up with lanyards, slack rigging was often held to be a factor assisting speed and many tales were told of the fortunes of racing yachts being radically altered when the standing rigging was slacked off. It seems there was no advantage in this practice, and logic demands that if standing rigging is thought necessary to support a mast it should be set up as tight as possible, though the present fashion of bending masts and rigs perhaps strengthens the theory of the old skippers. My own view is that standing rigging should be set up tightly, but not so tight that it might strain the boat's construction, which is quite possible with over-taut rigging screws. By the 1890s rigging screws had superseded deadeyes and lanyards in racing yachts and, though they lingered in cruising yachts and working craft for many years, and are still sometimes seen, the rigging screws' ease of operation and finer adjustment is greatly superior.

RIGGING SCREWS There are several types of rigging screw and the commonest form used in gaff rigged craft has a closed cylindrical body with a hexagon in the middle, for adjustment by spanner, and internally threaded ends; one to receive a right-hand threaded end-piece and the other a left-hand threaded end-piece. When the body is turned the end pieces are extended or withdrawn and thus tension in a shroud or stay can be varied. The end pieces usually have forked ends, fitted with a retaining bolt, or alternatively a threaded pin, the upper one attaching to the thimble eye in the lower end of the shroud and the lower to the chainplate eye. It is very desirable for lock nuts to be fitted to prevent the barrel from turning, when set up. If these are not fitted the screw must be wired to prevent turning by passing a wire through the end forks and the hole which is drilled through the body in way of the hexagon. Before setting up the rigging screws it is best to pack the body with a waterproof grease, to protect the threads.

Rigging screws are now made from galvanised steel,

stainless steel or bronze. To avoid chemical action and possible failure, care should be taken to select a type where all the parts are of the same metal. A rigging screw must be as strong as the wire it supports and a general rule is to divide the circumference of the wire by two to obtain the diameter of the screw's threaded end fitting. For example, a $1\frac{1}{2}$in circumference wire would need a $\frac{3}{4}$in diameter screw. A closer approximation can be made by working from the breaking strain of the wire and obtaining a screw of similar proof test; however, it must always be remembered that when anything parts at the shrouds it is generally the screws' threaded end which has broken, and it is best to allow an additional factor of safety of say 25 per cent. Care should be taken to avoid over-tensioning when setting up rigging screws. Each rig requires individual attention, and some experiment, to obtain the best results.

RATLINES Traditionally, ratlines or ropes seized between two or more shrouds to form a ladder up the rigging, were only fitted in a few English fore-and-aft rigged craft, except those setting squaresails. Most were arranged to be worked from the deck and, if anything needed doing aloft, a hand went up the mast hoops if the mainsail was set, or up with legs between the shrouds if it was not; and with a gantline or the topmast halliard he could reach the topmast head. However, where there was considerable work aloft, such as in large yachts and trading craft, one pair of shrouds on each side were usually fitted with ratlines.

Like deadeyes and lanyards these now have rather a theatrical appearance, but can be genuinely useful in larger craft, though they are often mistakenly added to small vessels, to the detriment of windward performance and appearance. Rope ratlines must be fitted in a certain way to remain reasonably taut and effective, and nothing looks worse than ratlines hanging slack or askew. They should be attached to the shrouds by a racking seizing or, if fitted over three shrouds, they may be clove-hitched to the middle one. If the shroud wire is of small diameter or smooth finish, the seizing should be done over insulating tape.

Hardwood battens are sometimes fitted as ratlines but they chafe the mainsail before the wind and their weight and windage are considerable.

FORESTAY Traditionally, the forestay was the last stay fitted to the masthead, at the hounds, with an eye similar to the shrouds. During the 19th century this arrangement was modified, in yachts and well developed small craft, to a larger eye which embraced the masthead and the yoke, diagonally, well above the hounds, and rested on top of the throat halliard bolt at the after side of the mast. The yoke also expanded the eye to allow the topmast to pass through it, if one was fitted. The splice was arranged below the crosstrees. This arrangement of forestay is still to be seen in the surviving Essex smacks. In craft with the mast stepped through the deck the lower end of the forestay ended in a variety of ways. A common arrangement in working craft until well into the 19th century (figure 21a) was for the lower end of the forestay to terminate in a deadeye, and three or four horizontal holes were drilled transversely through the upstanding wooden stemhead. A rope lanyard, with a Matthew Walker knot in one end, was rove through the forward hole in the deadeye and down through the forward hole in the stemhead, then up through the middle hole in the deadeye, continuing as before until the stay was set up tight, often with a tackle, after which the fall was taken in turns round the standing parts and seized. This arrangement is still seen in some working craft.

Another version commonly used in straight stemmed yachts and working craft (figure 21b), was to lead the wire forestay to the stemhead and down a score cut vertically in the forward side of the stem to a hole bored fore and aft through the stem. The forestay emerged at the back of the stem where it was seized back on itself round a heart block. A lanyard was rove through this and set up through an iron-bound heart block which was fastened to the bitts.

A later and improved arrangement was to fit a rigging screw to the lower end of the forestay and to set this up to an eyeplate securely fastened to the stemhead, or to a lug on top of the gammon iron, if the bowsprit was fitted on the centreline.

The commonest present arrangements in pole masted craft of moderate size are to shackle the forestay to a mastband and set up the lower end with a rigging screw, or in small sizes, by a simple lanyard .

Double forestays are a seamanlike arrangement useful when changing the size of staysail. It is best to fit a wooden spacer chock between the stays at the bottom to keep them apart, as, if they are too close together the sail hanks have a tendency to snap on both stays when sails are being bent, especially at night. They are best shackled to a mastband at the head and set up with rigging screws

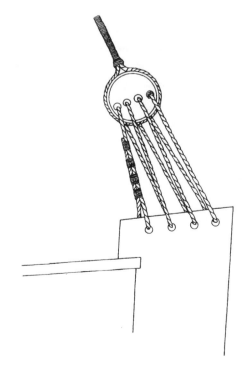

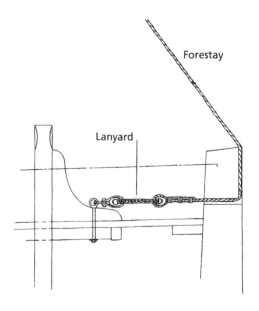

Figure 21. Traditional forestay setting-up arrangements; (a) forestay set up on a deadeye; (b) forestay set up inboard

Forestay

Lanyard

at the foot, attached to carefully designed eyes which must be well bolted to the stem head or the main forward structure of the craft.

Where the mast is stepped on deck in a tabernacle the forestay is also usually set up with a rigging screw or simple lanyard, and a tackle is hooked on when the mast is to be lowered. It is unusual and generally undesirable to leave the stay fall tackle set up to take the weight of the mast. However, this is sometimes necessary when passing a number of fixed bridges. It is bad practice to let the stay fall tackle carry the weight of the mast in a seaway, despite its being common practice in the English east coast sailing barges. The stay fall tackle should be well stretched rope which is not slippery. The number of parts required will depend on the weight of the mast, and whether it is to be raised by hand or the windlass. The fall should be carefully secured and stowed; a difficult matter in a small craft.

BACKSTAY RUNNER In most working craft and yachts a backstay was fitted, on each side, leading from the hounds to the deck or bulwark at approximately halfway between the mast and the rudder head, and commonly known as running backstays or runners.

Although there is little need for these with the wind forward of the beam, they do help support the mast against the pull of the headsail halliards, keep a tight forestay and jib luff, and relieve strain on the after shrouds.

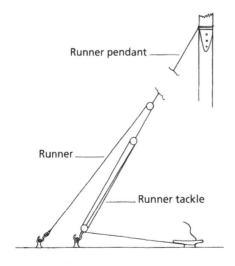

Figure 22. Backstay pendant, runner and tackle

The mast of a large commercial craft or cruising yacht should stand for a period without runners but if a runner breaks or suddenly fails in a hard breeze, it may damage or spring the mast.

In the yachts and commercial craft of the 19th century and early 20th century, the backstay was in two parts (figure 22). First a wire rope pendant having a single eye, as for the shrouds, at its head, which was passed over the masthead *before* the shroud eyes were fitted. The pendant's lower end had a single block either shackled or spliced to it for the runner. The length of the pendant is usually $^2/_3$ of the distance between the hounds and the deck, when set up. The second part was termed the runner, and in small craft this was of hemp or other rope, but in large vessels, especially racers, it was often of flexible wire rope. The total length of the runner should be $^3/_5$ the distance from the deck to the hounds, when set up. It should be as long as possible to permit overhauling when the boom is eased off, and it should be unnecessary to unhook the runner and tackle with the boom eased off a reasonable amount. The ends of the runner have thimbles spliced in them with a strong hook fitted at the end of the standing part, which is hooked to an eyebolt in the rail or on deck, so that the point of the hook is upwards to bring the strain on the throat of the hook and prevent it being distorted or broken.

The tackle which sets up the whole backstay is shackled to the eye at the other end of the runner. Often this tackle was arranged with a fiddle block above and a single block below, but sometimes the upper block was an ordinary double-sheaved type. In yachts without bulwarks the tackle was arranged to lead the fall out parallel with the deck to belay on a cleat. The usual arrangement is shown in figure 22. In sizeable craft the best option to this is to fit a flexible wire, or preferably synthetic rope tackle, leading the fall to a small winch giving singlehanded power to set up the runner effectively in any conditions, providing the fall is marked to prevent over-setting-up.

In pole masted yachts setting topsails, the upper end of the runner pendant sometimes ended in a span, the lower end of which attached to the mast at the hounds or the staysail halliard, and the upper end at the jib halliard position.

With the boat before the wind, in a seaway, with the

boom well off, the lee runner needs to be taken forward and secured at the shrouds to avoid chafe. To avoid having to go forward it is usual to seize a small block or bullseye to the after shroud and thread a light line through this; one end of which is seized to the runner pendant and the other led aft in reach of the helmsman who can clear it forward and belay to a small cleat. A plastic block and synthetic rope are best for this purpose.

Wire spans for runners are still sometimes used in small craft, though they are an inefficient fitting. The runner pendant wire is fastened by a large shackle to a span of steel wire shackled between two eyebolts which are fitted through the deck and beams. The wire and eyebolts should be of very strong construction to withstand the abnormal strains placed on the span. A rope tail is spliced to the span shackle and is used to set up the runner by drawing it aft along the span. The tail is belayed to a cleat. The span is of sufficient length and is so positioned that the runner pendant is at its after end when set up, and when slacked off runs forward to clear the boom and gaff. A variation of this arrangement uses a tackle to haul the runner pendant aft, but this is unusual. It is difficult to get a correct adjustment with this method, but the span can be marked with paint or a whipping at the desired position.

A variation of this method is the substitution of a strong deck-track and slide for the wire span. It is necessary for the track to be through-bolted to the deck and beams. There may be difficulty in getting the slide to shift forward when the runner is slacked off.

Many owners will prefer to fit one of the various types of runner lever which are attached to the deck, and by the throwing over of a comparatively short lever, slacken the runner sufficiently for it to clear the boom and gaff when the craft is close-hauled, though when the wind is freer the lee runner span has to be unhooked and taken forward. The runner is of flexible wire rope and must be arranged to lead horizontally to the base of the lever, and is fastened to its head. When set up the lever locks itself by having its head passing below the centre of the pivot pin, and the strain in the wire holds it securely. When the lever is slacked off the drift of the runner is twice the length of the lever and there must be sufficient drift to clear the sail when on the wind. With the wind free, the deck block should be released. Probably the best

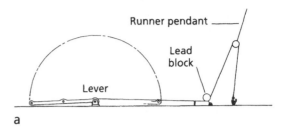

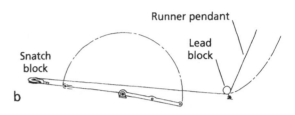

Figure 23. Improved running backstay setting-up arrangements; (a) backstay with runner set up by lever; (b) backstay with runner set up by lever via snatch block

arrangement is that shown in figure 23b; the runner is of flexible steel wire rope and is lead aft through a guide block to a snatch block aft of the runner lever, and then forward to the lever. When the lever is slackened the runner wire is lifted from the snatch block, giving ample slack for running free. The process is reversed when setting up. It is best for the snatch block to be about the after end of the cockpit, or adjacent to the helmsman.

Another method (figure 23a), is for the runner pendant to have a single sheave block shackled to its lower end. Through this is led a flexible steel wire rope whip, one end of which has a hook spliced to it which is hooked to a deck eyebolt. The other end leads down from the pendant block, through a guide block on deck, and is fastened to the head of the lever. With this arrangement, when the wind is free, the runner lever is slacked and the whip unhooked from the eyebolt, allowing the runner to run forward. However, it does not give so much slack as the snatch block method. When slacked off, these types of lever-operated runners can also be taken forward to the shrouds, as previously described.

TOPMAST RIGGING The topmast, or a pole mast extended to set a topsail, is supported by shrouds and backstays.

Craft of up to about 45 feet waterline length and moderate sail area should only require one topmast shroud each side. These are usually arranged to pass over a spreader and are set up to chainplates abreast the mast. Traditionally the upper ends of the topmast shrouds end in a single eye slipped over the topmast head and resting on a shoulder, or wood thumb. Alternatively, an eyeband can be fitted with the ends shackled to it. The lower ends are fitted with thimbles. The traditional method of setting-up is with a small tackle rove to an eyebolt at the sheer or at the channel. Where a fidded topmast was carried, the topmast shrouds were in two legs, the upper corresponding to the length from the topmast head to the top of the setting-up tackle when the topmast was housed, with its heel on the deck. As these two legs were, shackled together at the joint, the lower one could be stowed when the topmast was housed, and the upper set up with the tackle. In small craft with fidded topmasts, the topmast shrouds were often set up with simple lanyards, and rigging screws were normally used in pole masted yachts setting topsails.

The topmast shroud was made effective by being led over the crosstree, or spreader as it is now commonly termed. Wooden spreaders were usual, and had a notch in the outboard end through which the wire passed, and was retained by a clench nail or bolt. In large craft having great spread of crosstrees, a wire stay or standing lift was rigged between the mast cap and the outboard part of the spreader, which might be heavy enough to require support.

Where a large topsail is carried in craft rigged with a long fidded topmast, or in yachts with an extended pole mast in place of a topmast, an intermediate topmast shroud is usually fitted to support the masthead. This may lead to the topmast head of fidded topmasts, where a soft eye may fit over a shoulder or a pair of wooden thumbs, or may be shackled to an eyeband. The shroud leads down either through a hole in the outboard part of the crosstree, or through a wood or metal thumb fitted to the crosstree side. Where topmasts are housed, the shroud is in two parts, shackled together to facilitate re-setting up and stowing when the topmast is down, as previously described for topmast shrouds.

In some large vessels with a long masthead, a shroud is led from the mast cap, down to the spreader, being termed the cap shroud. Its function was to support the masthead against the pull of the peak halliard blocks.

When the topmasts were housed in fishing vessels, the topmast shrouds could be an embarrassment and a practice which seems to have been peculiar to smacks from Tollesbury, Essex, was to coil them up and secure them, hanging like giant earrings from the crosstree ends, making that village's smacks easily distinguishable in bad weather.

In pole masted craft the cap shroud is usually fitted adjacent to the peak halliard block or at a point about halfway between the crosstrees and the masthead. It may be attached with a soft eye or shackled to a band, but there is no need for it to be in two legs as for the housing topmast. The lower end of the cap shroud may be set up in any of the ways described for topmast shrouds.

Besides two topmast shrouds and the cap shrouds, large yachts with a pole masted Marconi topmast rig often had other, intermediate topmast shrouds supporting various peak halliard blocks, and leading down to separate chainplates, via the spreader. Another, rather impractical variant in these craft was the leading of the cap shroud down over a short spreader fitted to the mast about $^2/_3$ up the mainsail hoist, after the sail had been set. This gave added stiffness to the mast under the tremendous strains imposed by the ultimate height of rig and power of gaff racers.

The topmast forestay or pole stay is fitted between the topmast head and the bowsprit end to oppose the backward pull of the topmast when the topsail is carried to windward. In traditional craft a soft eye is slipped over the topmast head and rests on top of the topmast shrouds, which are held by a shoulder or thumbs. The lower end may either be fitted over the bowsprit end which is shouldered down to take a soft eye, or led through a block shackled to the cranse iron, and the fall set up to a cleat on the foredeck or to the bitts. In large craft a tackle may be used to set it taut. In yachts, the upper end of the stay usually has a thimble eye which shackles to an eyeband at the topmast head, the lower end being set up to the cranse iron by a rigging screw or a lanyard.

Some gaff yachts of advanced rig built between 1895 and 1925 had a jumper stay fitted on the forward side of the mast, just above the gaff jaws when the sail was hoisted, to relieve their thrust. The upper end of the stay was usually shackled to a mastband above the hounds, adjacent to the peak halliard, and led over a spreader or strut which was positioned on the mast fore and aft, above the gaff jaws, which was arranged to clear the staysail leech; then down to the foot of the mast above the deck, where it was set up with a rigging screw. Jumpers were usually only fitted in racing craft, being specially common in the large classes before and just after World War I.

PREVENTER BACKSTAYS These are now rarely seen as there are few yachts large enough, or setting sufficient canvas to need them. In gaff rigged yachts one preventer backstay was fitted on each side and supported the topmast head from going forward. In craft where the jib topsail halliard is fitted below the topmast head the preventer backstay should be attached at that point. The lee backstay had to be let go and the weather one set up when tacking or gybing. With a jib topsail set in a breeze any mishap in handling these stays might cause a broken topmast or, when gybing, a broken boom as well, if it fouled the lee preventer. These were known as shifting-backstays and were not usually fitted in commercial or fishing craft which did not carry extremely large topsails, or expect to carry their topsails in extremely strong winds. The upper end of the preventer had a single eye which was fitted over the shoulder or thumb at the topmast head, and the lower end terminated in a thimble to which was shackled the upper block of the tackle which set it up to an eyebolt on each quarter, usually the eyebolt to which the trysail sheets were attached. In large yachts the preventer pendant ended 6 or 7 feet above the deck and was set up with a tackle of two double blocks with the fall leading downwards to belay on a cavil if the yacht had bulwarks. If not, the fall was led on deck to a cleat.

The handling of these stays required judgment, especially in racing yachts where they usually were in the charge of the second mate. If they were not slacked off or set up correctly the topmast would break and, if gybing, so would the boom if it fouled the lee preventer. The power of the old racing yachts' rigs is well illustrated by a tragic accident on board the first-class cutter *Volante* racing at Harwich Regatta 125 years ago. She was running by the lee with a competitor close to leeward in a strong breeze when she accidentally gybed. Before the preventer and running backstays could be slackened the huge boom smashed against the backstays and flew to pieces, one of which transfixed a hand to the rail and killed him.

Probably the most difficult situation for a preventer man existed in the old 5-ton class Solent racers of the 1880s, which were long and narrow, six feet being an average beam for 40-footer. The deck aft of the rudder head was only three feet wide and on this narrow and heeling perch one hand had, when gybing round a mark, to handle the jib topsail sheets, the balloon foresheets, the preventer backstays and the spinnaker after guy all within a few seconds. There was room aft for only one man and as these yachts set 2,000 square feet of canvas (the sail area of a present 12-metre) his smartness can be appreciated.

BOWSPRIT RIGGING Long bowsprits are now unusual in yachts, but are still carried by ex-working craft setting traditional rig. Many of the larger types of sailing fishing and coastal vessels had very long bowsprits, sometimes dispensing with rigging of any kind, though most types had a bobstay.

A bowsprit is most vulnerable when a large jib is set in a hard breeze, or when seas are shipped in the jib through pitching in bad weather. Long bowsprits are stayed by wire rope shrouds, one on each side. A thimble is spliced in each end of the shroud which is shackled to the cranse iron at the bowsprit end and to a horizontal chainplate on the sheerstrake at the inboard end, or to an eyebolt fitted at the deck edge, through a deckbeam or the shelf, forward of the mast at a position giving good spread to the shrouds. In larger craft with bulwarks, an iron band with a forged eye was fitted round a bulwark stanchion for this purpose. In small craft, the shrouds can be set up with lanyards, preferably of synthetic cordage. Alternatively, rigging screws should be used as these are easier to adjust. In the traditionally rigged yachts and working craft, the shrouds were set up with outboard tackles consisting of a double fiddle block and a single block at the after end, the fall being belayed to a cavil or cleat. This was a clumsy method as the lee tackle dragged in the water when heeled in a breeze and an alternative

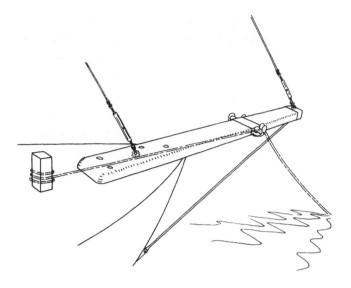

Figure 24. 'Plank' bowsprit

was to fit a chain to the wire shrouds, extending for the distance of the close reef of the bowsprit, lead through holes in the bulwark, and set up to ring bolts in the deck by an inboard tackle. When bowsprits were regularly reefed, or brought inboard in bad weather to ease pitching, additional eyebolts were fitted to enable the shrouds to be shifted aft as the bowsprit was brought aft, through the bitts, to new positions. This practice is now virtually extinct.

Narrow beamed craft often had the bowsprit shrouds spread with iron whiskers which had a jaw at the ends, in which the shrouds lay and were secured by a pin. The fitting of whiskers is detailed under Spars in chapter 1.

In bad weather traditionally-rigged small craft without bowsprit shrouds sometimes set up the bobstay fall and jib outhaul fall on opposite sides of the bow to support the bowsprit.

If a lengthy bowsprit is fitted, it is now usual to fit light rope netting slung under it, between the shrouds, to catch headsails when lowering and to make bowsprit work safer.

Increasing numbers of gaff yachts are being built or fitted with short bowsprits of rectangular section, often termed plank bowsprits (figure 24). Advantages of this type are

1. The ease of fitting to the foredeck by through-bolting to the deck beams, and to chocks between the beams.

2. Ability to walk on or stand on the bowsprit when necessary.
3. Ease of fitting fabricated ironwork for cable rollers and anchor stowage.

Plank bowsprits do not require shrouds but the bobstay is very important as their section has poor resistance to upward bending. A solid rod bobstay is often used, but chain is also effective with a substantial rigging screw. Some yachts, notably from the USA, have quite elaborate plank bowsprits which are often fitted with pulpits and lifelines for working at the headsails. Unless designed very carefully this results in a most inelegant appearance.

BOBSTAYS Bobstays may be either standing (fixed) or running. In small yachts and modern gaff rig, the bobstay is usually of standing type of either chain or wire rope shackled to the stem eyeplate and set up with a rigging screw to the bowsprit-end cranse iron. In very small craft this may be set up with a lanyard, preferably of synthetic rope.

Standing bobstays are chafed by the cable at anchor and the best solution is to sheath these with a plastic or rubber tube over the stay for a sufficient distance to cover the sweep of the cable. The lower end of the tube will rest against the stem and it should be free to revolve.

Sometimes double standing bobstays are fitted in large craft, especially in the USA, one supposedly acting as a preventer for the other. However, this is a doubtful practice as the tension in them will be uneven and, in bad weather could lead to one or other breaking. It is better to have one reliable stay, unless the inner bobstay is supporting a staysail, or second jib tacked down to the bowsprit.

Running bobstays may be of chain, wire or, in part, of rod, shackled to the stem eyeplate or clasp, and set up to the cranse iron with a tackle. The fall of this tackle is lead inboard and is usually set up to a cleat or the bitts, with the aid of the windlass. Sometimes a small winch is fitted, but this is unusual. Running bobstays are common in traditional craft and older yachts and are usually fitted to a long bowsprit giving power to bowse the bowsprit down to obtain a taut luff when the jib is set. They can also be hauled up to the stemhead when at anchor, to clear the cable, a lanyard being fitted about half way along the bobstay and set up to a cleat or the bitts. This means much to those sleeping forward, and in exposed anchorages may save the bobstay from excessive chafe.

It is now very rare to find a craft which reefs her bowsprit but where this is still done, the standing part of the bobstay should be short enough to allow the tackle to be bowsed down when the bowsprit is run in to its storm position. The martingale or bowsprit spreader is now rarely seen. It came into common use in yachts about 1893-4 when the new-fashioned and cut-away bows, such as are still fashionable, were introduced and the resulting poor lead of the bobstay caused the loss of a great many bowsprits in the big racing yachts. The martingale was of steel rod, fitted to the stem, often on a pivot bolt and led down to bisect the angle between the bowsprit and the stem profile. Its lower end usually had a U-shaped crutch over which the bobstay passed and was retained by a pin. The martingale was stayed transversely by wire shrouds shackled between its lower end and the outboard ends of the bowsprit spreaders.

BUMKINS These are still fitted to yawls and a few ketches, to sheet the mizzen. In small craft a stout-sectioned bumkin usually only needs a stay to the stern, shackled between an eyebolt in the stern centreline structure and a rigging screw fastened to the bumkin eyeband. In craft with very large mizzens, bumkin stays may be required to maintain transverse rigidity when the mizzen is sheeted hard in a breeze. In very large yachts with counter sterns, a martingale was also fitted to extend the spread of the lower stay.

CHAPTER THREE

R UNNING RIGGING

RUNNING RIGGING should be as simple as possible for ease of handling and to reduce wind resistance and weight aloft.

Blocks should have reasonably large diameter sheaves and good clearance for the rope, to reduce friction, which is improved by use of synthetic ropes and plastic blocks, which render more easily than natural fibre rope and traditional blocks.

Halliards should always be arranged to belay to pinrails, cleats or cavils, so that the fall will clear mast hoops and other gear, both for ease in working and to avoid chafe.

THROAT HALLIARD This purchase hoists the throat of the sail at the forward end of the gaff. The sail luff must be hauled as taut as possible and the throat halliard should have adequate power, remembering that the fewer sheaves used will assist quick lowering of sail. Traditionally the upper throat halliard block is shackled to a crane or short iron arm projecting sufficiently from the afterside of the mast to hang the block clear of the mast and allow the parts of the halliard to work freely. The lower throat block is shackled either to a strop rove through eyes in the gaff jaws or to a pivoting steel bar which is slotted through the forward end of the gaff and bolted through it. Sufficient drift or distance should be allowed between the crane and the gaff for the purchase to hang correctly when the sail is fully hoisted and ease the strain on the gaff jaws (see figure 14).

Usually one end of the halliard is shackled to a becket on one of the throat blocks with the result that the nip always comes in the same place on the blocks, when the sail is fully set. This is often overcome by reversing the halliard after a season's use. In large craft it

is practical to use flexible steel wire halliards, reeled on a halliard winch at one end and set up to a deck eyebolt with a tackle on the opposite side, to vary the nip. Rather than fit powerful throat halliards, some prefer to have a sliding gooseneck and use a tack tackle to set the luff taut. This can be done more easily with a loose-footed sail where the tack flows free of the boom. A tack tackle fall is hauled upwards, enabling the sailor to exert his full strength. The fall is made fast around the parts.

In large craft it was necessary when hoisting the mainsail for part of the crew, possibly 20 men in a racing yacht, to go aloft where some seized the peak and some the throat halliards, kicking themselves off from the hounds to string down with their full weight on the halliards assisting to raise the huge sail while the rest of the crew hauled at the halliard falls. This practice changed when lead sleeves were introduced at the heel of the mast and the halliard falls led through them to be hauled on, along the deck, by the whole crew.

Chain halliards, with special blocks, were sometimes fitted in large commercial craft to save wear, particularly in coasting ketches and schooners.

PEAK HALLIARDS The peak halliards are more used and abused than throat halliards and should be carefully arranged to give the most efficient lead to suit the shape of the sail and angle of gaff. Small craft with gaffs up to about 15 feet length require only one wire span on the gaff and often have the peak halliard blocks shackled or hooked direct to gaff eyebands or strops. The span should be of galvanised plough steel wire rope with a well-served eye spliced in each end, of suitable diameter to slip over the gaff, where they are held in place by small wooden chocks at each end, notched over the eye and screwed to

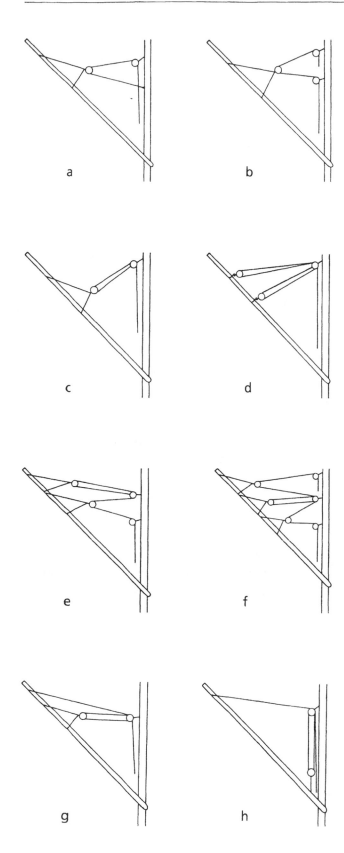

a

b

c

d

e

f

g

h

Figure 25 (a - h). Typical peak halliard arrangements

the gaff. A galvanised or bronze traveller, or a large shackle works on the span and the peak halliard block shackles to it. Various arrangements of peak halliards are indicated in figure 25. It is always preferable to lead the fall or hauling part of the peak halliard to the uppermost mast block as this keeps the gaff higher than if passed through the lower block (figure 25a). This may not seem theoretically true, but is so in practice.

Many owners fit double-ended halliards to avoid chafe, but this requires a further block on the mast (figure 25b); larger gaffs require more purchase in the peak halliards and double spans are commonly fitted in craft of about 40 foot length (figure 25e). The spans distribute the peak halliard's strain evenly on the gaff. Larger yachts had three spans (figure 25f). When designing new gaff rigged craft, the peak halliard should be arranged on the mast to pull upwards, at right angles to the gaff if possible.

Some gaff rigged craft used in sheltered waters have their peak and throat halliards combined in one and a typical arrangement is shown in figure 25h.

The power of all halliards can be imposed by using a whip purchase applied after the sail has been hoisted hand tight. Halliard winches can also be fitted to sweat up halliards.

THE MAINSHEET Mainsheet arrangements for gaff rig need not differ from those in common use for bermudian rig and a selection of arrangements is shown in figure 26. The boom block may be shackled to a steel boom-end eyeband; to a served wire or rope strop around the boom; to a boom span or spans in large craft or to a swivel boom-end fitting for roller reefing. In many cruising yachts it is an advantage to lead the mainsheet forward to belay before the helmsman.

The mainsheet must have sufficient power to sheet in the mainsail, full of wind, in strong winds and to gybe the vessel. A mainsheet winch is a great help when short handed.

Mainsheet horses are worthy of more thought than they usually get. A gaff rigged craft will not do her best to windward with the mainsheet pinned-in. The horse should extend to the side allowing the sheet to relieve the sail of twist. Some traditional craft such as bawleys even had the mainsheet horse extended slightly outboard with

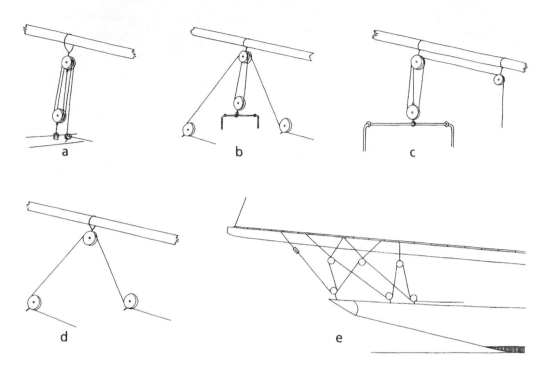

Figure 26 (a - e). Typical mainsheet arrangements

its ends returned by radius to the bulwarks for maximum efficiency, particularly necessary for a boomless mainsail.

TOPPING LIFT The topping lift supports the boom when the sail is lowered. Traditionally rigged English craft have a single topping lift, rove through a block at the masthead above the peak halliard blocks, with the fall set up to a sheerpole; often with a tackle shackled to a deck eyebolt to give additional power (figure 27a). American small craft frequently fit a topping lift permanently shackled aloft and set up through a block to belay at the forward end of the boom (figure 27b). A purchase may be added to this arrangement for additional power. A variant incorporates a boom-end tackle, and was much used in American working craft (figure 27d).

Double topping lifts are often fitted in larger craft, enabling the weather lift to be set up if the mainsail is girt across the other. They also help control the gaff when hoisting and lowering.

LAZYJACKS Lazyjacks appear to have been an American innovation designed to gather and control the sail and gaff when lowering or hoisting. Figure 28 indicates the usual arrangements which may be more elaborate.

STAYSAIL HALLIARDS The most common arrangement of staysail halliards is a single block with a single halliard rove through it and shackled to the staysail head (figure 29a). An improvement with two blocks, for sails over about 100 square feet, is shown in figure 29b with the advantage that the halliard end cannot get lost or come unrove. Other combinations of double or single blocks can be used to get a taut luff and a halliard winch is often

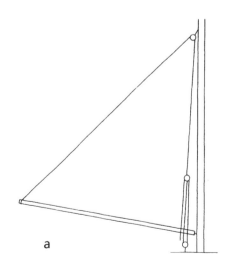

Figure 27 (a). Typical topping lift arrangement

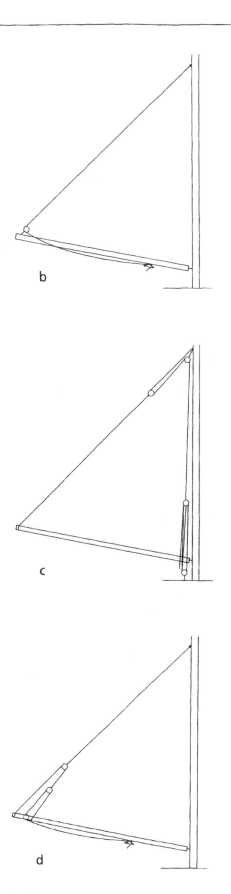

Figure 27 (b - d). Typical topping lift arrangements

used in modern craft. A tack tackle will set the luff taut and can be rigged as described for the main tack.

A staysail downhaul is always useful and can be of light synthetic line made fast to the head of the sail and led down through the hanks to a small block shackled to the forestay eye and the fall led aft. On casting off the halliard a smart pull on the downhaul whips the sail down the stay and secures it from billowing out while anchor work progresses.

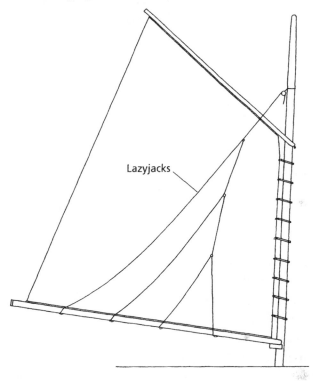

Figure 28. Lazyjacks

STAYSAIL SHEETS Single sheets are simplest but in craft with two or more headsails, the jib is usually sheeted first and the staysail needs powerful sheets to get the sheet in when full of wind. Nowadays sheet winches are commonly fitted in many yachts but some owners prefer traditional arrangements. A single purchase is often used (figure 30a). Its disadvantage is that the block on the pendant thrashes about in stays and could injure anyone on the foredeck. The alternative is to arrange the block to travel along the deck (figure 30b). The disadvantage then is that the block hammers on the deck in stays, or light weather. More powerful tackles can be arranged in this way aboard larger craft.

Headsail sheet leads should generally be positioned

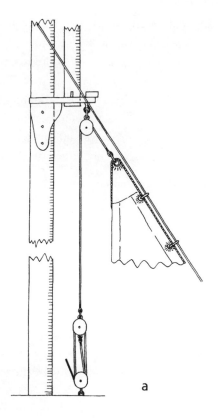

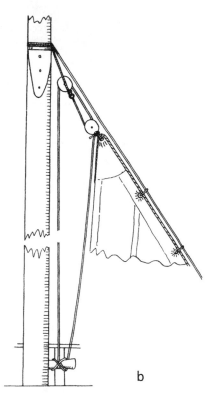

Figure 29 (a - b). Typical staysail halliard arrangements

on deck where a line drawn from $^2/_5$ up the headsail luff meets the deck if drawn down through the sail clew. This does not apply to very large light headsails where the sheeting angle will be much less. Adjustable sheet leads in the form of a bullseye or block travelling on track and

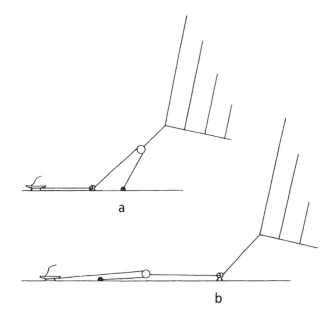

Figure 30 (a - b). Typical staysail or jib sheet arrangements

capable of being quickly secured in various positions are a great advantage, particularly if a variety of headsails are to be set. The transverse position of the lead is more difficult to determine but generally the widest possible spread of the leads is desirable in gaff rigged craft, especially the slower boats. Typical arrangements for headsails sheeted to a horse or fitted with a boom at the foot, are described in chapter 5.

JIB HALLIARDS The jib usually requires more power to hoist than the staysail, particularly if it is set flying and not hanked to a stay. A pair of single blocks are the minimum practical purchase to get a straight luff and double blocks are commonly used with sails over, say, 200 square feet.

A jib stands best when hanked to a stay, which has long been common practice in the USA. In Britain most traditionally rigged craft set the jib flying on a traveller working along the bowsprit with an outhaul and inhaul. However it is possible conveniently to combine the advantages of both methods by rigging the jib halliards

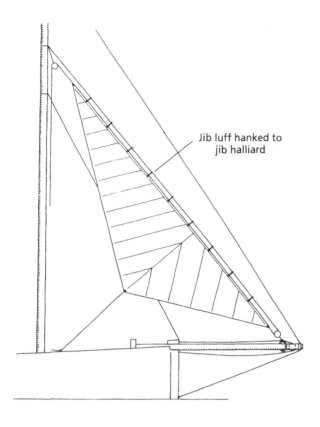

Jib luff hanked to
jib halliard

Figure 31. Combined self-tensioning jib stay and halliard

so that the luff of the jib is hanked to a stay formed by part of the wire jib halliard, which becomes self tensioning, over a block shackled to the bowsprit traveller (figure 31).

In traditionally rigged English craft with a fidded topmast the jib halliard is often a single rope rove through a single block at the jib head with the two ends led through vertical sheaves or cheek blocks on each side of the masthead encased in wooden cheeks. The 'nips' can thus be varied, one end being made fast and the other acting as the fall. The arrangement also allows the topmast to be struck down between the jib halliard parts without interfering with the halliard.

JIB SHEETS These may be arranged as detailed for staysail sheets. In small craft a single length of rope with a thimble seized in its centre often forms the jib sheets. This is bad practice as, if the seizing breaks, the sheets are instantly ineffective. Jib sheets should always be in two pieces of rope, with a thimble spliced at the end of each, well served, and shackled to the jib clew.

Many cruising yachts use the Wykeham-Martin furling gear on the jib, an indispensable gear for short-handed craft. The jib has a wire luff rope and shackles to a bronze swivel, incorporating a winding drum, at the tack. This lower swivel is shackled to the tack anchorage. A small synthetic cord passes through a guide on to the drum and is led aft to belay close by the helmsman. The jib is hoisted and its luff swigged-up taut, usually with a flexible wire halliard. The cord is wound on the drum with sufficient turns so that when the jib is rolled up or furled tightly on its luff rope, a few turns remain on the drum. When the sail is to be set the sheet is hauled aft, and a light tension maintained on the cord ensures its correct rolling on to the rotating drum. To furl the sail pull on the cord and release the sheet with slight tension to ensure a firm furl.

JIB TOPSAIL, BALLOON HEADSAIL AND MAIN TOPMAST STAYSAIL SHEETS These sails and their halliards and sheets can be rigged in various ways. They are usually single part in small craft but may need the power of a simple tackle in large vessels. As all are comparatively light weather sails, only minimum power is necessary.

BOOM GUY The danger of a boom gybing over accidentally is ever present at sea when running in a breeze and can be serious in large craft under many conditions. A boom fore guy is rigged from the boom just forward of the mainsheet attachment, to the after lee shroud, or an adjacent anchorage, and ends in a thimble which is made fast to a deck eyebolt or cleat by a few turns of light line. Properly rigged and used this guy makes a small fore and aft yacht almost as good as a square rigger, when running in a seaway.

DOWNHAUL Ensign halliards may be rove through a block at the gaff end and form a useful gaff downhaul when lowering. Alternatively a light line can be used, set up to the boom when not in use.

LUG SHEETS FOR SCHOONER FORESAILS The lug sheet foresail was frequently fitted to various types of schooners in North America and, between 1890 and 1914, to many large schooner-rigged racing yachts. A typical arrangement of the sail is shown in figure 32. The foot is boomed to just clear of the mainmast, which the clew overlaps. The boom is sheeted to a traveller on a deck horse and does not require tending in stays. The lug sheet must be handled each tack and this sheeting arrangement is not recommended as it can be dangerous with a small crew, and gains little useful area in small yachts.

Figure 32. Typical lug-sheet foresail in a schooner

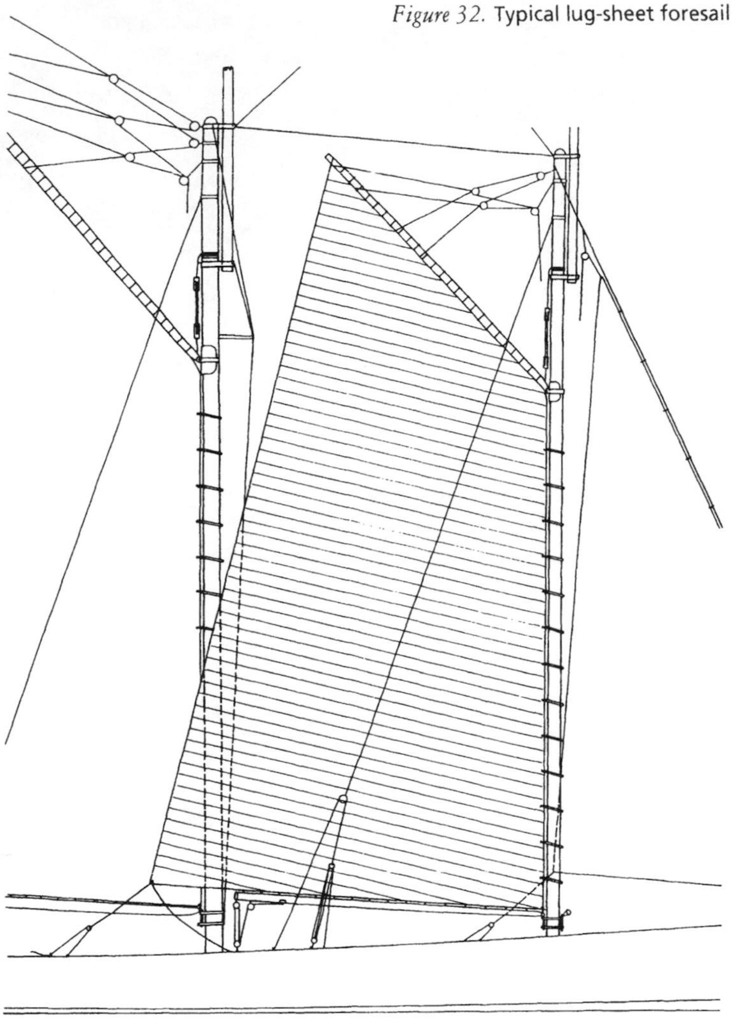

CHAPTER FOUR

GAFF SAILS AND BASIC SAILMAKING

THE BEST PROPORTIONS for a gaff mainsail to be set without a topsail are: luff 1·0, head 0·833, leech 1·73, foot 1·02. The angle of the gaff to the centreline of the mast should be 30°. The rake upwards of the boom should be about 6° depending on the height of the gooseneck, of the deckhouses, and the sweep of the sheer. If a topsail is to be carried then the angle of the gaff should be eased to about 42°. If the topsail is to be regarded as a serious part of the sail area in a new design, it needs to be of larger area than the usual fair-weather toy that most topsails have become in yachts, and the proportions of the mainsail to the topsail, and of the parts of those sails, must be specially considered. Laced footed

gaff mainsails may be cross cut, vertical cut, or diagonal cut, with a seam worked from clew to throat.

CROSS CUT OR HORIZONTAL CUT This is the best method of making a laced footed gaff mainsail (figure 33a). The cloths run across the sail at right angles to the leech which, in the case of cotton sails, results in a flatter set and better retention of shape than when the vertical cut is used. Also, since the seams run across the sail they offer little resistance to the windstream when going to windward, and eddying is reduced to a minimum, increasing efficiency. Many mainsails are now cut from synthetic sailcloth, and often wide panels are used, due to

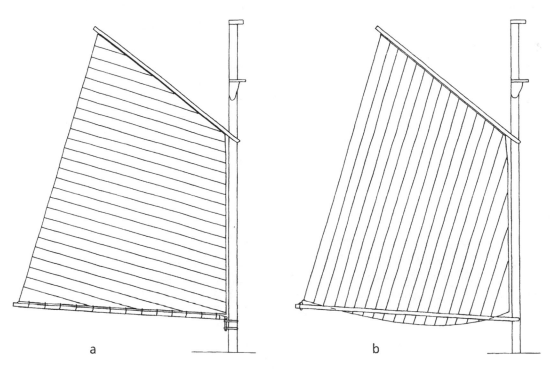

Figure 33 (a - b). Cross and vertically cut mainsails

THE GAFF RIG HANDBOOK

the very small amount of stretch of the material. Panel widths can be too great and lead to distortion of the sail, but an experienced sailmaker will restrict this. The principal disadvantage of a cross-cut canvas sail is that, if it splits, the tear will extend from luff to leech, rendering the sail completely useless, but this accident is unlikely with a sail of synthetic material, of correct weight. Another disadvantage is that, whatever type of sailcloth is used, the strain between the foot and the head of the sail is placed on the stitching of the seams. Battens are always required to get a good set to the leech and the reef points have to be fitted in the cloths, between the seams, with patches for reinforcement.

VERTICAL CUT Vertical cut is the oldest known method of making a sail (figure 33b). The cloths are fitted parallel to the leech, and stretch parallel to it, counteracting the tendency for the leech to go baggy in a canvas sail. Battens are not normally fitted in the leech of vertically cut sails and the leech cannot be rounded as it can in a cross-cut sail. Although it does not usually set so well as the cross-cut, the vertically cut is suitable for cruising yachts and working craft as it stands more rough use, and the strain between the head and foot lies within the cloths' weave, and in the strongest direction. A further advantage is the sewing of reef points in the seams which, as the cloth is doubled there, is stronger than the method used for cross-cut sails.

ATTACHMENT OF SAILS Traditionally, natural fibre cordage is used for seizing and lacing sails, but synthetic cordage of superior strength and with negligible stretch is now usual. The size of synthetic lacing can be smaller than natural fibre, saving weight and reducing wind resistance. Apart from not stretching or shrinking, it can be cast off more easily than natural fibre.

When bending a sail, the gaff should be hoisted to a convenient height and the throat of the mainsail seized to an eyebolt under the gaff, or through two eyes bored in the wood gaff jaws. The head of the mainsail may be attached to the gaff either by a continuous lacing or by robands.

The continuous lacing is now the commonest method, and it is usually spliced into the first eyelet aft of the throat, passed aft through the eyelets and around

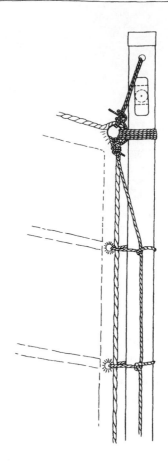

Figure 34. Lacing of mainsail head to gaff

the gaff in a series of marlin hitches (figure 34). The head of the sail is stretched along the gaff with moderate tension and the lacing end set up through a hole in the end of the gaff, or around a wood thumb or eyebolt. The main disadvantage of this method is that if the lacing parts, the whole head of the sail unroves, which could be serious in bad weather. Also the lacing requires occasional setting up due to stretching and slacking in its length.

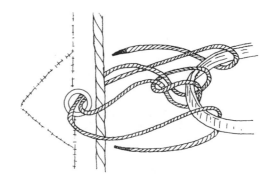

Figure 35. Lacing mast hoop to mainsail

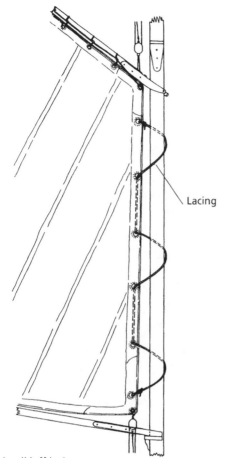

Lacing

Figure 36. Mainsail luff lacing

copper nails rivetted on rooves. Alternatively, they can be built up in glued laminates. The inner face should be well rounded and they are usually varnished to give a smooth surface. A little waterproof grease on the inner face helps working. Wood hoops can, of course, be fitted as replacements to a standing mast, and clenched in place, which is why they are sold in an unfinished state. Plastic hoops are now manufactured in the USA where they are especially used with aluminium alloy spars.

Each luff eyelet or grommet is seized to the hoop with spunyarn or synthetic cordage, allowing sufficient slackness for the hoop to lie at right angles to the luff rope. When fitting out it is important to remember that mast hoops are fitted to the mast before it is stepped, for it is surprising how often this is forgotten.

Luff lacings are used in the working craft and yachts of several nations, notably the Dutch and Chinese. An efficient lacing can be made from the more slippery types of synthetic rope which are spliced to the upper grommet of the luff and passed around the mast and through the grommets as indicated in figure 36, being secured to the lower grommet or to the boom.

The tension of a luff may be adjusted with a lacing, and the luff brought closer to the mast or slacked off, to improve the set of the sail or avoid eddies. If rigged as shown it will not jam, and it is cheaper than hoops, though not so convenient for climbing aloft.

The tack may be secured to the boom by either lacing it to an eyebolt, or an eye on a boom-end band, around a thumb on the boom or, in working craft, through two holes in the boom jaws. The lacing should be slack enough to allow the sail to lie vertically when set. The commonly-seen practice of lacing the tack round the gooseneck is bad as it could easily be chafed through. A good arrangement is to shackle the tack to the boom eye, or use a spring clip.

The foot of the sail may be laced to the boom as described for the head. A system often used in American yachts is to fit a wood batten along the top of the boom, raised on small wooden chocks (figure 37) and glued and screwed to it. The foot eyelets are laced to this either by individual robands or with a continuous lacing which has the advantage of being more easily adjusted because of the reduced friction compared with a lacing passed round the boom.

The traditional and more reliable attachment is by robands or, literally, rope bands which were generally used in commercial craft and large yachts. After seizing the throat, the head of the sail is made fast to the gaff with individual seizings through each eyelet. This is the better arrangement for large sails but the head has to be stretched out to its full extent before lacing commences, as longitudinal adjustment is not possible afterwards.

A track and slides could be used, provided the track is well fastened to the gaff and the outhaul kept taut.

The luff may be held to the mast by either hoops or a lacing. Hoops are the traditional British method and are generally used elsewhere in larger commercial craft and yachts (figure 35). They may be made of wood or plastic. Steel hoops have been fitted, but score the mast severely and should not be used. The diameter of hoops should generally be $1\frac{1}{4}$ times the diameter of the mast at the boom. Wood hoops are usually made of either Canadian rock elm, ash, or young oak, steam bent to the approximate diameter and clenched to the exact size with

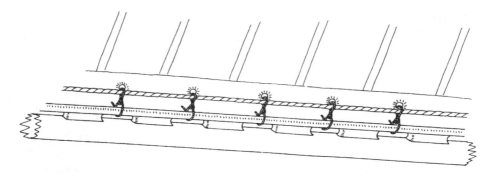

Figure 37. American mainsail foot lacing to wood batten

Strangely enough, the modern methods of track and slides, or the footrope fitted in a groove, are little used with gaff rig, though both would be practical.

The clew attachment of a gaff sail always needs frequent adjustment and, although a seizing round a wood thumb or through a hole in the boom is traditional, the use of a large slide and track with an outhaul passing over a sheave is desirable to adjust tension, especially in larger sails.

In craft whose owners are not bound by traditional appearance, the set of gaff sails can often be improved by fitting battens in the leech. These are now usually of plastic which will not break easily, and as they are fitted in slip-in pockets, give little trouble compared with the old wooden tie-in type.

Point reefing is the most common method used with the gaff rig, but roller reefing is a practical alternative with a laced-foot mainsail, especially for deep-sea work, and was often fitted in commercial craft with small crews in the later days of sail. A continuous reef lacing is still sometimes used. It was common in gaff-rigged racing yachts, but is not a convenient method for a cruising yacht. Eyelets or grommets are fitted across the sail in place of reef points and a lacing rove through them and under the foot of the sail in a series of marlin hitches.

It is possible to obtain some variation in the fullness or flatness of a laced-footed mainsail by using a roach reef. The sail is cut with excessive roach or fullness in the foot and is fitted with a row of reefing, eyelets so that a lacing can be rove through them and under the footrope and the 'reef' pulled down to flatten the sail. This method was reputedly first used in the American sloop yacht *Regina*, designed by A. C. Smith and built in 1876.

If a sail is arranged for point reefing, rows of points

or nettles as they are occasionally termed, are sewn across the mainsail at the intervals of the desired reduction of area. At the ends of each row a large cringle is worked in the luff and leech. When a reef is taken in the tack and clew cringles are pulled down to the boom with ropes known as reef pendants, also sometimes called reef earrings. The craft is first put on the wind with the mainsheet slacked, the topping lift set up, and the throat and peak halliards slacked away to suit the area of sail to be reduced. The tack pendant is pulled down first and secured against a wood thumb at the forward end of the boom. Alternatively, the pendant is fastened to the luff cringle and is passed down through the tack cringle and then aft to belay to a cleat on the boom. A good substitute for the tack pendant is to fit the same number of spring clips as there are reefs, to the forward end of the boom on short lanyards. These can be quickly clipped into the luff cringles, saving much time and avoiding a conglomeration of ropes.

The clew pendant is next pulled down. The traditional method of reeving this is for a length of rope having a wall or stopper knot in one end to be rove up through the appropriate hole in the wooden reef comb or bee block on the side of the after end of the boom (see figures 38 and 17), through the reef cringle in the leech, and down through the corresponding hole in the reef comb on the opposite side of the boom, to pass forward and belay to a cleat on the side of the boom. In small craft this gives sufficient purchase to haul down a reef, but in larger vessels a reef tackle is fitted; usually a single purchase secured to the forward end of the boom, against a thumb, and to the end of the reef pendant to haul it taut. Care must always be taken to ensure that the after end of the sail is not chafed by being nipped between the

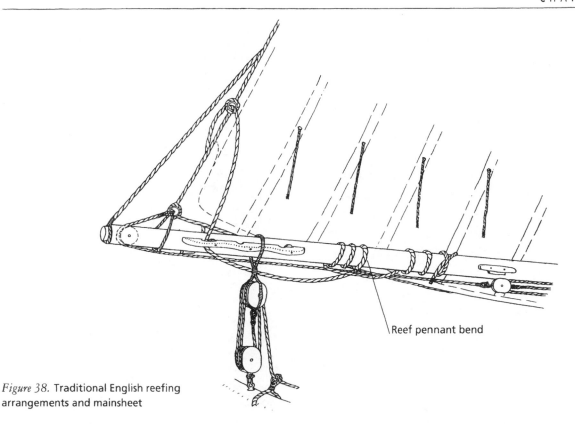

Reef pennant bend

Figure 38. Traditional English reefing arrangements and mainsheet

pendant and the boom or reef comb. This method has the advantage of bringing the reef cringle down on the centreline of the boom, which is important where large booms are fitted. Small craft may splice the reef pendants to the clew cringles and lead them down on alternate sides of the boom for each reef, through the reef combs, to belay as already described.

The pendants take considerable strain, especially at the clew, and should be of best-quality rope. Flexible synthetic rope is very good for this purpose and its slipperiness minimises the friction of this reefing arrangement. The first and second clew reef pendants should always be rove ready for use, and the others kept in a handy place as reefs are often wanted quickly.

When the pendants are secured, the reef points are tied. These are sewn to the sail and are of sufficient length to allow them to be tied together around the neatly bunched sailcloth when a reef is taken in. In craft having large mainsails, the points of the third and fourth reefs are usually considerably longer than the lower points to allow for the amount of reefed sail they secure. As each reef is taken in, it is tied over the one already made, so that when the weather improves the reefs can be shaken out gradually. The points are tied with two kinds

of knot; the reef knot or the common bow, as used for tying shoe laces. The bow knot is commonly used by the Essex fishermen and yachtsmen, but the reef knot is probably in general use throughout the sailing world. It is important to ensure that each pair of points are tied to each other and not to one of another reef. In bad weather mainsails have been torn by the foul tying of reef points. Reef points should always be tied around the footrope but *never* round the boom. The footrope will adjust to the various tensions of the points, but the boom cannot, and the sail will be distorted in a hard breeze. This is an important and seamanlike action which is now often disregarded, to the detriment of the sails.

When the reef is tucked in, the peak and throat halliards are set up and the topping lift slackened, before the craft is put on course. When shaking out a reef, the topping lift is set up and the throat halliard slacked slightly. The reef points are untied and the luff pendant cast off, the clew being let go last, the halliards adjusted and the topping lift slackened off before trimming the sheet.

If a reef lacing is used, the pendants are secured as previously described and a continuous lacing is passed through eyelets corresponding to the reef points, around the reefed part of the sail and under the footrope, in either a spiral manner, or in marlin hitches. The ends of the lacing are made fast on the standing parts.

Various types of roller-reefing gear are available and the commonest used in gaff rigged craft has a worm wheel mounted on the gooseneck, turned by a crank handle, which engages with a toothed pinion mounted on the forward end of the boom, which revolves around the gooseneck. This pattern is self-holding and can withstand considerable force applied to it in bad weather. The gooseneck must be very strong as it takes the wringing force of the sail. This type of gear is often referred to in England as Appledore reefing gear, having been used extensively in craft from that area, and in the Bristol Channel generally. However, it has been claimed, with some justification, that the original of this type was devised and made at St Helens, in the Isle of Wight. Roller gears are very useful in craft having a gaff sail of over about 500 square feet and a small crew. The sail can be reefed on almost any point of sailing, provided the weight of the boom is just taken by the topping lift. However, there are some disadvantages. A reef cannot be rolled in at anchor without the sail being set. Unless the boom is carefully designed with a large diameter after end and a suitable taper of the diameter throughout its length, the clew will often droop. The mainsheet can only be attached to the extreme after end of the boom and to a swivelling fitting. This restricts the length of the boom to close above the lower block of the sheet. It is preferable that the foot of a gaff sail to be used with roller reefing gear should have its footrope fitted in a groove in the top of the boom or alternatively, attached to a sunken track with slides.

LOOSE FOOTED GAFF MAINSAIL The loose footed mainsail must be of either vertical or diagonal cut. It may be attached to the spars as already described for the laced-footed sail, except that the foot is only attached to the boom at the clew, and the tack to a lanyard from the deck. The foot has considerable round or roach and, as it flows freely away from the boom, this type of sail, when cut by an experienced sailmaker, has more power for a given area than the laced-footed sail, even though it may appear not to set so well because of the wrinkles often seen spreading out from the clew. Its efficiency lies in the

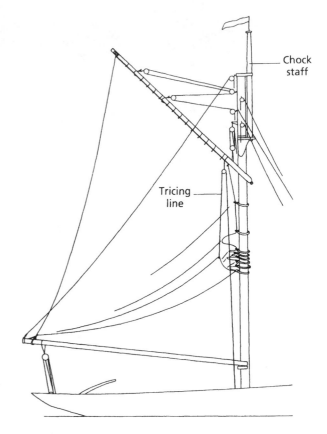

Figure 39. Tricing-up mainsail tack

correct flow which can be obtained and varied to suit wind strength by tensioning or slacking the clew outhaul (see figure 38). In a breeze the sail can be flattened by heaving on the outhaul or in light weather this can be slacked to induce a more rounded aerofoil section or as sailors say, throw more draught in the sail. Such adjustment was regularly practiced by the professional users of the loose-footed sail, especially in fishing vessels and the large racing yachts which used them until the late 1880s. A gaff rigged boat designed by the author for personal use was fitted with both types of gaff mainsail and the difference in their power was remarkably in favour of the loose foot.

A traditional feature of the loose-footed mainsail is the ability to trice up the tack to improve the helmsman's vision or quickly reduce the mainsail's power, without reefing; this was especially useful when fishing with trawls or dredges. In small craft a light tricing line is rove through a block suspended beneath the gaff jaws. This is made fast to the tack cringle and the fall is belayed to the boom (figure 39). To truss up, as the

Essexmen say, the tack line is cast off and the tricing line set up, bringing the tack up the mast to the right height, spoiling the set of the sail, and reducing, its efficiency. In large craft a simple tackle was rigged in place of the tricing line to give sufficient purchase to hoist a heavy sail. A gaff mainsail set with the tack triced up and the peak slacked away or settled or rucked down, is described as scandalised; a manoeuvre frequently used when bringing up to a mooring, lying at anchor without lowering the mainsail in quiet weather, when fishing, to quickly reduce sail in a squall, or when gybing in heavy weather. Although these practices were once widespread, tricing up is detrimental to the set of a gaff sail and it is not a recommended practice for pleasure craft.

A further advantage of the loose-footed sail is the ease of reefing. The sail can easily be rolled up to the points which do not have to be passed between the foot-rope and the boom. Of course roller reefing cannot be used with this sail.

The loose foot throws great strain on the tack and clew attachments and these should be stronger than would be required with a laced-footed mainsail. The clew is particularly important as it will require frequent adjustment. The traditional clew arrangements were a rope grommet spliced through the clew cringle and around the boom, with sufficient slack for it to slide along. A rope outhaul was also spliced into the clew cringle and passed aft over a sheave let into a slot in the after end of the boom, and led forward to belay to a cleat on the side of the boom or sometimes made fast round

the reef combs. This arrangement was common in working craft.

A superior arrangement is the use of a large slide running in a U-section track on top of the boom. The clew cringle is shackled to the slide and the outhaul rope is spliced to the cringle and led over the sheave in the boom end as before. Provided the track is adequately secured this gives excellent adjustment to the set of the mainsail.

JIBS In traditional English practice, a jib is set flying from a traveller on a bowsprit but in modern craft, they are often hanked to a stay. Jibs are not reefed and when it is necessary to change sail area, they are handed, and a different sized jib is sent up. A typical outfit of jibs for a gaff rigged cruising yacht includes a working or average sized jib, often known as the No. 1 jib, and set in ordinary weather; a smaller or No. 2 jib, for heavier breezes; and a storm or spitfire jib for heavy weather. Most yachts and working craft also carry a large jib for light weather.

The working and No. 2 jibs are now almost always of diagonal cut, (see figures 40b and 40c), the cloths below the mitre being best worked at right angles to the foot and those above it at right angles to the leech. They are cut fairly high in the clew and are now usually made of synthetic sailcloth. Theoretically, the clew of the working jib should not overlap the forestay, but an overlap is usually necessary to obtain an efficient area.

The traditional cotton or flax jib was originally of

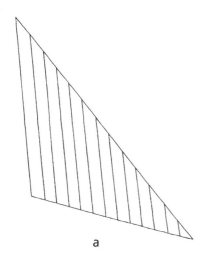
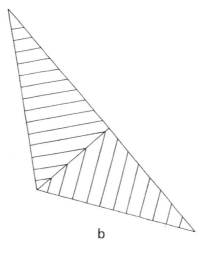
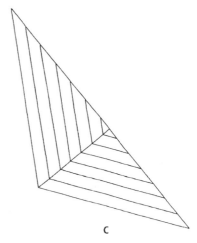

a   b   c

Figure 40. (a-c) Jib cuts

vertical cut but after about 1894 the diagonal cut became general. The luff rope was of galvanised wire, well parcelled and served, and the clew rope, head and tack tails were of Italian hemp. In those cotton or flax sails still made, the luff rope is now usually of plastic coated galvanised steel wire, or of pre-stretched synthetic rope.

The weight of jibs varies with their size and the size and type of craft setting them. As the storm jib is only used in bad weather it must be of equal weight and strength to the trysail, which is set at the same time. The traditional storm jib was diagonally cut, made of flax, cutch tanned for preservation, and hand sewn for strength. The foot and leech were roped all round and galvanised iron thimbles were often fitted. In yachts, this sail spends most of its time stowed and needs to be inspected frequently for deterioration. Nowadays, most storm jibs are made of Terylene. They are also roped all round and, one hopes, spend most of their lives in the sail locker.

Working craft found much use for their spitfire jibs, as they called them. The boats were at sea in all weathers, winter and summer, and generally feared calms more than ordinary gales. Sailing fishing vessels frequently set them for balance or steering sails, or to reduce speed when trawling or dredging in fresh winds.

The smaller jibs are usually fitted with a wire pendant at the tack to keep them clear of a sea which often split the low-cut jibs of the past, especially when pitching to windward in bad weather.

FORESTAYSAIL, STAYSAIL, OR FORESAIL The forestaysail, to use its full name, is hanked to the forestay and sets between it and the mast. In craft with a jib and staysail the forestay is naturally at a steeper angle, but in sloops with a single headsail, especially those with bowsprits, the forestay has a slacker rake and the forestaysail becomes a very large sail, often cut low in the foot and difficult to sheet in a hard breeze. The true terms of the sea and seamanship have suffered much recently from the sloppy attitude of some yachting journalists and sailing people. The single headsail of sloop rigged craft is now often wrongly referred to as the jib.

The forestaysail is one of the most powerful sails in the fore and aft rig, and is best cut diagonally from sailcloth of equivalent weight to the mainsail. Present practice is for the luff rope to be of plastic coated wire or pre-stretched synthetic rope; the sail is roped at the clew and has rope tails at the head and tack. In cutters the staysail is usually cut flat to set in the draught of the jib, but in sloops they are sometimes very full, almost like the foresail of a spritsail barge, with the luff bellying out at a considerable angle to the forestay, yet the sail pulls like a horse, if cut by a knowing hand. Generally, a well cut staysail sets in a gentle aerofoil curve.

The head and leech of a staysail should not be too close to the mast or it will tend to backwind the mainsail. In modern bermudian-rigged craft with most of their sail area in overlapping headsails, this does not much matter, but in a short-handed cruising yacht under working sails it can seriously injure windward performance. Short battens are sometimes fitted in the upper part of the leech of staysails which have to be perpendicular. This generally overcomes the fluttering which would otherwise occur.

WORKING STAYSAILS In cruising yachts the working staysail is best cut reasonably high in the foot to reduce the pressing effect, and preferably trimmed with a pair of sheets to obtain a good set. A row of reef points is often fitted, corresponding in depth to the mainsail's first reef. In traditional craft the working staysail is cut with cloths parallel to the leech and it is often sheeted to an iron horse across the deck, just forward of the mast, the leech being almost perpendicular. This makes it an awkward sail to set well as the leech is pulled out of shape with the sheet pulling vertically on it and distorting the foot. A single row of reef points is usually fitted.

Staysails were originally attached to the stay by rope grommets spliced in place, the sails being left stowed on the stay. Lacings may also have been used. Sister hooks were used early in the 19th century and were supposed to have been devised by James Husk, the Wivenhoe, Essex, yachtbuilder. They were also known as cliphooks or calliper hooks, but frequently required mousing to keep them closed.

The piston hank, today's universal attachment, was supposed to have been invented by Captain John Into, a noted American yacht-racing personality who devised them for use in racing yachts in 1905.

The tack of a staysail is often set on a short pendant

which shackles to the deck or to the stemhead, allowing the foot of the sail to be clear above water coming aboard. In large craft, tack tackles were sometimes fitted to bowse the luff taut. The sheeting arrangements for staysails which are not fitted with a boom are generally as already described for jibs.

If the staysail is to be sheeted to a horse across the foredeck, this should be of robust construction, carried as far outboard as possible and bolted through a beam or reinforcing chocks under the deck, and fitted with deck collars, crushing washers and traveller stops in the form of circular washers welded to the horse at the outboard ends to prevent the traveller jamming. Bronze, galvanised mild steel, iron (if you can get it) or stainless steel are the best materials. The traveller should be strong and of ample diameter or a large shackle may be used. A simple tackle is usually used as the sheet, shackled to the traveller and clew. Plastic blocks are ideal, with synthetic rope parts secured on themselves when adjusted. It is best for the horse to have a pronounced curve forward in plan view, to slack the sheet as the craft goes through stays. An arrangement used by the Dutch is shown in figure 41. If a track is used instead, this should be of heavy, double fastened, internal type with a slide of substantial size and loose fit. Bronze track is preferable but aluminium alloy will work provided it is well lubricated.

Staysails are sometimes fitted with a boom which is sheeted to a horse or track across the foredeck. The staysail is either laced to the boom along its foot, secured to it at tack and clew, or at the clew only. The arrangement is usually fitted to save handling sheets when tacking but is of doubtful value when the yacht is sailing on rapidly varying courses as it is necessary to go forward to tend the sheet, and at such times the staysail is often left inefficiently set.

Lacing the foot to the boom through eyelets along the foot is a bad arrangement, though it was common in large American sloop yachts until about 95 years ago. Securing the sail to the boom at the tack and clew is little better as its natural curve is spoiled and the leech makes eddies, however far the sheet can be got to leeward. With both these arrangements it is usually impossible to get the staysail fully down at the luff, as the boom is usually fitted to pivot on the forestay.

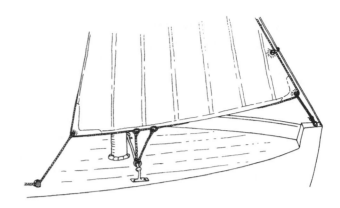

Figure 41. Dutch overlapping staysail on horse

The best arrangement for a boomed staysail is to fit the boom to a large slide working in a short, heavy track on the foredeck or the bowsprit, the forward end of the boom being fitted with a gooseneck, and the after end attached to the clew of the sail with the sheet working on a horse or track. This allows the sail to take its natural curve, and to lower completely as the boom end rides forward in the track and relieves the tension across the sail's belly. This refinement is not often seen in England but is common in the USA and elsewhere.

Small or storm staysails are seldom set by English cutters but large light weather staysails were common in commercial craft, as they were and are in yachts. They were generally referred to as balloon staysails or reaching foresails and, although now seldom seen, are well worth having in any craft. The sail is set on the forestay in the same way as the working foresail but for sailing off the wind the sheets, which are single parts, are led outside the shrouds. The critical dimension of these sails is the length of the foot, and the most efficient length for an average cruising yacht is $1\frac{1}{2}$ times the distance between the mast and the forestay at deck. They are very powerful sails off the wind and, cut with a slightly hollow leech, for windward sailing also, the sheets are led *inside* the shrouds. They are best diagonally cut, and the construction is as already described for the working staysail, the weight depending on the size and the material. In many ways these sails anticipated the genoa jib, now universal in yachting. Fishing smacks carried large reaching foresails for working in light weather since the early part of the 19th century, and probably earlier, and many used them to windward. An early recorded use of a very long-footed staysail for windward work in a racing yacht was in 1885 when Captain John Carter of Rowhedge, Essex, had one specially made for the racing cutter *Genesta*, with which he raced for the America's Cup.

JIB TOPSAIL The jib topsail is a triangular or jib-shaped sail of light cloth set from a single part halliard at the topmast head or masthead, with its luff hanked to the topmast forestay and the leech parallel to the masthead. The tack is made fast to the bowsprit end or stemhead by a tack pendant of galvanised steel wire, preferably plastic coated, or alternatively prestretched synthetic rope. A light downhaul should be fitted, running from the jib topsail head, through the hanks, to the bowsprit end, where it can either be made fast or rove through a small block and led aft, to the foredeck. The downhaul is important, especially in short-handed craft where the least possible amount of bowsprit-end work is desirable. The sheeting arrangements may be complicated by the spread of the shrouds, especially the topmast shrouds. Often it is necessary to lead the sheet through a bullseye fitted on a short lizard or lanyard, seized to a shroud at a height determined by trial, by which a fair lead can be obtained. This sail should be sheeted as hard as possible, and in large craft it is sometimes necessary to put a purchase on it, or lead it to a winch. The jib topsail is a light weather sail which will stand to windward if the wind is true and the sail is correctly sheeted. In a gaff cutter it is usually set to balance the area of the topsail. Being set so high it is a very powerful sail, raking wind out of the sky, and in the large racing yachts of the past it was nicknamed the topmast breaker. In those days the jib topsail sheets were handled by the cook and stewards, and though they were also good sailors they usually had a hard job to get the sail in. In small yachts it is something of an affectation for light days, but adds character and fun on a calm day, and adds appreciably to speed on passage — but watch the masthead.

Generally two sizes of jib topsail were set in the large racing and cruising yachts: the short luffed baby and the true jib topsail which had a luff about half the length of the topmast forestay. However, a larger variant was known as the yankee jib topsail believed, as its name implies, to have originated in the USA. This is a much larger and even more powerful sail, known in the larger yachts as the long roper because of its lengthy luff, which frequently stretched from the bowsprit end to the topmast head. The clew was cut high, and overlapped both the jib and staysail. It was also carried in light breezes and, even then, the cook and stewards had to put some beef into its sheeting. A sail of this type is worth having in today's small craft as it can be carried to windward in light breezes, or over the staysail with the staysail stowed but a topsail set.

SPINNAKERS Spinnakers have been used on board gaff rigged working craft and yachts for about a century and a half and, provided they are of suitable type and moderate size, are useful sails for cruising.

The now commonplace parachute spinnaker is a most efficient sail but is only useful where it can be set clear ahead of the forestays and, as few would now wish to handle it from a bowsprit, its use in gaff rigged craft is restricted, especially as it requires a large crew to set and trim it. The most practical spinnaker for small and moderate sized gaff rigged yachts is the triangular shaped type as used in the working craft and racing yachts before 1914. This sail is set from a halliard at the hounds, or at the masthead in pole-masted rig, and is boomed out on the opposite side to the mainsail, when running. It is triangular in shape and vertically cut of light sailcloth, lightly roped at head, clew and tack, and fitted with a headboard. The foot is roped.

However, a sail of this type can often be sheeted in on a reach, or even a close fetch, when it is used as a balloon jib. The Essex smacks and other working craft carried such a sail which was known as the bowsprit spinnaker. This could be set as described and was often boomed out when set at the bowsprit end. A sail of this pattern is most useful in light weather, but should be cut fairly flat, might require roping or taping on luff and clew, and would certainly need ample reinforcement at the tack and clew.

There is a wide variation in spinnaker handling arrangements for gaff rigged craft. Wood or aluminium alloy tube are the obvious alternatives for the boom, but its attachment to the mast is more difficult than in bermudian rig, because of the mast hoops, luff lacing, and gaff jaws working up and down. The traditional method is to fit a pair of wooden jaws or crutch to bear against the mast at a height to suit the set of the sail. A better arrangement is to fit a gooseneck cup on the fore side of the mast and this can usually be fitted on the same mastband as the main gooseneck. The inboard end of the spinnaker boom is tapered to fit into this conical cup and the tension on the guys helps to keep it in place, but a pin is usually provided to make sure. If a mainsail luff lacing is fitted it might be possible to fit a track on the foreside of the mast and use a modern-pattern spinnaker fitting. On long runs a kicking strap aids control of the boom and could be used with advantage as it helps minimise chafe and wear, apart from improving the set and efficiency of the sail.

In its most primitive form the spinnaker was, and sometimes still is, a large headsail with its clew boomed out with a long boathook or sweep. Regulations against this practice were laid down in the racing rules of the Colne and Blackwater smack races held during the 1780s, and it was obviously in common use among working craft for many years before that. Prior to the 1860s, commercial craft and large racing yachts set light squaresails when running. These had long yards and were hoisted by a halliard at the hounds and set in addition to the staysail and jib. In 1852 the small yacht *Leo*, owned by McMullen, the noted Victorian amateur cruising yachtsman, set a triangular spinnaker from the topmast head, boomed out to weather in addition to his ordinary canvas and, no doubt, similar sails were set by many other craft years before the supposed invention of the spinnaker in 1866. What is nearer the truth is that a triangular running sail set from the topmast head and boomed out was first set in yacht racing, in 1865, by the cutter *Niobe*, owned by William Gordon, and during that season it was called a Niobe. The following year the yacht *Sphinx* came out with a slightly improved version of the sail which, as her south coast crew called her the Spinks, was referred to as the Spinker. From that the term was, apparently, gradually corrupted to Spinnaker which is the term still in use.

In 1874 John Harvey, the noted Wivenhoe yacht builder, and his partner Mr Pryer, patented the 'Shadow sail' as an 'improvement' on the spinnaker. This complicated but powerful gear was intended to be set when a yacht was running under all ordinary canvas. It had a gaff which was sent aloft on halliards and socketed into a gooseneck at the masthead, and a boom similarly socketed at the deck and set up with a topping lift. Then the 'Shadow sail' was sent out to the peak and throat of the gaff and its clew hauled out to the boom end. The yacht was now, in effect, running under two mainsails of equal area, but this was not enough for Harvey. A jib-headed topsail was then sent up to set above the gaff, which was restrained by an after guy and a fore guy. The first and only yacht fitted with this curious rig was the large racing schooner *Sea Belle*, designed and built by Harvey and commanded by Captain Wadley of Wivenhoe. She set it once when racing, and after that hair-raising experience the captain ordered it to be returned to the yard.

A watersail is nowadays something of an affectation but it was widely used in various craft until a century ago, and in some working craft early in this century. Some cruising yachts still set them as a bonnet or addition to the mainsail. The sail is rectangular, the length being that of the boom, and the depth is usually not more than two to three feet in small craft. It should be made of light sailcloth and has a row of eyelets at the upper edge, with tack and clew cringles for setting and sheeting with light line. It is usually set laced to the boom but, in some cases, it is laced to a row of eyelets across the lower part of the mainsail. A spare jib can be used as a watersail .

TRYSAIL The trysail is principally used when there is too much wind for a fully reefed mainsail. It is also a substitute if the mainsail is seriously damaged and in the past it was often used as a passage making sail in racing yachts or a bad weather sail in fishing vessels. For clawing to windward in very bad weather the gaff trysail sets better, and is a better driving sail than the jib-headed trysail now often fitted to yachts of all rigs.

The term trysail originates from the old term to lie a-try, or lie to the storm wind under close sail making little if any leeway or headway. The traditional form is the gaff trysail which used to be made of very heavy hemp or flax canvas but is now more often of synthetic sailcloth, generally cut vertically but sometimes with a diagonal seam. This quadrilateral sail is laced to a gaff but has no boom (figure 42). A special trysail gaff of

Figure 42. Gaff trysail: (a) trysail gaff; (b) parrels to mast with toggles; (c) trysail sheets; (d) eyebolts; (c) mainsail stowed; (f) low crutch; (g) tack tackle; (h) topmast housed

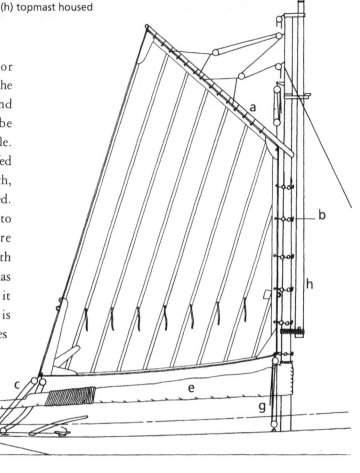

suitable length is often carried for the sail or alternatively, the mainsail is unbent from the gaff and the trysail set from it, but this is naturally a long and difficult job in bad weather at sea, and could be disastrous, so a separate gaff should be carried if possible. The trysail area should be about that of the close reefed mainsail, but one deep reef should be fitted. The leech, foot and luff are roped and all parts strongly reinforced. Cringles are worked in the luff, which may be laced to the mast, preferably by synthetic rope or, more traditionally, secured to it by wire or hemp toggles with ash parrel beads, rove through the cringles. The head has robands or stops rove through each eyelet ready to lace it to the gaff if a special gaff is not carried. The clew is specially reinforced to take the strain of the sheet tackles and, in large trysails, a relieving cringle is often worked in the leech, about 10 inches above the reef cringle. The sheets are shackled into this while the reef is being taken in. The trysail is set after the mainsail is handed and stowed with boom lashed to the quarter. The peak and throat halliards are cast off from the main gaff and shackled to the trysail gaff. The trysail jaws are secured to the mast by a parrel line with ash balls, and the luff is toggled to the mast. The trysail sheets are in the form of two tackles rove from the clew to strong deck eyebolts in the quarters. These sheets should utilise the strongest blocks and synthetic rope. A simple purchase of two single-blocks is sufficient for small craft but greater power will be required in larger vessels. The two tackles are required to attempt to control the boomless sail and their falls should be belayed to strong cleats, cavils or posts, as they take tremendous strain in bad weather. Care should be taken to have a clear lead from the clew to the quarter blocks, and if the sheets chafe on the stowed mainsail or its spars, spare canvas or other anti-chafing material should be frapped round them. The trysail is set in conjunction with the storm jib and, if lying-to, the helm is usually lashed slightly to weather.

SAILMAKING

The story of sails and sailmaking is very interesting but too long and involved for a detailed study. However, some understanding of the sailmaker's art can be gathered by considering the traditional manner of making a gaff mainsail.

To make a typical vertically-cut gaff mainsail a sailmaker requires the exact dimensions of the spars and is usually told the type of cut and weight of sailcloth desired, though both should be agreed with him by the owner. The principal dimensions required to make a mainsail are from the throat eyebolt on the gaff to the boom at gooseneck, or jaws eyebolt; from the throat eyebolt on the gaff to the peak lacing hole or thumb; from the boom gooseneck or jaws eyebolt to the centre of the clew sheave-pin at the boom; the diagonal measurement from the throat halliard eyebolt to the

Figure 43. Parts of a gaff loose-footed mainsail: (A) head; (B) luff; (C) foot; (D) leech; (e) peak piece; (f) throat piece; (g) tack piece; (h) clew piece; (i) reef patch or fly piece; (j) reef cringle fly piece; (k) head rope; (l) luff rope; (m) tack tail; (n) clew tail; (o) clew rope; (p) head tail; (q) head thimble (r) throat thimble; (s) tack cringle; (t) clew cringle; (u) reef cringles; (v) reef points; (w) head lacing eyes or grommets; (x) head tabling; (y) luff tabling; (z) foot tabling; (za) seams; (zb) luff grommets or eyes; (zc) leech tabling

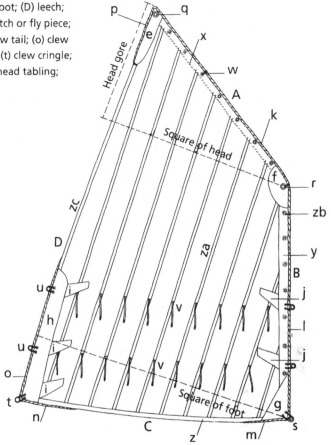

centre of the clew sheave-pin, on top of the boom, to determine the height of the boom at the sail clew; the dimensions of the leech from the centre of the clew sheave-pin on top of the boom to the peak lacing hole in the gaff, in a straight line; the diagonal from the peak lacing hole to the boom gooseneck or jaws eyebolt, which fixes the angle of peak of the gaff.

Often a plan supplied to a sailmaker will be redrawn by him, and probably altered to suit his own experience and practice, particularly regarding the angle of peak and height of clew. Precise stretch allowances cannot be given by the designer or owner as these depend on the type, quality and weight of the sailcloth used, and on the cut and size of the sail. These allowances are generally impossible to calculate and are the result of experience which is part of a sailmaker's knowledge.

The sailmaker sets out the squares of the head and foot on the sail plan by drawing lines across the mainsail at right angles to the leech to throat and tack as an aid in computing the gores of head and foot (see figures 43 and 44). The gore is the total of the gores of each cloth plus the overlap of the seams. When set out on the plan the

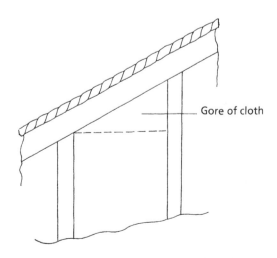

Gore of cloth

Figure 44. Gore of sail cloth

number of cloths and amount of material required can be calculated.

Work can now be commenced by marking the sail's final dimensions and shape on the loft floor, full size. The bolts of sailcloth are unrolled on the floor and the quality of the weave is examined before work commences, particularly of flax or cotton canvas which should have an even selvage, free from tightness. Methods of working vary, some sailmakers commencing at the tack and others at the leech, which method is now described. The clew gore is cut first, making allowance for leech and foot tablings. Next the leech cloth is cut, with allowance for leech and head tablings. The succeeding cloths are each cut from the preceding one, after allowing for the overlap or eating in of the seam in each cloth, until the tack is reached. The amount of flow in a mainsail will vary with type but for a loose-footed gaff mainsail the forward $1/3$ of the sail width is the most rounded. The amount of round or flow is achieved by cutting the cloths of the after part longer than those in its forward end, the allowance varying with material and the fullness or flatness of the sail, but typically might vary from $\frac{3}{8}$in in the forward cloth to possibly 3in in the after one, for a flax or cotton sail. This extra length is worked evenly in the seams and the practice is generally termed adding in the slack.

After the cloths are cut, the seams are marked out, broad seams being necessary at the largest gore, such as near the tack in a loose-foot, vertically cut sail. The use of broadseaming to counteract excessive stretch in the cloth is dependent on experience and may be used.

When the cloths have been laid on the loft floor and the seams marked, they are sewn together in pairs, then in fours and then eights, until the whole mainsail is stitched together. Machine sewing is now almost universal but some sails are still specially hand-stitched, at greater expense, but they are stronger than machine-sewn sails due to the smaller needle used, with a thicker double twine while the machine uses a single twine to make its lock-stitch. Sailmaking stitching is reckoned at the number of stitches per 'quarter' of 9in, or a quarter of a yard. The number can vary depending on the weight of sailcloth used. Flat seams are now always used for yacht sails but in the past a round seam was sometimes used for flax sails, but never for cotton, which would not flatten or rub-out after sewing.

When sewn, the sail is spread and the tablings turned. The luff, head, leech and foot tablings of cross-cut sails are cut off and placed on top or on the starboard side of the sail so that the threads of flax or cotton canvas lie in the same direction as those of the sail and these are then stitched. In vertically cut sails the leech tabling is turned over. The sail corners are reinforced; at the clew a sheet hand and clew piece are stitched on; a peak piece at the peak; a throat piece at the throat; and a tack band and tack piece at the tack. Reef patches, also termed fly pieces, are sewn to the sail as reinforcement in way of the reef cringles at leech and luff. Then the grommets formed from several yarns from a strand of boltrope are worked into the head of the sail for the head lacing, and into the luff for hoops or luff lacing. Similar grommets are also made for the throat, tack, clew and peak.

Traditionally, a hemp boltrope dressed with Stockholm tar was fitted to cotton or flax sails but lightly tarred Italian hemp rope was usually fitted to yacht sails. Terylene or other synthetic rope is now commonly used with synthetic sails. The boltrope is sewn on the port side of the sail, the stitches passing between the strands. In heavy commercial sails the boltrope was sometimes fitted all round the sail, but in yacht work and most smaller working craft it is usual to rope only the head and luff, carrying the roping a short distance down the leech from the peak in a head tail, and along the foot from the tack in a tack tail. The leech roping is fitted from above the upper reef cringle, round the clew, to finish with a tail, on the foot. The boltrope tails are tapered for appearance by gradually reducing the number of strands until the rope is pointed, with no ragged ends visible. A whipping is put on the boltrope, leaving sufficient length for the desired taper. The strands are unlaid and the yarns unravelled and thinned out with a knife until the taper is achieved. Natural-fibre ropes are treated with beeswax before the yarns are rolled up, the strands relaid, and roped to the sail.

The tack, clew and reef cringles are then worked. Each is made from a single strand of bolt rope, formed and stretched over a fid and the thimbles fitted in them. Head and throat thimbles are worked and are usually fitted of gunmetal, brass, galvanised steel or now often of plastic, usually nylon.

Finally, the reef eyelets and points are fitted. In a vertically cut sail the eyelets are pierced through the double cloth at the seams, in a line parallel to the foot of the sail and at a level measured from the bottom eyelet of the leech and luff cringles, so that the maximum strain of the sail when reefed falls on the cringles and not on the reef points. The points are passed through small grommets in the seams and are either sewn to the seam below the grommet, or a crowsfoot is worked in the point and each part sewn separately to the seam. The sail is then completed, ready for delivery. The parts of a jib are shown in figure 45.

The introduction of synthetic sailcloth during the past 40 years has led to much change in sailmaking methods and, in many instances of gaff mainsails, an apparent lack of understanding of the materials, judging by the number of badly puckered sails to be seen. Paradoxically, the best makers of synthetic cloth gaff sails are probably those who have experience of cutting synthetic racing sails. Such men accept and have wide knowledge of materials and are accustomed to thinking

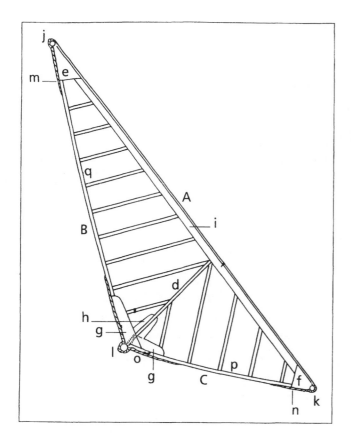

from basic principles, and have few prejudices. The result is usually an excellent gaff sail. Terylene is now the popular British synthetic sailcloth, and Dacron its American equivalent. Terylene is manufactured in various qualities but the one favoured for many British gaff rigged craft has a red-brown hue and a stiff and shiny surface. After weaving, this type of cloth is finished by passing between hot rollers and being coated with a resin which fills the weave and improves its wind-holding efficiency, although it makes such sails very slippery and difficult to stow without creasing. Some new forms of Terylene (polyester) sailcloth are not finished in this way and retain the supple feel of cotton. It is much used for racing sails and seems preferable where ease of handling is required and cost is not the prime consideration.

Figure 45. Parts of a jib: (A) luff; (B) leech; (C) foot; (d) last or diagonal; (e) head piece; (f) tack piece; (g) clew pieces; (h) tongue; (i) luff or stay tabling; (j) head thimble; (k) tack thimble; (l) clew cringle; (m) head tail; (n) tack tail; (o) clew rope; (p) foot tabling; (q) leech tabling

CHAPTER FIVE

GAFF TOPSAILS

A TOPSAIL IS USEFUL in a gaff rigged craft to increase the sail area aloft, particularly in light airs. It also extends the luff of the mainsail when sailing close to the wind. Provided the topsail is regarded as a working sail and is arranged accordingly, it is worth having; otherwise it becomes a toy, and possibly a nuisance. In cruising yachts it could be regarded as the first reef; though in many working craft, and some yachts, a jib-headed topsail was often set above a reefed mainsail.

Topsails should be cut to set flat, with a diagonal seam from the clew, and if made of Terylene or Dacron, will avoid stretch. The luff roping, halliard, sheet and tackline should be of pre-stretched Terylene.

Before setting a topsail the mainsail must, of course, be set and if a topmast is fitted it should be well bowsed forward at the head as it will pull upright with the topsail set. The weight of the boom should be taken on the topping lift while the topsail is being set. The topsail sheet is attached to the clew by a topsail sheet bend (figure 46) and when hoisting or lowering a topsail a strain should be kept on the sheet to prevent it taking a turn round the gaff end, but it should not be fully set up until the halliard is belayed. The tack of a topsail should be set to windward of the gaff and its peak halliards, to suit the longest courses to be sailed on that tack. This relieves the strain on the topmast head when running in fresh winds and prevents the topsail bellying out and bending the topmast, even if it is supported by backstays. Generally, topsails should be set and taken in to windward to assist control of the sail, though large topsails were often

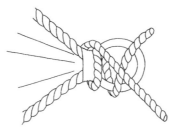

Figure 46. Topsail sheet bend

lowered to leeward in some craft, depending on the weight of wind and the skipper's preference.

The jib-headed or triangular topsail is the simplest to handle and sets best if the head and foot are of equal length, though this is not always possible. Thimbles and reinforcements are fitted at the head, clew and tack. The single part halliard reeves through a sheave in the topmast head or, less commonly, through a block at the topmast head, and is shackled or bent to the sail's head, the fall belaying at the foot of the mast. The topsail sheet is shackled or bent to the clew and leads through a block at the gaff end either shackled to a band or, more commonly, to a strop fitted on a thumb. In some craft cheek blocks are fitted at the gaff end for the topsail sheet, but this usually causes considerable chafe. The sheet leads down and forward from the gaff-end block, through a single-lead block or jewel block, hanging on a pendant from the throat of the gaff, and the fall is belayed to the forward end of the boom or, in large craft, to a deck eyebolt just abaft the mast. A very large topsail requires purchases on the halliard, sheet, and tack but this is unnecessary in small craft, except that the tackline may be set up with a purchase to get a taut luff. The head is set as close up to the sheave as possible, and the clew should be cut to allow a short drift of sheet between it and the gaff-end block, to haul the sail flat. The tack should be cut to set below the forward end of the gaff as this helps to set the sail flat on at least one tack. The luff of a jib-headed topsail is cut parallel with the topmast from the head to the masthead cap and is cut away from the

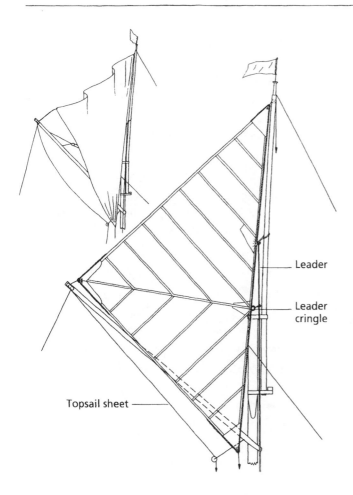

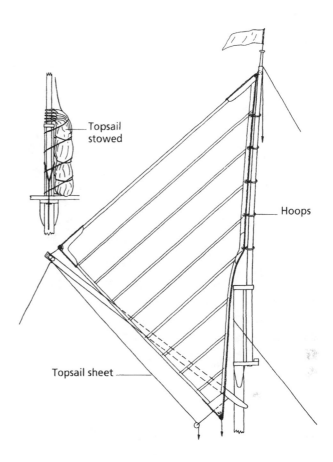

Figure 47 (a). Jib-headed topsail with luff cringle on leader or jackstay *Inset.* topsail luff set up before sheeting

Figure 47 (b). Jib-headed topsail hooped to topmast *Inset.* hooped topsail stowed aloft to masthead

mast below this to avoid chafe on the mainsail halliard and topping lift blocks. The luff may be unattached to the topmast, which results in a badly setting and inefficient sail, except when running. The luff is usually connected to the topmast by hoops, a lacing, a leader or jackstay (figure 47a).

If the luff is hooped to the topmast the arrangement of the halliard, sheet and tackline will be as previously described. The hoops may be of wood, steel or plastic, having a diameter of about $1\frac{1}{2}$ times that of the topmast at the mast cap, and should be seized to eyelets in the topsail luff with rather slack seizings to allow the sail to set smartly. If hoops are fitted the topsail must be stowed aloft, requiring going up the shrouds to gather, furl and stow the topsail to the masthead; an unpleasant task in a breeze or at night. Although some larger working craft, particularly in America, carried hooped topsails, the method is not recommended for small craft (figure 47b).

Topsail luff lacings are now unfashionable but were once commonly used in racing yachts and many other craft to set the topsail luff smartly. Their use requires work aloft when setting and taking in the topsail. The halliard, sheet and tackline are rigged as already described and, when the sail is hoisted the luff is laced to the topmast commencing at the head and working downwards with a strong line passed around the topmast and through cringles on the luff (figure 48a). A well-cut jib-headed topsail, properly laced, will stand as well as the mainsail and can be set over a reefed mainsail. Alternatively, strops and toggles could be fitted instead of a lacing, as described for trysail luffs.

Rope topsail leaders were commonly used in fishing and coasting vessels and were formed by splicing a large eye around the topmast and leading the fall to the deck. The diameter of the eye should be about $1\frac{1}{2}$ times the topmast diameter. The fall should be long enough to allow the eye to rise up the topmast to the height of the thimble spliced in the topsail luff, through which the leader is rove before hoisting the topsail. As the sail goes up it carries the leader with it and, with the halliard and tackline belayed, the leader is set up to bring the topsail luff in to the mast. The leader may be of fibre or wire rope or, better still, of pre-stretched synthetic rope. A single leader is sufficient for small craft but larger vessels may have two or more fitted. Leaders also help to control a topsail when lowering.

The jackstay can be used where an eye is spliced to fit the topmast diameter opposite the luff thimble, which is threaded on it as previously described and the jackstay fall set up, often to the shrouds to keep it clear of the mast hoops. When the topsail is hoisted the luff thimble is held close to the mast immediately under the eye. Or it is possible to fit a wire jackstay from the topmast head to the deck close abaft the mast, to which the topsail luff is attached by hanks. The stay can be set up by a rigging screw.

A variant of this which was uncommon in English craft but which has been used successfully by the author in cutters up to 17 tons is to fit a pre-stretched synthetic

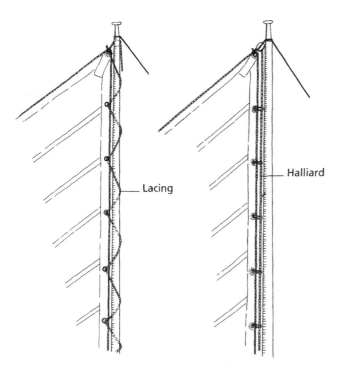

Figure 48 (a). Topsail luff lacing
(b). Self-tensioning topsail luff set on halliard

halliard with its fall threaded through large piston hanks seized at right angles to the topsail luff, about 2ft 6in apart. The sail is hoisted with its luff on the halliard, which is belayed close to the mast, after which the topsail sheet and tack are set up. When the topsail fills with wind the sagging strain on its luff is resisted by the halliard tensioning itself, and the rig will stand quite close to the wind (figure 48b). Large jib-headed topsails were often send up in stops to assist handling.

There have been attempts to evolve jib-headed topsails which set on mast track and the best solution was devised by R.A. Pinkney in association with W. Martineau for use aboard Mr Pinkney's elegant gaff cutter *Dyarchy*. The length of *Dyarchy's* topsail luff equals that of the mainsail, making a snug rig with the topsail handed. The mast has a long, pear-sectioned masthead forming a topmast which has a groove on its after side to take the large topsail luff rope. A small traveller is permanently inside the groove, through which the wire topsail halliard runs freely. The halliard runs over a sheave in the topmast head and is made fast to the topsail by a threaded fitting which makes it a continuation of the luff rope. When the halliard is hauled, the traveller remains at the bottom of the groove, keeping the halliard in the right place until the head of the sail enters the bottom of the groove. The head of the sail then picks up the traveller as it is hoisted and when it has entered the groove it cannot jam or come out, is under control when hoisting and lowering, and sets on the wind like a bermudian sail. Provided the wind is forward of the beam the sail can be set easily, but with the wind aft it may be necessary to luff or for a hand to go aloft to start the lead of the sail into the groove. This topsail can be hoisted on any point of sailing and Mr Pinkney has written of his satisfaction with it.

Yard topsails were common in working craft and yachts of all types until forty years ago, and the recent revival of gaff rig has brought them out again. The yard topsail passed through similar transition to the mainsail from a horizontal rectangle to an almost vertical triangular sail. Early 19th century topsail yards were almost horizontal but gradually increased in angle of peak until, towards its end, they became almost vertical extensions of the topmast. The old-fashioned square-headed yard topsail aids speed running or reaching, but is

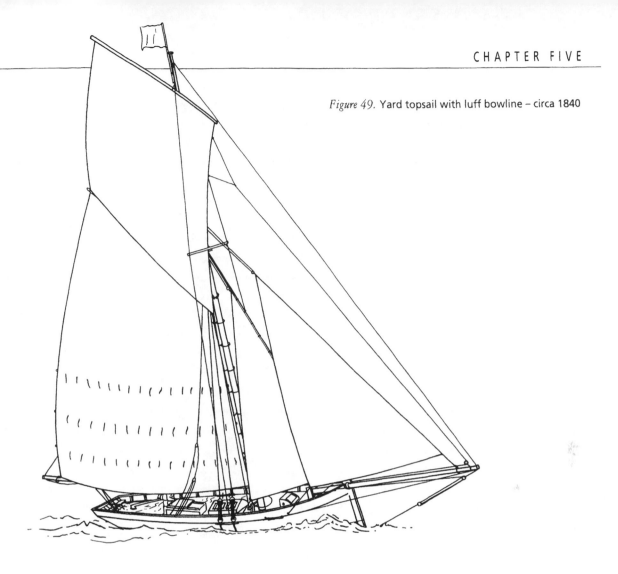

Figure 49. Yard topsail with luff bowline – circa 1840

useless closehauled as it shakes badly and is better handed before a long beat to windward. This type of topsail requires a very heavy and stiff yard to set it correctly and it adds weight aloft, but has the minor advantage that, if the topsail is stopped to the yard before hoisting, it can be broken out after hoisting the yard to the topmast head. However, these sails are usually difficult to take in and should not be carried at night.

A clewline is a useful aid when hoisting or lowering. The line is bent to the heel of the topsail yard and reeves through the topsail clew thimble which should be made purposely large and preferably of nylon to reduce friction, and back on the other side of the topsail to reeve through a small block seized to the heel of the topsail yard, with the fall lead to the deck. When lowering the sail the topsail sheet is slacked and the clewline hauled on, clewing the topsail up to the yard and spilling the wind from it, which greatly assists getting it down.

A yard topsail's halliard, sheet and tackline are arranged as for a jib-headed sail, except that the halliard

is seized to the yard direct, or shackled to a strop around it.

The luffs of yard topsails were seldom lashed to the topmast in early 19th century vessels. They were usually set rather like a lugsail, with about $1/_3$ the length of the yard before the mast. In contrast contemporary jib-headed topsails were often laced, usually being set for lengthy windward work. During the 1840s attempts were made to improve the set of racing yachts' yard topsails on the wind by tautening the luff with a form of bowline (figure 49), but this rather complicated gear had a limited vogue.

Jackyards to extend the clew of a topsail and increase area began to appear in racing yachts the 1850s, and seem to have originated as a means of increasing the size of a topsail to obtain sail area unmeasured by the rules. There is nowadays much misuse of the term jackyard topsail. The yard extending the luff of a topsail above the topmast is the topsail yard, and in vessels with short or high-peaked gaffs the length of topsail foot may be increased by a short yard called a jackyard. This yard of about $\frac{1}{3}$, or rather more, than the length of the foot is laced to the after end of the topsail foot, extending it beyond the gaff end (figure 50a). The topsail sheet is bent to the middle of the jackyard. Jackyards are normally only fitted on topsails having a topsail yard, usually in racing yachts or workboats, for racing or light weather work, but some racing yachts set jackyards on jib-headed topsails (figure 50b). In America the jackyard was known as a club and such topsails called club topsails. It was also common practice in American racing yachts to lash the heel of the topsail yard to the mast to assist it to stand rigidly.

Jackyard topsails were usually set when the yacht was at anchor or gently jilling about preparing for racing, in both cases head to wind, and were taken in on the lee side of the mainsail even if the yacht had to come about to bring the topsail tack to leeward of the gaff. Large racing yachts had two or three sizes of jackyard topsails and eventually the light-weather ones became so large that racing rules restricted their area. Typical was the largest jackyard topsail set by the British racing cutter *Valkyrie III* of 1895; the topsail yard was 65 feet and the jackyard 55 feet long, the topsail luff 102 feet, the foot 85 feet and the clew 56 feet. This topsail area was 2,380 square feet; considerably more than the total sail area of a present day 12-metre. Sails of this size were fairly common in the large yachts and required smart handling, particularly by the mastheadsmen whose job was to see that everything went up or down clear and quickly, and that the topsail luff was well laced and everything drew well. There were only two of them aloft in a big cutter and they were amongst the smartest of the clever sailors who manned these craft and for their skill, and risk, they received an extra shilling every week – in those days when yachts cost many thousands of pounds but men were paid in shillings.

In large vessels the topsail yards were stowed on deck and the topsail bent to them with stops after the earrings had been hauled out and seized taut. In small craft the yards are often stowed with the sail. To set a yard topsail the halliard and sheet are bent to the yards, or preferably shackled to rope strops fitted at predetermined positions to obtain a good set for the sail. The topping lift is set up

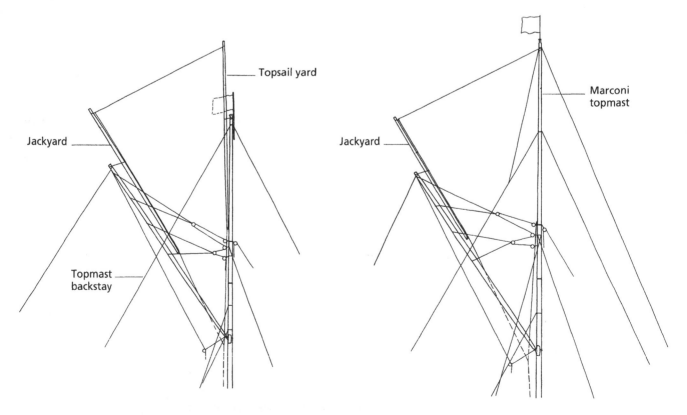

Figure 50 (a). Jackyard topsail, 1890-1914 *(b).* Marconi topmast with jackyard on topsail – 15-metre *Istria*, 1912

and after a tackline is bent on the halliard is hauled on and the topsail goes up inside the topping lift. Large craft often had a tripping line on the lower end of the topsail yard and this was also used to guide it when going aloft.

A lazyjack can be used to keep a large yard topsail from blowing away from the mast during hoisting. This is a short line with one end made fast to the tack cringle of the sail and the other end passed round the main or peak halliard, if these are belayed close to the mast. The halliard forms a jackstay and the lazyjack a traveller, which is cast off when the sail is aloft ready for trimming. When hoisting, a slight strain should be kept on the topsail sheet to prevent it fouling. With the halliard set up the tack is bowsed down hard. In large yachts the mastheadsman guided the topsail up clear aloft, and lashed the heel of the topsail yard to the topmast, before lacing the topsail luff to the mast. The topsail sheet should be hauled out last and the sail should stand flat on the wind, when the topping lift is slacked.

Large jackyard topsails were often cut with a mitre seam from the clew to the luff, and any jackyard topsail requires very careful cutting to ensure that, when set, there is a gap between the gaff and its foot, so that a tautening strain can be kept on the sheet.

Large racing yachts sometimes used tripping lines to lift the foot of the topsail over the peak halliards to reset it to weather, when the yacht was head to wind for a few seconds in tacking and, at the same time the heel of the topsail yard was sprung a'weather by one of a pair of tripping lines rove from the heel of the yard, through blocks on the gaff and down through jewel blocks to the deck. In a fresh breeze it might be necessary to start the topsail sheet a little.

Topsail halliard bend

When making very long tacks or racing, the foretopsail of a schooner should be shifted to windward. As the vessel goes about the sheet and tack are eased and the sail clewed up with a clewline. A hand aloft clears the tack and sheet over the main topmast stay, and when head to wind for a few seconds the tack must be bowsed down quickly and the sheet hauled out smartly after it. Handling this sail is smart work of which few amateur crews are capable.

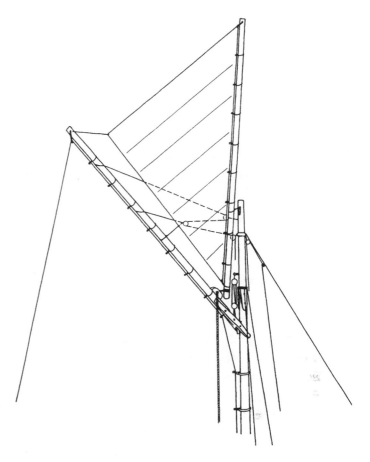

Figure 51. Cornish yard topsail

The Cornish topsail is a convenient type for small, pole masted yachts having the now usual short masthead (figure 51). It is a triangular sail having its luff extended well above the masthead by being laced to a yard hoisted by a halliard which is rove through a sheave fitted diagonally in the masthead so that the yard clears the peak and throat halliard blocks. The sheet is rigged in the usual way but the heel of the yard fits immediately above the gaff and is kept in position by a tackline called a timminoggy which passes up one side of the mainsail, through a hole in a wooden chock fastened to the top of the gaff just abaft the jaws, and down the opposite side of the mainsail. The timminoggy is made fast to the heel of the yard and sometimes several holes are bored in the chock to vary the position of the heel of the yard to adjust the set of the topsail.

CHAPTER SIX

THE BOOMLESS GAFF SAIL

THE BOOMLESS GAFF SAIL is now little used. Its principal advantages are the ease of area reduction and furling by brailing and the resulting clear deck space with the sail brailed to the mast and standing gaff due to absence of a boom. However, such sails flog badly in stays or, more particularly, when getting under weigh or bringing up, head to wind, and the mainsheet needs careful tending when running, as a gybe with a slack sheet can be extremely dangerous, and it must be kept fairly taut on the wind to minimise flogging in stays. Generally, a slightly heavier mast and gaff are needed compared with a boom mainsail, and the length of gaff is often increased to bring the leech almost parallel to the

luff to maintain sail area. The gaff halliards are rigged as usual for a gaff mainsail but large vessels using the rig often rove chain halliards when the gaff remained standing for long periods. The head is laced to the gaff and the luff hooped to the mast, as for other gaff sails.

Usually only a single throat brail is used. A pair of small single blocks are hung from the gaff jaws and a light, supple brailing line, preferably now of Terylene rope, is led from a cleat at the foot of the mast, up through one block and across one side of the mainsail, through a large (and preferably nylon) cringle about $2/3$ the distance down the leech from the peak, back across the other side of the mainsail, and down through the other single block to belay at deck. A sail of up to 500 square feet area can be brailed up by this method.

The mainsail's weight should be slightly increased above normal sailcloth weight to compensate for the chafe of the brail, which cannot be fitted with any form of chafing gear. Point reefing is also usually fitted with this rig for reducing area for windward work. The mainsheet of a boomless gaff sail has to be arranged to flatten the clew and leech, and the commonest practical arrangement, as used in the English bawley boats, is powerful and effective, if somewhat alarming to the uninitiated when going about. The single part sheet is shackled to a single lower block working on an iron horse across the after bulwarks; up through a single block hooked into the clew cringle; down through the traveller block; up to the upper single block hooked into the first reef cringle, and down to belay to a long pin through the traveller block. The sheet blocks on the sail can be shifted up the reef cringles for reefing, after the reef is tied, or alternatively, the sail can be reefed down to the foot.

The boomless sail is an inefficient shape when

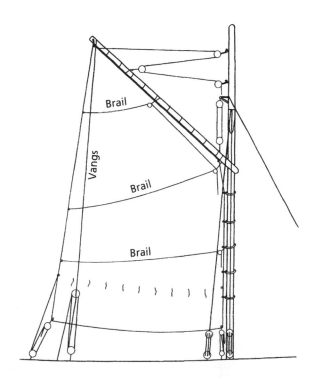

Figure 52. 17th century boomless gaff sail with vangs and brails

would be described as small bawleys.

Stephen Frost of Leigh is credited with starting the local shrimping trade in 1830 which brought a living to the port for over a century. It grew quickly with the Victorian popularity of seaside holidays and by 1872 became the principal fishery of Leigh, employing about 100 vessels, the largest being transom-sterned bawleys, 32 feet overall. When shrimps were scarce some Leigh fisherman sailed to Harwich to join the fleet of smacks and bawleys working out of that north Essex port. Sometimes, in bad seasons, Leigh craft worked on the south coast using Shoreham as a base, and occasionally sailed to Holland, Belgium, France and the Channel Islands in search of shrimps to trawl or shellfish to dredge. The sailing bawley as finally evolved and built at Harwich, the Thames, the Medway and the Swale, was powerful in hull form and rig. A typical boat built during the 20 years prior to 1914 varied between 35 - 42 feet length overall; a 35-footer having 12ft beam and 5ft or 5ft 6in draft. The hull had a straight keel, straight

A Thames hatch boat. Engraved by E W Cooke 1828. Note vangs to gaff, sprit mizzen and rudder yoke

running and in light weather a spar should be used to bear out the clew which is otherwise difficult to keep quiet, with consequent frequent risk of gybing in a sea.

A topsail can be set above the boomless gaff sail in the usual manner, but can also remain set while the mainsail is brailed, in light weather. This is often useful when temporarily anchoring, or running down a narrow, wooded river where only the topsail can get an air.

The rig appears to have become popular in fishing and trading smacks and yachts of the mid-17th century, which set boomless mainsails spread by a half-sprit, or standing gaff, which was hoisted by throat and peak halliards but generally remained aloft in all but very bad weather and at anchor (hence the term standing gaff).

The sail was hooped to the mast and fitted with up to four brails which furled it close to the mast and the gaff, which was fitted with one, and sometimes two pairs

of vangs (figure 52). Many of these craft were rigged as sloops, with a staysail, and a jib on a well-steeved fixed bowsprit. Some also set a square topsail on yards crossed above the hounds, and the braces of these yards were sometimes led to the gaff, putting additional strain on it.

This rig was the origin of the cutter and the half sprit almost certainly originated in Holland where for centuries a sprit had spread the mainsail. When the gaff was introduced to spread the head of the sail as effectively but with less than half the weight and a gain in stability in bad weather, the term half-sprit was used. It has now almost disappeared from nautical language.

Charles II's yacht *Mary II* of 1677, the first English-built craft to be called a yacht, was of this rig which, with the dimensions of 66ft 6in x 21ft 6in x 7ft 6in draft, was typical of many contemporary Dutch and English smacks and coasting and inland waterway hoys;

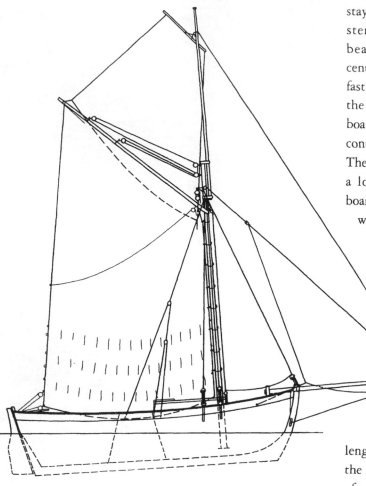

Figure 53. Peter boat built at Strood, Kent, about 1820
Dimensions: 27ft 6in x 9ft 6in x 4ft 2in draft aft

staysail and jib and a sprit mizzen stepped close to the sternpost, resulting in yoke and line steering. The beautifully formed hull had little draft and no centreboard but was beamy, well balanced and reputedly fast – hatch boats being much used for match sailing on the Thames in the 1820s, with cutter rigs. The hatch boat's mainsail often had no brail but the gaff was usually controlled with vangs and the luff was laced to the mast. The boats were decked, with fo'c'sle accommodation, and a long well amidships, hatched over with portable boards, hence their name. Hatch boats were often fitted with wet wells for the catch and besides the Thames, often worked along the Kentish and Essex shores. They died out by the 1850s as the Peter boat type grew rapidly in size, evolving into the bawley.

Peter boats were originally pointed stern, clench-built, rowing fishing boats with a wet well amidships, used in the Thames fisheries. They developed to set a sprit mainsail and staysail, without a bowsprit, and there are references to craft of this type at 'Lye' in 1540. By the 1840s Peter boats were often 28 – 30 feet in length, decked forward for accommodation and working the same waters as the hatch boats and later, by a change of name, became the bawley; a word which has always puzzled nautical pundits but probably merely meant a Peter boat with a shrimp boiler aboard. Detailed plans exist of a Peter boat built at Strood on the Medway about 1820 which on dimensions of 27ft 6in overall, 9ft 6in beam, 4ft 2in draft aft and 2ft 6in forward, set a boomless cutter rig identical with that of the later bawleys of which she was obviously a prototype, differing only in having a fish well amidships (figure 53). This type was refined in hull form and the speed of this example was noted by an enterprising Admiralty which recorded her lines and sail plan, noting at the foot that she was 'particularly marked for her superior sailing'. Peter boats of this size and rig had probably been sailing the Thames for many years. Some were of greater draft than this Strood boat. The *Favourite* worked by the Young family of Greenwich in the whitebait fishery during the mid-19th century had a 6ft draft on 27ft length. A watercolour drawing of a similar boat, also owned by the Youngs, shows them to have been what

though her crew of 20 and armament of six guns was unusual.

From this rig the moderately long-gaffed loose-footed boomed mainsail was evolved and came into general use during the 18th century for its improved windward ability compared with the boomless sail; a truth which has since always been recognised but in some craft was ignored in favour of the convenience of rapid sail reduction of the boomless mainsail by brails, to suit certain types of work and crew economics. Thus it was that the boomless gaff sail survived in English commercial use until the late 1940s, particularly in the Thames Estuary.

One of the principal types of Thames fishing craft was the hatch boat; a double-ended clench-built yawl, rigged with a half-sprit mainsail with topsail over, would be described as small bawleys.

Stephen Frost of Leigh is credited with starting the local shrimping trade in 1830 which brought a living to the port for over a century. It grew quickly with the Victorian popularity of seaside holidays and by 1872 became the principal fishery of Leigh, employing about 100 vessels, the largest being transom-sterned bawleys, 32 feet overall. When shrimps were scarce some Leigh fisherman sailed to Harwich to join the fleet of smacks and bawleys working out of that north Essex port. Sometimes, in bad seasons, Leigh craft worked on the south coast using Shoreham as a base, and occasionally sailed to Holland, Belgium, France and the Channel Islands in search of shrimps to trawl or shellfish to dredge. The sailing bawley as finally evolved and built at Harwich, the Thames, the Medway and the Swale, was powerful in hull form and rig. A typical boat built during the 20 years prior to 1914 varied between 35 - 42 feet length overall; a 35-footer having 12ft beam and 5ft or 5ft 6in draft. The hull had a straight keel, straight stem slightly rounded at the forefoot and a transom stern with slight rake but often surprisingly delicate shape. The sheer was very high forward and swept down to little freeboard aft, in the manner of the finer-lined Essex smacks. The hull had a very hollow entrance with bold 'shoulders' about the mast and usually considerable shape in the bold hull sections which, with the generous beam, made the bawleys rather stiffer sail-carriers than the smacks in smooth water. Construction was heavy and cheap, with sawn oak frames, beams and centreline structure, pitch pine bottom planking and pine topsides, decks and spars. Low bulwarks and a heavy oak wale belted the sheer. All the six or seven tons of iron ballast was carried in the bilge and ceiled over. Below decks they were arranged with accommodation for four in the fo'c'sle, which usually extended aft of the mast and was entered by a hatch from the deck. The crew slept on locker tops and cooked on a coal stove. The hold occupied the remaining aft space. In a shrimping bawley it was divided from forward, into net and gear store space, room for the shrimp boiling copper, and general storage of the catch and a steering well. The hold was served by a long, narrow and tapering hatch, with wooden covers for bad weather leaving wide sidedecks for working trawl or stowboat gear and sorting the catch. A handspike

Leigh bawley *Gladys* in working trim about 1922.
Tack triced up for vision

windlass with pawl post was fitted across the foredeck, and many had a simply-geared hand winch or wink mounted on a post amidships to haul the trawl warps.

The boomless cutter rig was the bawleys' most distinctive feature. The mast was rather short but a long fidded topmast was stepped and in a 35 foot boat the topmast head would be about 48 feet above the deck. Standing rigging comprised two shrouds a side; the foremost in line with the mast and the after 2ft 6in or so further aft, with single topmast shrouds passing over wooden spreaders and set up midway between the lower shrouds. A forestay led to the stemhead and the two fish-tackles for handling nets could be, and occasionally were, set up as backstays in heavy weather. The bowsprit stood 17ft 6in outboard, without shrouds, but had a running bobstay and a topmast forestay.

The sail plan was tall and comparatively narrow but efficient and handy to those accustomed to it. A 35-footer would set a 500 square feet mainsail (referred to by some bawleymen as the trysail); a 200 square foot topsail; a 95 square foot foresail (staysail) and a 117 square foot working jib; the rig totalling 912 square feet. A spitfire jib or a large light-weather jib could be set to suit weather conditions and a jib topsail and balloon foresail could be set in light weather, when a bowsprit spinnaker was often set to the topmast head and used boomed out for running, or sheeted in as a balloon jib. The jib sheets had no purchase and led in through holes in the bulwarks, then forward to make fast to the bitts but the balloon foresail sheet was led aft, outside everything. The gaff averaged 25 feet in length and was hoisted by smack-style peak and throat halliards. The leech was almost parallel to the luff and the boomless foot of the sail very short. The luff was hooped to the mast and the main tack could be triced up with a tackle and a single brail was fitted, running from a block at the throat, round the leech and back again to another throat block, and down to the deck. Both truss and brail were in constant use when working gear, to regulate speed, and two or three rows of reef points were also fitted. The mainsheet was arranged as already described and a bearing-out spar was used when running in light weather. The foresail (staysail) sheeted to a horse across the foredeck and the fall of the halliard was always made fast to the neck of the lower block to retain it when lowering in haste. The topsail was often diagonally cut and the luff laced, hooped, or leadered to the topmast. In light weather the topsail often remained set above a brailed mainsail, for ease of working or when brought up for short periods. In winter the bawleys often struck their topmasts and some removed them altogether. When the mainsail was lowered it stowed along the gaff which was stowed with the peak on deck, to one side. With their square forefoot and long keel, bawleys looked after themselves when sailing or hove-to. The tiller had a pin rail across the after coamings (called the 'old man' at Leigh) and with the foresail just a'weather these boats

held their course for miles while the crew were culling, cooking and packing the catch. If hove-to, most bawleys would fetch directly up to windward in any weather. Originally all their sails were of flax canvas, but later many had cotton. Eighty years ago a suit of sails for a bawley cost about £80, ordered by the owner, direct from the sailmaker, usually F.A. Turnidge of Leigh or Pennick of Harwich, both of whom specialised in the rig.

A bawley took about six months to build, and construction was usually in the open. Over many years the majority were built by Aldous of Brightlingsea, Essex, most of them being creations of the yard foreman, a Mr Foote, who produced shapely and well-built craft. Besides the Canns of Harwich, Heywood of Southend was a clever designer of bawleys and sailing cockle boats always designing by drafting. Many fast sailers, principally for Southend, were launched from his yard near the gasworks on Eastern Esplanade. Frank Parsons, one of his apprentices, also built a few bawleys at Leigh, where Bundock Brothers were also building the boats. Gill at Rochester, E. Lemon of Strood, Fiddle of Gravesend, Stone of Erith and Peters of Southend each built a few. Lemon's craft formed the bulk of the Medway fleet belonging to Chatham and Strood.

In 1890, 86 bawleys sailed out of Leigh which also had 32 open cockling boats, and in 1913 there were still 72 sailing bawleys there.

Originally, the Leigh shrimpers worked a type of small primitive trawl. This evolved into a triangular net whose mouth was spread by a beam towing along the bottom and extended by a vertical post in the centre, which had a light horizontal beam, making a quadrilateral mouth. As the size of bawleys gradually increased they towed more nets until, by the 1890s, a 35-footer worked four nets, two boomed out from the hull. Eventually this mass of gear was replaced by a single ordinary beam trawl with fine-mesh net. The net was recovered by hand, the mass of shrimps being shaken out on deck, often filling it to the rails, and while the bawley had another tow her crew sieved and sorted the catch. The catch was boiled in the hold copper and spread on a drying net spread on light battens before packing into pad baskets, labelled for the train. A few bawleys worked from Ostend at various times and the bawley rig was adopted by the pointed stern, cutter rigged Belgian smacks which also shrimped from the port.

The bawleys and Peter boats also worked Peter nets, or stop nets, for flatfish and seines for smelts or whitebait, which were more usually caught with the stow net (described for the Essex smacks, but having a very fine mesh).

Whitebait fishermen from Greenwich worked the river until the 1880s when pollution drove them downstream to settle at Leigh, where the Young family continued the whitebait fishery with bawleys until the 1950's. Five Leigh bawleys joined the Colne smacks scallop-dredging down-Channel in 1883 but had to give it up as they were not good seaboats.

Harwich, in north-east Essex, was the other principal home of the bawley and about 60 worked from there, owned mainly by the Goods, Dennys and Smith families. Between May and September they were shrimp trawling off Walton and Clacton, in the Wallet, inside the Gunfleet, off Felixstowe and the Cork and Cutler shoals. From September to May the whelking season kept them working off the Ridge off Harwich, and under the Pye Sands. Many laid up in winter, particularly when the demand for whelks, used as bait for long-line fishing, declined after 1914.

The Harwich boats had especially shapely hulls, particularly those built by John and Herbert Cann who in about 1890 took over a yard established in 1868 by their father, George Cann, who had moved to Harwich from Brightlingsea. About 30 men were employed, principally building sailing barges and barge boats, of which one was completed each week for stock. About 12 bawleys were built and they included the *Helen and Violet* and *Doris*; the fastest of the type. Most were designed by John Cann who carved skeleton half-models before laying off. His bawleys were very shapely with raking transoms, rounded forefoots and splendid hull sections. John had tremendous enthusiasm for his work, rising at 5 a.m. for an early start. At very busy times he sometimes went to bed in his boots to save dressing time in the morning.

In 1897 Canns' built the 39ft 6in bawley, *Alice Matilda* on speculation and she sold for £100, complete with all gear. They also built at least one of the Leigh centreboard sailing cocklers. Cann's ceased building shortly after the First World War. The brothers Norman

also built a couple of bawleys at Harwich, one of them, *Wings of the Morning*, won several bawley races at Harwich regatta until 1914. Harwich owners specialised in bizarre names and *Auto-da-Fe* was built by Vaux at the Naval yard where they principally built schooners and sailing barges.

Bawleys raced annually at various periods at Leigh, Southend and Harwich Regattas. At Leigh they sailed for money prizes and a Championship Pennant made and presented by Mr Turnidge, the local sailmaker.

In 1910 Cann's built *Doris* for William Lucking of Leigh, who won the Leigh Race with her from 1910 until 1913. She was an extremely fast boat, cleverly sailed, and is now under reconstruction and restoration to her original condition. After 1919 her owner could not afford new racing sails and Arthur Felton's *Helen and Violet* also built by Cann in 1906, invariably won the races, which continued until 1928. That year, stiff though bawleys were, *Helen and Violet* was caught under an enormous yard topsail by a hard squall near the Chapman Light which hove her down until water ran down the cabin hatch and her keel was visible, until she was eased. Another Cann-built bawley, owned by T. Emery of Leigh, won a challenge match at Harwich against a local bawley.

About 20 bawleys were owned at Southend, where the Regatta also featured a bawley race and there was great rivalry with Leigh. Many pleasure-tripping boats working off the Southend foreshore were rigged bawley-fashion, sometimes with a mizzen added. The smaller ones generally resembled the cocklers but a few larger, such as *Volunteer*, were open boats of bawley size and proportions.

A few bawleys were used for boarding and landing pilots in the Lower Thames until the 1920s, and also for marking wrecked vessels and generally tending their raising, for which they were painted green and carried special lights.

The Leigh cockle galley was another type which developed with a rig identical to the bawleys. These very shoal-draft boats had strong bottoms to stand grounding on the sands of the Thames Estuary while their crews raked up cockles from the extensive natural beds. They worked all year round, the grounds varying from the 'Main' (Dengie Flats) between the Rivers Blackwater and Crouch, which took two tides to reach through

Havengore or through the Whittaker; the Maplins, Shoebury, and on the Kentish shore; Sea Salter, the Pollard, Leysdown Flats and the Grain Sands.

When dried out three or four hands raked up the catch by hand into a landnet, where it was washed and put into a pair of baskets carried by a yoke across the shoulders, and tipped into the hold. The catch was sailed home, sometimes to be re-laid for a short time in Leigh Creek for 'purging'. However, sewage polluted the layings and regulations required the cockles not to be relaid, and to be boiled in steam instead of in coppers, as they are still.

The cockle boats had to be fast to sail to the grounds before the water left and to sail home as quickly as possible, if necessary with a heavy load, but also be capable of sailing well loaded or light as they carried no ballast. Until 1900 old warships' boats were reputedly converted for this work, from which the term cockle galleys is supposed to derive. They were rigged with a boomless mainsail and a foresail. About 60 baskets filled these old-time galleys but in 1901 Franny Noakes had the larger, carvel-built *Lilly and Ada* built by Heywood of Southend. Like all the sailing cocklers she was undecked aft of the cuddy-fo'c'sle; the well being ceiled for the catch and partitioned aft to form a steering room. *Baden-Powell*, *Shah*, *Jane Helena* and *Florence Edith* quickly followed and established the new type which set full bawley rig. Typical dimensions were 28ft x 9ft x 2ft 6in, increasing to 32ft, and many were fitted with a shifting bulkhead down the centre of the hold to control the slippery cockle cargo of up to two tons weight. The fo'c'sle had limited headroom and overnight accommodation for four men. Size increased until about 1912 when Heywood built two 34-footers. The sail areas of one, *Shamrock*, were mainsail 416 square feet, topsail 168 square feet, foresail (staysail) 83 square feet, working jib 86 square feet, total 753 square feet. They also carried the same additional sails as the bawley. About a dozen worked the cockle fishery under sail and were smartly kept and handled. They also raced as a class at Leigh Regatta until 1911.

The bawley rig, with addition of a standing lug mizzen stepped on the transom, was used by many Lower Thames cruising yachts of the 19th century. *Waterwitch* of about 1830 was typical and was sailed by her owner, and

his brothers, to Holland, France, the Baltic and all round the British Isles. Her rig was tall and she won many racing prizes. It is interesting that she had a cockpit and a coachroof cabintop, both often regarded as American inventions of a later date.

The transom sterned fishing boats of Faversham and Whitstable also carried this rig, sheeting the lug mizzen to a long outrigger, and were reputed to have carried it for herring drifting. They certainly set enormous yard topsails and emphasise the fallacy of attempting to catalogue types too closely.

The boomless gaff mainsail was popular for coastal tripper boats. Margate and Ramsgate, Kentish ports on each side of North Foreland and much frequented by day trippers, evolved a type of pleasure yacht up to 70 feet length and resembling the small Margate fishing ketches. The Margate *Moss Rose* was typical with a clench-built counter-sterned hull and a full ketch rig with a jibheaded topsail over a boomless mainsail which would not sweep overboard the hundred or so 'sixpenny sicks' crowding the thwarts of the huge open-well amidships. Four of these pleasure yachts, cutter and ketch rigged, also worked from Ramsgate harbour. Slightly shallower and lute sterned boats with boomless mainsail ketch rigs but standing lug mizzens worked off the beaches of Eastbourne and Brighton, one being the immortal original *Skylark*.

In Dorset the 20 foot, transom sterned, open clench-built Weymouth watermen's boats, locally called stuffy boats, used the rig without topsail but with a triangular mizzen sheeted to an outrigger and with the staysail tacked down to a short iron stem bumkin.

Some of the larger east coast sailing barges used the half-sprit rig with brails, as did several of the smaller ketches working out of Langstone, Chichester and Portsmouth harbours, and the Solent ports.

Boomless mainsails were also set by the pilot dandies working from Lowestoft and the Falmouth pilot cutters, in winter and after amalgamation killed competition. At the other extreme, the Norfolk and Suffolk wherries set a single boomless mainsail in an exaggerated, freshwater form.

After the mid-19th century boomless gaff mainsails became widely used in vessels drifting for herring and mackerel, and it replaced the traditional lug rig in many

Lowestoft fishing vessels. Ketch-rigged trawling smack LT 136 in foreground and dandy *Thankful* built at Yarmouth 1894. Note dandy's boomless mainsail and differing topsails typical of these types. *Ford Jenkins*

of the larger English, French, Dutch and Belgian drifters. This change of rig was related to the improved distribution of fish by railways which, with the organisation of the fishing industry, led many English owners of herring-drifting luggers to take up profitable fish-trawling after the herring season and, the lug rig being unsuitable for this, the drifters soon set a boomless gaff mainsail on the foremast, leaving the standing lug mizzen sheeted to an outrigger, and setting a tall, narrow staysail to the stemhead, and a variety of jibs on a long reefing bowsprit. The mainsail had no brails but several rows of reef points, and sometimes a bonnet or laced piece on the foot was added. They were pole masted craft which gave them a rather squat appearance compared with the lofty trawlers, but in light weather they set a large yard topsail above the mainsail.

Gradually size increased to 65 feet or more, and the mizzen was changed to a gaff sail, loose-footed but with a boom and the mast heavily raked forward, which dispensed with the cumbersome outrigger and enabled the mizzen to be topped up out of the way in crowded harbours. A yard topsail was also set on the mizzen as a final development of the rig. The main (or foremast) was stepped in a tabernacle and frequently lowered when riding to nets, which might set for two to three miles when fully set.

It was a handy rig for a short-handed and hard worked crew, and spread rapidly among the great herring fleets of Yarmouth and Lowestoft and also the clench-built Yorkshire 'yawls' from Filey, Scarborough and Staithes.

This dandy rig reached a spectacular peak in the huge, 90 feet Boulogne and Fecamp, herring drifters of 70 years ago. Many of them were built at Lowestoft or Yarmouth; splendidly powerful vessels, gaily painted, scouring the Channel by the score, each with up to 6 miles of drift nets set, hauled by a steam capstan and crews of 20 to 30 hands. Despite their great sail plans, with bowsprits often 38 feet outboard, they were clumsy sailers and usually towed out to sea, two and three at a time, their crews clustered forward, singing themselves out of a Boulogne harbour solid with sailing and steam drifters and coated with a pall of smoke and coal dust, backed by the lusty clangour of a successful herring season.

The West Country inshore line-fishing vessels known as hookers also used the boomless gaff mainsail. There were two types; the smaller cutter rigged of 30 to 35 feet, and the larger yawl or dandy rigged and between 35 to 40 feet long. Their heyday was just before 1900, when hookers built by F. Hawke of Stonehouse, England and Hines of Mount Batten, Williams of Millbay and Borlase of Turnchapel were to be found tumbling about off Burgh Island or Rame Head, line fishing for whiting, herring and mackerel, or boultering as it was locally called. This type of boat worked from Torquay, Plymouth, Looe and Polperro, where they were called gaffers. The type were also built at Porthleven and Mevagissey and Tom Pearce of Looe was a noted builder. In 1893 he launched the cutter-rigged *Dayspring*, for Captain C. Lambert, who worked her from Plymouth until the 1930s. Her deep, lean and slack-bilged hull was 31ft overall x 9ft 8in beam x 6ft draft and the long straight keel, heavy sawn-frame construction and plumb stem and transom belied the speed she attained in her work and local regattas. Hookers were flush decked and usually had a large open wall aft of the mast, with a handspike windlass across it, amidships, to handle long lines and other gear. The crew berthed in the fo'c'sle which was sparsely furnished with locker seats and a coal stove. The mast stepped well inboard, West Country fashion, and the cutter rig spread generous area, the mainsail having two rows of reef points. A yard topsail set on a 30 foot topsail yard timminoggy fashion, which appears to have originated in craft from this area.

The hookers and gaffers had great family relationships with the Mounts Bay luggers, Falmouth punts and oyster dredgers and particularly with the Polperro gaffers, so called to distinguish them from the local luggers. The gaffers were generally about 30 feet in length, but some were as small as 25 feet, and rigged like *Dayspring*.

Their seaworthiness has become well known as the 29 foot *Lilly* built by Ferris of Looe in 1898 was converted into the yacht *Moonraker* for the late Doctor Peter Pye and has voyaged all over the world.

Although smaller, the Polperro gaffers were more extreme than the Plymouth boats in that their open cockpit extended forward of the mast, which stood supported by a tabernacle and main beam; and from the

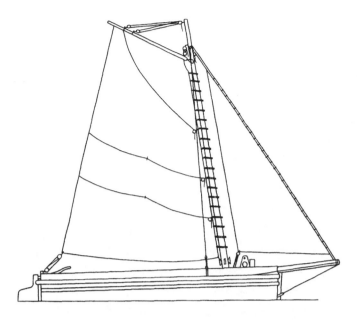

Figure 54. Sailing cargo lighter of New York Harbour, circa 1900. Dimensions: 96ft x 41ft x 9ft draft

masthead purchase served the hatchways. Its length was commonly about 103 feet above deck, and the heavy bowsprit reached 37 feet outboard, being fitted with a bobstay and shrouds. The staysail hanked to the forestay and was fitted with lazyjacks. The boomless mainsail was hooped to the mast and the almost horizontal gaff was about 37 feet long and fitted with a downhaul. Two brails were fitted; one across the middle of the sail and one between it and the foot, rove through blocks on the luff and belayed at the mast foot. The sail sheeted by a single purchase to the transom and one shroud was fitted on each side. No topmast or topsail were carried and the lighters were tiller steered. A large hand winch for working cargo was fitted on the foredeck and two small cargo hatches left ample deck space.

The crew of two men and a dog lived forward and earned a hard living lightering freights to and from ships in the port to outports and railheads, providing an erratic service in light weather but rarely able to afford a tug.

The last vessel built with the rig known to the author was the 43ft 6in Leigh motor cockler *Theodore E.M.* launched by Dan Webb & Feesey at Maldon, Essex in 1946, and setting a staysail to the stemhead and a traditional bawley boomless mainsail on a tall pole mast. She temporarily carried it as a practical form of east coast 'yacht'. She would not go to windward under the rig but, as a steadying sail in a seaway or emergency canvas in case of engine breakdown, combined with convenient stowing with the mainsail made up to the gaff which can be stowed hoisted well up the mast and the peak on deck, it would be hard to beat for motor yachts of workboat type, and this is probably the likeliest future use of the oldest form of English gaff sail.

20 foot topsail yard of the timminoggy they set a large jib topsail in light weather to the end of the exceedingly long bowsprit.

The boomless gaff rig was widely used for the foresails of North American schooners but its use as a standing gaff, with brails, seems to have been very limited. The best example is the New York harbour lighters; bluff-bowed, very beamy, flat-floored sloops lightering cargoes around the port from about 1800 until about 1905 (figure 54). Typical dimensions were 96ft x 41 ft beam x 9ft draft unladen. Stem and transom were vertical, the keel straight and sheer flat. The mast stepped well forward and raked heavily aft so that a

THE CATBOAT

THE CATBOAT may be broadly described as having one fore and aft sail set on an unstayed mast stepped in the extreme bow of a shallow hull which has great beam for its length, and is usually fitted with a centreboard. Traditionally, a gaff sail was used, laced to the boom and hooped to the mast. The rig was the most economical possible, being set by a peak and throat halliard and fitted with a topping lift and sheet. A forestay was usually set up to the stemhead, and a vang was sometimes fitted to control the gaff when lowering.

Holland may well have been the original home of the cat rig, as it was used in at least one of the many Dutch traditional, inland waterway types having the usual bluff bowed, beamy, shallow, light displacement hull with leeboards, and about 30 feet length. The sail was laced to the mast, had a loose foot, and a curved gaff. A forestay and one shroud each side were fitted and the rig was in use until just before 1914.

The single sail rig with a short gaff and the mast stepped well forward was used in at least three types of commercial keelboats sailed in North America during the early 19th century: the Bahama sharpshooter; the Narragansett Bay, or point boats of Rhode Island; and the double-ended fish carrying boats of Eastport, Maine; all of whose rigs seem to stem from the earlier shallop rig. However, it is difficult to trace the catboat's ancestry to any of these. It is probable that the shallow draft, beamy, centreboard catboat first appeared in the New York area during the 1830s and then, at later periods, became extensively built and sailed from Long Island Sound, the bays along the south shore of Long Island, Narragansett Bay, Cape Cod, and in New Jersey, especially Barnegat Bay. As these craft were built for fishing and working throughout the year the sail areas and proportions were

modest but, as the type was adopted for pleasure sailing, especially for racing, in various areas, it developed exaggerated features which, paradoxically, led to a decadent influence on the working types.

The mast could not be effectively stayed by shrouds due to the small width of deck at its step and, consequently, wood masts of large diameter were fitted. These relieved strain on the rig by bending in strong winds, but their weight caused excessive pitching even in a small sea and although centreboard catboats were in limited use for coastal fishing, they were really developed for working in shoal, sheltered waters.

Despite the apparent simplicity of the rig, great experience is required to design a fast, balanced catboat which will handle well under all conditions of wind and sea. There must be perfect balance between sail and hull as there is no possibility of adjusting it with a headsail. Important factors are the proportions of the hull and sail, position of centreboard, sail shape, freeboard, fineness of the hull forebody, breadth of stern, and the shape of the bilge and topsides.

The cat rig is ideal for windward work in smooth water, and in these conditions will lie closer to the wind than any other, although in light weather, a sloop's greater sail area usually enables her to beat a catboat of similar size. Other advantages are the speed of getting under way compared with other gaff rigs, and minimal sail and rigging costs. However, there are several disadvantages with the rig, even when used in smooth waters. The limited forward deck space makes hoisting and lowering the single large sail difficult, especially in a breeze. Most catboats are hard on the helm in any weight of wind and must be reefed early to make them steer properly. Running in a hard wind often requires the peak to be lowered horizontally and the topping lift set up, and these ropes are usually led to belay near the helmsman. Even when reefed this makes hard work at the helm and there are instances of catboats becoming unmanageable under these conditions. Great care is also necessary when gybing in strong winds or the rig may be damaged. Cats were steered with tillers until the 1880s when wheel steering became commonly fitted; probably being first needed in the grossly overcanvassed pleasure catboats of that period, though excessive weather helm can be eased by slightly raising the centreboard. It is not always easy to make a cat pay off when getting under way, especially when in a chop of sea or in a confined anchorage, despite backing the boom, adjusting the centreboard, and reversing the helm. The extreme forward position of the mast step, coupled with the shallow hull, lack of shrouds and weight of the wooden mast, often leads to leaky garboards at the forefoot, but this can be minimised with proper construction. Unless moderately canvassed, catboats are not easy to reef and, when this is done, they are usually poor performers in the sea which builds up. These features, with the types' low freeboard and general wetness in a seaway, restricted cabin headroom and the obstruction of the cabin by the centreboard case, usually restrict their use to limited coastal cruising.

However, several long voyages have been made, notably between 1901 and 1905 when the 27 foot catboat *San Toy* was sailed the length of the Atlantic seaboard between New York and Key West, Florida, and 'lived through much wicked weather outside', according to her owner, W. Pearson. He described her as a 'wholesome, able, fast little ship, a perfect home afloat, I have lived aboard her every day winter and summer for the past four years.' She had a bold sheered hull with 12 foot beam and 3 feet draft, excluding the centreboard .

Most modern catboats have a general resemblance but in the past they differed greatly in form and the proportions of rig. These characteristics of various areas gradually disappeared when the type ceased to be used for fishing. During the 1830s centreboards were beginning to be fitted in American working craft and the large numbers of inshore fishermen working the vast bay approaches to New York evolved a shallow draft, beamy centreboard boat for oystering, line fishing and

Figure 55. Catboat *Camilla* built 1876 by T R Webber, New Rochelle, New York.
Dimensions:
27ft 5in, 27ft 7in waterline x 12ft 6in x 3ft hull draft.
Sail area 800 sq ft

wildfowling. The waters then were unpolluted and flanked by rural shores. These broad beamed, plumb stemmed craft had very fine bows with hollow waterlines, a midship section which was a very shallow vee, and high bilges, which were hard in working cats and slacker in those built later for pleasure sailing (figure 55). They ranged from about 17 feet to 30 feet length, with large centreboards of about $1/_3$ length of keel, which, when raised extended above the sheer, where they protruded from the case in a triangular hump almost touching the boom. The deck had a rectangular or oval shaped cockpit amidships, and steering was by a long tiller, with the rudder hung on the transom. A typical boat was 21ft x 9ft x 1ft 6in draft, extending to 5ft with the board lowered. These early New York catboats were fitted with dual rigs.

The common summer rig was as a sloop, with long boom and bowsprit, a fairly long gaff set almost horizontally, and considerable sail area to suit the average light summer weather conditions of the area. In autumn, winter, and early spring most of these working boats removed the bowsprit and staysail and shifted the mast forward to an alternative step just inside the stem, setting the sloop mainsail as a single sail or cat rig, as it became known. Many small boats of the type, under 20 feet length, set the cat rig permanently. The mainsail was laced to the boom and attached to the mast by hoops. It was hoisted by throat and peak halliards, or, in some boats, by a single halliard. The early cats seem to have used a single topping lift, but a peak vang or downhaul was fitted to control the sail when lowering. The boom extended beyond the transom and the mainsheet attachments were spread over its after end, the lower sheet block being shackled to a traveller on a horse across the transom. Three reefs were usually fitted and, due to the length of the boom, the reef pendants were kept rove in working cats. In sudden squalls the peak was dropped. The catboats of the New York area were not very seaworthy craft and some were capsized in strong winds. Due to extreme rig and hull form they often steered badly, usually accentuated by improper adjustment of sail and board when built. However, they seem to have ben better balanced when rigged as sloops with the mast stepped further aft.

Larger centreboard catboats were developed along the south shore of Long Island, especially in the Great South Bay and Moriches Bay stretching for about 65 miles inside the narrow shingle spit of Fire Island Beach, which enclosed a strip of water often 5 miles wide but averaging only a fathom depth. The villages along the north shore of these bays were the homes of fishermen and boatbuilders who loved fast sailing craft, many of which were built at the small town of Patchogue. Protected shoal water and fresh Atlantic breezes led to extensive use of catboats averaging 35 feet length. As many parts of the bays were only 3 feet deep, they were designed with a very hollow entry and run to keep up to windward when the centreboard was raised .

Samuel Wicks and his son Edward, of Patchogue, were noted builders of these catboats, and sloops of similar form, and their business and background were typical of many small yards building sailing craft in the USA. Samuel Wicks was born in 1835, and seven years later was holding on to clench-up boats built by his father. Apprenticed as a boatbuilder at 10, he subsequently also served his time as a ship-caulker, then a separate trade. He commenced building at Patchogue early in life and quickly gained a reputation for good work. In common with most of his contemporaries, Samuel designed by carving a scale half model of soft pine layers or 'lifts' of thickness equivalent to level waterlines through the boat's length, screwed or dowelled together. When the desired shape had been carved the principal mould stations were marked across its back and struck square across the faces of the lifts when they were taken apart. These offsets were measured, laid off on the mould loft floor, and faired. The moulds were erected on the keel and five or six ribbands or battens were bent round them. The shapes of the sawn frames were lifted from the ribbands at the correct intervals, and the boat was framed up and planked. Samuel Wicks specialised in building catboats for working the bays' oyster and clam fisheries in winter, and for party sailing with visitors in summer. These cats were about 35 feet long and had large, long-gaffed sails with four reefs to suit the strong bay breezes. The hulls were plumb bowed and had little sheer but the transoms were heart shaped, and raked like the sterns of pilot schooners or clippers, with well rounded quarters. The stock of the long-bladed rudder passed through a trunk to the deck. The deck was carried all round the boat, which had a long foredeck and a large cockpit, left open forward for fishing but with a trunk cabin covering most of the after part, leaving a small steering-well aft. A portable summer cabin top, with canvas side curtains instead of coamings, could be added over the forward cockpit for party sailing. The fishing cats were sailed by two men or, occasionally, by one. The local pleasure catboats were similar but often had open cockpits.

Samuel Wicks' son joined the business which also built many oyster sloops for owners in Long Island sound and elsewhere, besides many small, fast yachts. The family was proud that it could design, build and rig a complete boat on the premises, as the sail loft of George Miller, Samuel's son-in-law, was over the boatshop.

Their neighbour, Gilbert Smith, commenced building similar craft at Patchogue when he was 32, without having served an apprenticeship, and built about six craft annually, using the same methods as the Wicks.

Catboat builder Samuel Wicks' boatshop with Millers sail loft above at Patchogue, Long Island 1898.
Local fishing catboats in foreground

Although he built numbers of fishing catboats, he was best known for racing cats and sloops which carried the local form to extreme delicacy, with the addition of finely formed counter sterns. They had beautifully-setting sails made by his wife Miriam, but were flat sheered and wet in a small sea. His son Asa followed the trade, and this truly family business produced about 400 craft, mostly catboats, during 60 years. During the 1880s the 23 foot catboat *Coot*, designed and built by Smith, made a 1500 mile singlehanded voyage from New York to Beaufort, North Carolina, and return, sailed by C. P. Kunhardt, the noted American small craft historian. Her waterline length was 20ft 9in and she drew 2ft with the board up, and 5ft 6in with it lowered. Sail area was 390 square feet and she displaced 5,000lb, 1,000lb of which was inside ballast.

Catboats were also used in Barnegat Bay, New Jersey, in oyster and other fisheries. Between 1890 and 1903 these cats held races at Beach Haven which were the principal regattas in the area, attracting entries of 40 or more catboats between 30 to 35 feet long. Into this happy annual event sailed the 32 foot Herreshoff-designed pleasure cat *Merry Thought*, owned by a Philadelphia yachtsman, which in two years ruined the racing by her superior performance, largely due to her finely cut sails and superior gear which the fishermen could not afford to imitate. Although a yacht she was an exceptionally good seaboat and well finished, costing over $5,000 when launched.

Transom-sterned working catboats also fished from various places along the coast south of Barnegat Bay and in the Gulf of Florida. The type was also tried, unsuccessfully, in the San Francisco Bay fisheries.

A more seaworthy type of catboat was developed for the alongshore fisheries from such localities as Newport, Rhode Island, Buzzards Bay, Martha's Vineyard, Nantucket, Falmouth, Plymouth, Cohasset, Hingham and Boston. These catboats were of similar basic form to those from the New York area, already described, but had a harder bilge, greater depth, freeboard and sheer, and were sometimes built with short counter sterns.

Probably the best known type of eastern catboat were those from Cape Cod, principally those hailing from Chatham and Cotuit, later known as Osterville. The

Catboat *Thelma* of Vineyard Haven, swordfishing 1916. *John Leavens*

fishing catboats working from Chatham, on the exposed eastern shore of the Cape, were necessarily the most seaworthy of all the catboats. They had bold high bows, considerable sheer, and moderate sail area. Cotuit, or Osterville, is situated on Nantucket Sound, an area having an average depth of six feet and subject to strong tides and winds, demanding boats capable of easy handling and quick turning among the many restricted fairways. The most noted builders in the area were the Crosby family, two of whom, in 1857, began building fishing catboats at Cotuit, where an annual average of 35 cats were launched from their tall, narrow-windowed boatbuilding sheds by the shallow creek. Their six sons

followed in the business and one of them, Daniel Crosby, is credited with first altering the long established beam/length ratio of catboats to develop the 'two beamed cat', as it is now known; a powerful, some would say brutish, craft half as broad as she is long.

Generally, the Cape Cod type are characterised by a hard bilge, rather upright transom stern, and a very large transom-hung rudder. They ranged from 18 to 25 feet length and were used for inshore handline fisheries, lobstering, oyster and scallop dredging, wildfowling, and even the carriage of light goods. Some larger working catboats were built for party sailing, the American equivalent of the English tripper boats. Some of these ranged up to 40 feet length but 30 to 35 feet was average. A typical 26 foot waterline party cat built in 1902 set 828 square feet in her sail and had 11ft beam and 2ft 6in hull draft. She could accommodate about 20 people for day sailing.

These eastern fishing or party cats were all generally similar in arrangement; the mast stepped in the bow and decked all round with a coachroof cabin top (or trunk cabin as it is termed in America) forward, which, even in fishing cats, extended to amidships, enclosing a cabin with two bunks and stowages which was divided on the centreline by the large centreboard case. Aft of the cabin a semicircular cockpit, usually watertight but not self-draining, gave protected working space, or room for a party. Stone or iron ballast was fitted under the sole and, in some working cats, notably those in the lobster trade, wet wells were fitted on either side of the centreboard case, often with access through hatches in a bridge deck which was fitted across the boat at the forward end of the cockpit. The mast had no shrouds and the usual boom and gaff sail was fitted, but of moderate area compared with those of the smooth-water cats. Two or three reefs were fitted and the pendants carried rove, ready for instant use. Lazy jacks were common and the sails were invariably laced to the boom. Some of these catboats carried a forestay set up to a plank-type bowsprit which was usually used for convenient anchor handling; but some boats set a small staysail to aid in steering and staying.

Keel catboats were built in Narragansett Bay and Block Island. These developed from the double-ended stone-ballasted boats of the area which were rigged as shallops, with two long-luffed short gaffed sails, the foremast being stepped in the bow. This rig is now called a cat ketch but was more properly known as Periauger rig. Boats of this type and rig survived at Block Island until 1900. The boats became modified in Narragansett Bay, especially at Newport, Rhode Island, where a keelboat with a single tall gaff sail was built. By 1830 a typical boat was 24 feet long, of moderate beam, had a small Y-shaped transom, drew 4 feet aft, and was not fitted with a centreboard. The keel had no rocker but the stem and sternpost were well raked and the forefoot rounded. Ballast was of beach stones. A small cuddy cabin was arranged forward and a very large cockpit took

up half the boat. The tall mast stepped just inside the stem and was 38 feet long in the 24-footer. The boom was 26 feet long and the gaff 8 feet, peaked at 50°. Peak and throat halliards were fitted and a topping lift and a forestay were the only other rigging. The luff had mast hoops but the sail was loose footed, with a clew outhaul grommet. Two point reefs were usually fitted.

The most unusual feature of these boats was, for a cat rigged boat, the proportions of the sail. The 24 foot boats' sail dimensions were 28ft luff, 7ft head, 24ft foot, and 35ft leech. The area was 460 square feet and the rig was powerful, effective and easy to handle compared with the long gaffs and enormous areas of the later catboats. The proportions of these sails were almost those of the mainsails and foresails of the fast American pilot boats from the Chesapeake, Delaware and New York, discussed in another chapter, and were also reproduced in the racing catboats of the 1920s, whose Swedish rig mainsail had a slightly longer gaff, but were otherwise remarkably like this old rig. These Narragansett boats were not referred to as cats but were called point boats, from the point of land in Newport harbour where most of them were built by several small boatbuilders. Many point boats could be fitted with a plank bowsprit to set a foresail when passage making, and this was slipped over the stemhead. The bobstay was an iron rod which hooked into an eye on the side of the stem at the waterline, and the whole of this gear was easily removable.

The point boats were gradually replaced during the middle-19th century by shoal draft centreboard versions which eventually adopted the long-gaffed cat rig, the beam/length ratio increased, and the keel became a flat plank. The reason for this change is not obvious.

Many catboats of this type were built, several by the Herreshoff brothers of Bristol who became noted yachtbuilders. Their ability showed in cats such as the *Trojan*, launched in 1872, a 26ft 6in boat which was remarkably well balanced and would keep a course with the helm lashed. In common with many eastern cats she set a staysail on a short bowsprit, which was fitted with a diagonal club or sprit, a rather doubtful method of flattening the sail for windward work.

Larger pointed-stern craft similar to the point boats and clench-built were used between Cape Cod and the western end of Long Island Sound for menhaden fishing during the mid-19th century, primarily for carrying the catch to shore canneries. They used purse seine nets which had been developed at Portsmouth, Rhode Island about 1845, and which were worked by similar boats. Most carried the same cat rig as the point boats but some, possibly in the sound had triangular or leg-of-mutton mainsails. In 1864 two Rhode Island firms opened canneries at Bristol and Blue Hill, Maine, and probably took this type of boat with them. Shortly afterwards a cannery opened at Lubec, Maine, on Passamaquoddy Bay, further east, and by the 1880s similar but deeper, carvel planked, cat rigged boats were working a growing fishery which spread, with the type of boat, to Eastport; and ultimately to the Canadian side of the bay, where many boats were built at St Andrews. They became known as Quoddy boats, after the bay, but were locally called pinkies which points to their alternative origin from the old pinky schooners which had comparable hull form; deeper and of greater displacement than the Rhode Island type. As well as purse seining some drift-netted sardines, and others carried away the catch. As well as the short-gaffed cats

Una boat *Vigia*

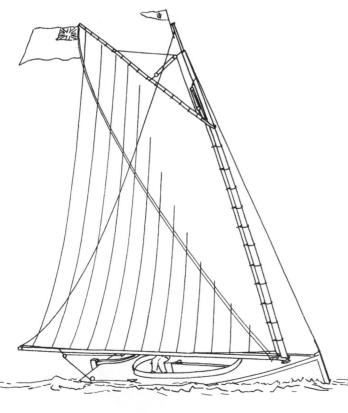

many were rigged as sloops. But with either rig they were often handled by one man and the main halliard was rigged with a single part to ease work. An average length was 35 to 38 feet and they were arranged with a two-berth cabin forward, with a small coachroof, large fish hold amidships, and a small steering-cockpit aft. All ballast was inside, usually of stone but sometimes of iron or iron ore. They spread along the Maine coast, fishing alongside the transom and counter sterned Maine sloops, and were much used for live lobster carrying, particularly in Penobscot Bay, and reckoned drier and abler boats than the sloops.

The water-carrying boats of Gloucester and Boston harbours were among the larger catboats. These were keelboats of about 40ft length and, 5ft draft, flush decked, with large wooden tanks within the hull and a small cabin aft for the two hands. Many of them were built at Essex, Massachusetts, then noted as a building place for New England fishing schooners. Under short-gaffed mainsails of large area, the waterboats proved the handiness of the cat rig by working about the crowded harbours and wharves, lying alongside ships and schooners to discharge with a hand pump. Shrouds and a forestay were often fitted, and double topping lifts and elaborate lazyjacks were common.

During 1853 a 16 foot New York pleasure catboat was shipped to England for Earl Mount-Charles who bought her when visiting Bob Fish's boatshop at Pamrapo, New Jersey. She was of the local shallow centreboard type, and he named her *Una* because of her single sail hoisted by a single halliard and trimmed by a single part sheet. She was sailed for a season on the London Serpentine without attracting much attention, but when she was taken to Cowes her unusual simplicity, handiness and speed caused a sensation and, within a year the Solent had a number of locally built Una boats, as the type came to be called in England, all copied from Bob Fish's creation.

The Una's single halliard arrangement is worth describing. The standing part was made fast to the gaff about a quarter of its length from the peak, led through a double block shackled to the mast, down to a single block at the gaff jaws and up through the double block and down to the deck, where a lead block led the fall aft to the helmsman. Double topping lifts were fitted. Her mainsheet arrangement was equally simple, one end being shackled to an eyebolt on one quarter and led through a single block on the boom, through a block on the opposite quarter, with the fall belayed in the cockpit. The surviving English Una boat *Vigia*, retains this practical sheeting arrangement which is both efficient and effective for estuary sailing.

The popularity of the English Una boats was confined to sheltered waters and, even there, in Britain's average weather, they could be wet and wild, especially when running or going to windward in a small sea. Their lines, arrangement and rig closely followed the original *Una*, and most were designed and built by C. Corke of Cowes, including, the typical *Vigia*, launched in 1872 and still sailing as a yacht, owned by Peter Sainty of Arlesford, Essex. Her dimensions are 23ft 9^3/$_4$in x 8ft 4in x 2ft 11in x 5ft with plate lowered and her hull and deck remain as originally built, but a light cabin has been added over her oval-shaped cockpit and she is now rigged as a bermudian sloop. She is as fast as ever and remarkably close winded in smooth water. Her original mast step and bowsprit are still fitted.

Sometime during the 1880s *Vigia* migrated from the Solent to the east coast and apparently led to the building of Una boats there, especially on the Orwell where Gildea & Co. of Ipswich designed and built the similar sized *Wenonah* in 1886, and four years later Orvis and Fuller of the same port built the very similar *Tern*. These, with *Puritan* and other Unas, raced and day sailed in the surrounding rivers and estuaries, notably between 1893-98, after which they dispersed, and *Vigia* seems to be one of two survivors, the other being the *Emma*.

The hulk of another Una boat laid at Wivenhoe until recently and was referred to as the old Bubfish. It is possible that this was another product of the builder of the original Una, but more probably she may have been built in Germany where the Una type also achieved limited popularity, but were known as bubfish boats.

CHAPTER EIGHT

THE SLOOP

THE MODERN SLOOP is a single masted, fore and aft rigged craft having a mainsail and a forestaysail hanked to the forestay, and is the commonest type of rig now in use throughout the world. The sloop's origins are complicated; prior to the 19th century seafarers often classified craft by hull form and characteristics rather than rig, and the present term sloop denoting rig, derives principally from the Dutch Sloep, used in the 16th century to describe a ship's open rowing boat. As the size of ships increased, so did that of the sloops and, to reduce labour of pulling, they were rigged with one or two masts, and sails of various types. Gradually, large numbers of sloeps were built for coastal and inland use in Holland; a common rig being a sprit mainsail and a forestaysail hanked to the forestay.

The type spread to England and other countries bordering the North Sea, Channel and Baltic, and finally evolved with a gaff mainsail. Writing in 1780, Falconer, a noted authority, defined a sloop as 'a small vessel furnished with one mast, the mainsail of which is attached to a gaff above, to the mast on its foremost edge, and to a boom below, by which it is occasionally shifted to either quarter. It differs from a cutter in having a fixed steeving bowsprit and a jib stay, nor are the sails so large in proportion to the size of the vessel.' The single headsail of the sloop was properly called the forestaysail, as it hanked to the forestay which gradually became more frequently set up on a bowsprit, a standing spar which could not be run in as the forestay supported the mast. Later sloops often set two headsails, in the manner of a cutter except that both the forestay to the stemhead, and the outer forestay to the bowsprit end, were permanently set up, and the bowsprit remained a standing spar. The forestaysail was hanked to the forestay and the jib,

instead of being set flying as in a cutter, was hanked to the outer forestay or jib stay as it became known. These were the principal differences of rig between sloops and cutters, but there were differences of design and arrangement.

English sloops of that period had greater beam, were shallower and slower than contemporary cutters, and were extensively used for coasting and fishing, and occasionally on ocean voyages. The early 18th century timber trade between North America and England was principally carried in sloops, for political reasons. Sloops shown in Chapman's wonderful *Architectura Navalis Mercatoria*, 1769, were predominantly Baltic craft, much coarser in form than the English cutter also illustrated. The sloops' masts were little forward of amidships, and the rig illustration (figure 56) shows the sloop with an enormous bowsprit and jib-boom, setting, besides the sails already mentioned, a flying jib and a square topsail, with a topgallant over; altogether a saucy rig.

The English merchant sloops of that period had greater beam and were shallower than contemporary cutters. These tough, hard-worked little ships were extensively used for fishing, and in trade; carrying coal, grain, and other homely cargoes coastwise, but were liable to be suddenly ordered on deep water voyages. William Stanton of Deal, who became a noted Channel pilot, recalls in his fascinating journal the voyages of the Ramsgate sloop *Speculation* during 1817, which seem typical. After a passage from Shields to Bridport, Dorset carrying coal, she sailed round the land for a similar cargo from Milford Haven to Whitstable, Kent, and from there in ballast to Portland, loading stone for Dover Harbour. A mid-Atlantic voyage to St Michaels, Azores, followed to load oranges but, these not being ripe, she took a cargo

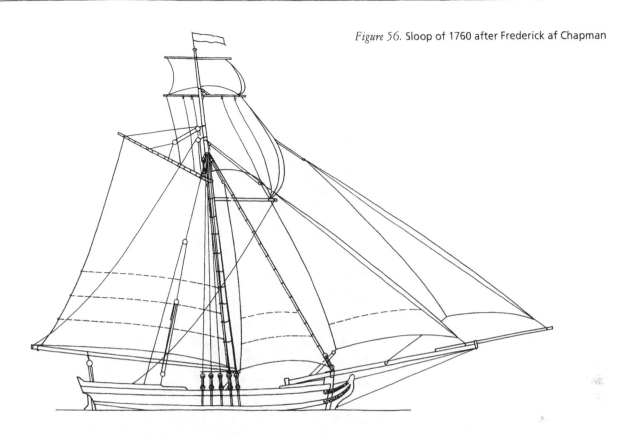

Figure 56. Sloop of 1760 after Frederick af Chapman

from Fayal to Gibraltar, returning with rockstone, and arrived back at the Azores weather-beaten and with gear in bad shape, to load her oranges for London. Dogged by severe gales the battered sloop worked back to the Channel and, after several misadventures made Ramsgate, where she discharged.

The sloop rig survived commercially in England until recent times. Until the 1920s commercial traffic on the Rivers Humber and Trent and their maze of tributary waterways was principally carried by simply rigged, bluff bowed barges called keels averaging 60ft x 15ft x 8ft depth and loading about 100 tons. The stern was only slightly less full than the bow and a large hatchway occupied most of the deck. Leeboards helped them work to windward under a large squaresail, though later keels also set a square topsail and sometimes a topgallant.

With the aid of numerous winches they were smartly worked by two men who lived aft. Most were built of wood, some of iron or steel, in small yards all over Lincolnshire and East Yorkshire. A smaller type without leeboards for upriver work were locally known as ketches.

The Humber sloops developed from these keels at the end of the 19th century, as handier and faster cargo carriers more suited to the lower Humber and its estuary. They traded well up the Humber and Trent, but seldom went far up the canals. Freed from the restrictions of confined waterways, the sloops were larger than the keels, carrying an average of 160 tons. Above water the early sloops strongly resembled small coasting billyboys and were heavily built of wood, with lumpy-looking round sterns, considerable hull depth, and shaped bilges. Continuous waist-high bulwarks were fitted, with a tiller port at the stern. The main cargo hatches, with well cambered hatch covers, left only narrow sidedecks, and a smaller cargo hatch was arranged forward of the mast which stepped in a wood tabernacle or lutchet. The tiny fo'c'sle was a store and the handspike windlass occupied the extreme bow and was exceptionally strong for frequent anchoring with heavy gear. The crew of two men and a boy lived in the aft-peak cabin, which was sometimes home for the skipper's family who worked the sloop and seem to have been skilful and courteous sailors.

Heavy wooden leeboards hoisted with tackles enabled the sloops to turn to windward quite well under their rig of loose-footed gaff mainsail, boomed staysail, and jib set on a long bowsprit which could be steeved up

Humber sloop *Esme*. Note steel hull, leeboards and staysail boom

out of the way in confined waters or berths. The jib was hanked to a stay and a jib topsail was also set, hanked to the top-forestay. Some carried fidded topmasts and set thimble-headed or headstick topsails. Many later sloops set a yard topsail to the pole masthead and for running, squaresails were often carried by large sloops.

Steel sloops were introduced early in this century, with the advantages of greater cargo space, a tight hull, and rather less maintenance. A typical sloop built on the Humber by W. L. Scarr had dimensions of 68ft x 17ft 3in x 8ft 3in depth and loaded 170 tons. They were very bluff-bowed but had a finer run to a stern which was full at the deckline. The hull was arranged as for the wooden

sloops but the main and fore hatches were combined and the mast stepped in a tabernacle supported by a heavy hatch beam, which assisted rapid cargo working. Three shrouds each side were set up with deadeyes and lanyards. Few of these steel sloops carried bowsprits, jibs, jib topsails or topmasts and the typical pole masted rig is shown in figure 57. The mainsail area was 1200 square feet, the boomed staysail 260 square feet, and the relatively forward position of the sail plan is interesting. A powerful hand-winch for raising or lowering the mast was sited at the forward hatch coaming or hedledge, and the halliards were led to winches (called rollers) fixed about the mast, above the coamings. The staysail halliard led to a barrel on the main halliard winch and the leeboards' tackles were usually led to winches to ease the work of the crew. Only a footrail was fitted around the sheer, with guardrails around the after deck.

These sloops ventured coastwise; to the Wash fairly frequently and Bridlington, Yarmouth, Lowestoft and the Thames occasionally; but they were poor seaboats, really at home in the fierce tides of the Humber and Trent, where George Holmes caught their character in his delightful drawings.

The sloops were famed locally for their speed, and the best were said to beat a steamer from Hull to Goole – provided they had a fair wind and tide – in the days when steamers averaged nine knots.

The Humber sloops raced annually at Barton Sloop Regatta for many years, and as late as 1928 there were six starters. The last two Humber sloops working under sail were *Ivie*, unrigged in 1949, and *Sprite*, cut down a year later; both owned by James Barraclough, the Hull coal traders who had 15 sailing sloops working in 1939.

Sloop rigged barges called flats were also used in the Lancashire and Cheshire rivers and coastwise. They developed from the small, transom sterned, flat bottomed barges, which traded on the Rivers Mersey and Irwell after about 1725, when the cotton trade's expansion led to improvements in navigation. They were also used on the River Weaver, Cheshire. At the beginning of the 19th century the single squaresail rig changed to a spritsail and forestaysail sloop, and the cargo capacity increased to the greatest size which would fit the locks. Some flats were gaff rigged by 1827 when the *Elizabeth* was launched by Dobsons, on the Weaver, her dimensions of 62ft x 17ft x 6ft depth being typical of the early 19th century flats, as was her transom stern.

As trade increased throughout the mid–19th century, several hundred flats were built and in regular use principally for carrying salt from Winsford to Liverpool, with coal as return cargo. In later years many were built and owned by firms whose goods they carried, and scantlings were heavy to withstand rough usage,

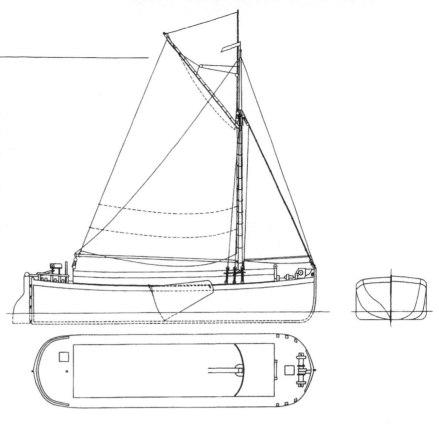

Figure 57. Humber sloop, 1908. Steel hull. Dimensions: 68ft x 17ft 3in x 8ft 3in draft

including taking hard ground when loaded. Framing, beams and forward planking were of oak, bilge strakes of 4in rock elm, bottom planking of pine and the side and deck planking usually of pitch pine. The 52-foot flat *Avon* built at the Cross yard, Winsford in 1868 was typical of the double-ended form finally evolved after enlargement of the navigation locks in about 1860. The flat-bottomed hull had a round bilge of small radius and the stern was rather finer than the bow, with slightly raking stem and sternpost. Rigged as a gaff sloop, the mast was stepped in a tabernacle just over $\frac{1}{4}$ length abaft the stem. Two shrouds were fitted on each side and the forestay had a stemhead tackle for lowering when passing bridges or going alongside ships. The forestaysail hanked to the forestay and shackled to an iron horse across the foredeck. The mainsail had a short luff and a high peak, and was loose-footed with a moderately long boom and three rows of reef points. Separate peak and throat halliards, topping lift and a peak downhaul were fitted, and the mainsheet led through three single blocks, the lower attached to the main hatch coaming. In later flats the halliards and topping lifts were often led aft to belay by the helmsman and some craft had crab winches for the halliards. The windlass straddled the foredeck with the forward cargo hatch abaft it, and a small hatch to the forecastle store between it and the stem. The main cargo hatch extended from immediately abaft the mast to the bulkhead dividing it from the tiny after cabin for the crew of two. Frame heads were let through the deck at intervals, to form bollards. The large rudder was tiller-steered and excessive weather helm appears to have been common. No leeboards seem to have been fitted and how they got to windward is a mystery, though one suspects the strong local tides did most of the work.

Later flats ventured coastwise to the Dee and north Wales ports, or north to Preston, Barrow and Fleetwood; adventurous voyaging for the river skippers, one of whom described his voyage to north Wales from the Mersey to an enquiring public bar as 'Nowt to it lads. Tha' needn't be particular to a handful of pints (points). Let her go as much one way as t'other and tha'll make as good a fetch as most of them'.

Larger flats up to 73ft x 18ft x 7ft 6in were built for these trades, many rigged as cutters, but later converted to ketch or jigger rig, as it was known locally, while a few such as *Goldseeker* were rigged as schooners with fixed masts. The introduction of steam flats and development of railways diminished this trade and many smaller flats became coaling lighters at Liverpool, eventually being unrigged, though about 20 were still sailing in 1931.

The sloop type spread to north-eastern America with Dutch colonisation, generally retaining its form and rig of sprit mainsail and forestaysail, but sometimes setting the double masted rig favoured in the shallops, which were also adapted to the new waters. By the end of the 17th century American sloops had crystallised to single masted craft setting a short-gaffed mainsail, forestaysail and jib. As they developed for coastal trading, fishing, and eventually ocean voyaging, the size increased, the gaff lengthened, and up to four headsails were sometimes set, especially in the deep-sea sloops, which also carried square topsails and often a lower course for passage making. The ability of these American deep-sea sloops was demonstrated by the 98 ton *Union* which, in 1794, sailed from Newport on a trading voyage lasting $22^1/_2$ months, during which she circumnavigated the world and made good weather of it, averaging 130 miles daily run.

However, the more suitable ships and brigs soon took over such heroic voyaging and, as the deep-sea sloops died out the coastal and inshore sloops came to represent what is now regarded as the typical North American sloop rig; developing shoal draft, beamy hulls with gaff mainsails and single headsail and, often, a jib-headed topsail, arranged for convenient handling and frequent turning to windward in inshore conditions.

The trading and passenger sloops of the New York area, particularly those working as packets on regular voyages between New York and Albany, up the Hudson or North River, were among the largest types of American working sloops, and showed considerable Dutch influence. Conditions required good windward ability for sailing up-river among steep hills and mountains which created frequent, fierce squalls, in which sloops were sometimes lost. About 100 sloops sailed the Hudson in 1790. Their mainsails had a long luff laced to the mast, a short gaff, and the foot laced to the boom. The large diameter mast stood without shrouds and stepped well forward. A large forestaysail was set, often with club or boom, on a forestay extended by a short bowsprit. Some had the throat and peak halliards arranged in one halliard, for ease in letting-go when luffing to heavy squalls when the mainsail dropped between three or four sets of lazyjacks, until the wind eased. Three lazyjacks were usually fitted to the foresail with a yoke around the forestay, which hoisted them with the sail.

Two men could sail one of the early sloops, which were generally fast and of good form, but others were purely carriers, often clench-built with apple bows emphasising their Dutch origin. Many carried hay and straw fodder to New York from country farms, bustling along with specially small mainsails set above huge deck stacks, reminiscent of the English spritsail barge.

The 125 ton *Illinois* typified the largest form of the Hudson packet sloops at the height of their passenger activity in 1825, before steamers spoiled the trade. This 75-footer had 20 feet beam, no centre board, and drew 12 feet loaded. Her topsides were painted in strakes of contrasting bright colours, as were most of her contemporaries. She was rigged with mainsail, staysail and a jib-headed topsail which became popular in the type in about 1840. Her mainboom was 90 feet in length and gybing meant hard work at the tiller. The poop contained two after cabins, two staterooms and a large

Figure 58. Hudson River sloop.
Hull dimensions: 73ft overall x 66ft waterline x 24ft x 7ft draft
ex. centreboard

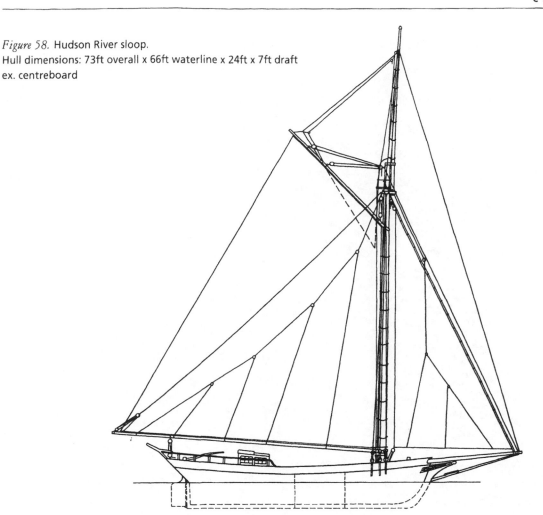

passenger cabin. Forward of this was the hold, divided into pounds by shifting boards to suit the various cargoes of produce and grain. The crew of four hands berthed in the fo'c'sle-cum-galley, where all food was cooked on a huge brick fireplace, with brick chimney, mantlepiece and hearth. Autumn harvest saw the sloops deep loaded with grain, the passenger cabins full of butter casks, meat carcasses and goods, and even a sheep pen round the quarterdeck.

Many of the 19th-century sloops were designed and built at Nyack, New York, by William Dickey who died in 1890 aged 80. The sloop shown in figure 58 is typical of his work and of the type. Hull dimensions were 73ft overall, 66ft waterline, 24ft beam, and 7ft draft, increased by a centreboard. The round-sectioned flat-bottomed hull had a long, easy run and easy entry which filled to the rounded, flaring bow above water. The small mainhatch was abaft the mast, and a smaller one at the

after end of the centreboard case which extended through the hold between them. The sail plan had developed with a fidded topmast and jib-headed topsail, shrouds were ratlined and mast hoops held the main and topsail luffs. Separate throat and peak halliards were rove, but she retained tiller steering, a raised quarterdeck, flat transom stern with windows, and stern davits for the yawl boat.

Such sloops were still built during the 1870s but the handier and economical schooner rig had grown in favour and the last Hudson sailing traders were schooners which retained the hull form of the old sloops into this century, working in the brick trade before 1914.

Until the 1880s working sloops enjoyed racing against similarly sized yachts in the New York area and the Larchmont Yacht Club gave races specially for them. At least two, *Admiral* and *Commodore*, were enrolled in the fleet of the Brooklyn Yacht Club and regularly raced in its sloop class, when not dredging oysters in the lower

New York Bay clam dredging sloop *Anna V* of Highlands, New Jersey, circa 1905. Note lazyjacks

bay. They were owned by Captain Joe Ellsworth of Bayonne, New Jersey; one of a family noted in local fishing, and for designing and sailing yachts. These New York professional yachtsmen combined winter and autumn oystering with successful summer yacht racing, in similar fashion to the English professionals from the Essex Colne and Blackwater Rivers. Captain Ellsworth had an encyclopaedic knowledge of local waters, and sailed in the Yankee *Puritan* which beat *Genesta* for the America's Cup in 1885, largely through his advice. His wide oyster interests included ownership of the 60 feet long *Admiral* built in 1872, and *Commodore* launched a year later and having typical dimensions of 45ft hull

length, 41ft 4in waterline length and 12ft 6in beam. There was little difference in the form and rig of these sloops and their yacht contemporaries except that until the late 19th century, oyster sloops had an open hold with no sidedecks and their jib, or foresail, had elaborate lazyjacks leading to a travelling toggle which hoisted up and down the jib stay. Their two hemp shrouds each side were set up by deadeyes and lanyards, fitted with ratlines, and well served against chafe. The long and heavy fixed bowsprits were square at the stem, octagonal outboard, and round at the end.

The dredging sloops from Keyport, Atlantic Highlands, Belford, Jersey City and Perth Amboy were typical of hundreds of fishing and cargo sloops once working the bay approaches to New York, and the coasts, bays and sounds between Cape Cod and the Delaware. Carvel planked and of moderate draft, their waterline lengths varied between 28ft and 40 feet, and the hulls were full bodied, very beamy, and stiff; sailing faster than their full bows indicated. Below water the entry was fairly fine, the run long, and the keel straight. The freeboard was low, the sheer slack, and a clipper bow with trail boards set off the heavily raked, flat transom with its well rounded quarters. The rudder stock was plumb and tiller steering general. The rig was moderately proportioned and a topsail often set. The laced-footed mainsail had three rows of reef points and the gaff was well peaked. The large working foresail had a club laced to its foot, or was spread with a boom pivoted on the bowsprit. Standing and running rigging followed usual practice; deadeyes and lanyards set up the shrouds, double bobstays were fitted and lazyjacks were used on both lower sails. Topmasts varied in height, often being rather short and the mainsail and foresail were regarded as the working canvas.

The crew of two lived aft under a trunk cabintop, and the after deck was sometimes raised, quarterdeck fashion, inside the low bulwarks which ran all round the roomy deck. The hold was amidships and the small fo'c'sle held stores. Many carried their rowboat across the stern in curved wooden davits protruding horn-like from the quarters, similarly to the Hudson sloops.

The New York Bay sloops raced annually at the end of the 19th century for such practical prizes as a pig or a sack of coal. In 1911 the race was revived over a 12-mile triangular course off Atlantic Highlands, New Jersey, for money prizes. Nine sloops competed, many sailed by Scandinavian immigrant crews, and the 35 feet *Lillie B* took the championship pennant in close racing. Shortly afterwards auxiliaries were fitted in most of them, though some remained fully rigged into the 1920s.

Captain Joshua Slocum's *Spray* became the best known of this type of sloop because of her epic voyage round the world in 1898-99, described in the genial captain's book *Sailing Alone Around the World*, which most cruising yachtsmen have read more times than they can remember. The lines and sail plan of *Spray* are given in the book and her dimensions of 36ft 9in overall

x 14ft 2in beam x 4ft 2in depth, spartan arrangements, and heavy rig are food for thought when considering simple offshore cruising yachts.

Small sloops were widely used along the Maine coast for hand lining, tub trawling (long lining), lobstering, carrying light goods, and for party sailing with summer visitors. Until the 1890s the most popular type ranged between 16ft and 25ft length, with a draft of 3ft to 4ft 6in, increased by a centreplate. The sheer was bold, beam about a $1/3$ of the length, and by the end of the century some had clipper bows balanced by a short counter stern, though many had plumb bows and transom sterns, resembling the New England wherry, from which these sloops seem to have evolved. Stone ballast was stowed inside and they were reasonably stiff and safe, being decked all round with a large cockpit and sometimes a small cuddy cabin forward. In these early sloops the mast stepped about $1/4$ or $1/3$ of the length abaft the stem and its height equalled hull length. Shrouds were unusual except for carrying light-boards or if a topsail was carried, but a forestay and bobstay were always set up to the standing bowsprit. The gaff was sometimes hoisted by a single halliard, and the boom extended beyond the transom and was attached to the mast by jaws. The larger sloops sometimes set a topsail and were often rigged with staysail and jib.

This type of Maine sloop has become known as the Muscongus Bay sloop after the area where many were built, but they were also built all along the coast, mainly by fishermen-owners who turned boatbuilder during, the harsh winter, constructing sloops, wherries, dories, and double-enders for their own summer use and for sale to less skilled fishermen. Among noted builders were Alan Ratcliff, and Freeman Ratcliff and his son Paris; who lived at the mouth of the Wessawesskeag River near Rockland, and turned out many fast, seaworthy and well balanced sloops which were in great demand along the coast. They designed by model, from which offsets for moulds were lifted, without lofting. Having set up the keel, stem and sternpost, a few sawn frames were erected and fairing ribbands bent round and adjusted to the desired shape, to which other frames were fitted. Keels were of oak or beech, sawn frames of oak or hackmatack crooks, sheerstrakes of oak, and planking of pine or cedar. Many sloops were clench or lapstrake planked, and carvel

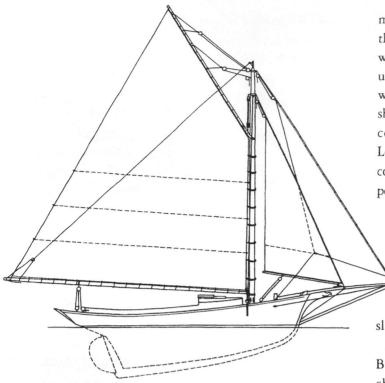

Figure 59. Maine lobster sloop, circa 1900.
Hull dimensions: 28ft 6in overall x 23ft 10in waterline
x 9ft 6in x 5ft 4in draft. Sail area 623 sq ft

most two men. The mast was now stepped about $^1/_5$ of the length inboard, so that the boat would pay off easily when hauling a string of lobster pots (the English term used in New England before trap became common usage) without any headsails set. The rig still often had no shrouds and the grown mast was heavy; a 28ft 6in sloop commonly stepped one 9in in diameter at deck. Lobstering increased in importance and a large deep cockpit or standing room was needed to carry dozens of pots, and the sidedecks were narrow with a low coaming to facilitate handling. The cockpit drained into the bilge but the sloops were quite seaworthy and their owners had great faith in their windward ability in bad weather. Accommodation with bunks for two and a coal stove was arranged under a low trunk cabin-top forward, with access through a sliding hatch (figure 59).

Large numbers of sloops were owned in Penobscot Bay and Isle-au-Haut. Their crews were colourful characters, generally dressing in suits of butchers' frocking to stand the work of pot hauling. They were always on the lookout for a good salvage and much of this seems to have been done by sloops when vessels were wrecked on the rocky coast.

Most sloops had a vertical flat bar or knife fitted on the underside of the tiller, which engaged in a curved rack or comb when dropped by the helmsman. When hauling pots the sloop luffed up, dropped her staysail, and with tiller set in the comb she tacked and gybed while the pot buoy was hauled and the pot came aboard for cleaning and baiting.

Bremen, Long Island was a most active sloop-building centre on Muscongus Bay with the boatsheds of the Priors, Carters, Collamores and McLains turning out many sloops; 22 in 1900. Charles Carter was typical of these Bremen builders most of whom were self-taught shipwrights. He built a sloop each winter for 26 years, working each one long-lining all summer and selling her in the autumn. His wife sewed and roped all the sails, which were cut by Charles, using a meadow as a floor. Oliver, Edward and Cornelius Morse were building centreboard sloops on Morse's Island by the mid-19th century in a yard inherited from their father John Morse, who died in 1835.

In 1874 Wilbur Morse, a young fisherman, built

planking was locally termed set work or smooth skin. Occasionally strip planking was used, edge nailed and set in lead paint. The sloops built in these small boatshops or by fishermen owners were usually of good materials, but the concentrated building of later years often resulted in plank fastenings of poor quality galvanised dumps, leading to early nail-sickness.

Sailmaking was often a community effort by the builders and their wives; the seamed-up cloth was laid out on a convenient meadow where the desired sail dimensions had been marked out with wood pegs, and was pinned and patted into shape under the critical gaze of the village experts, before being machined and roped. Exceptional peak was often given to the sails of lobster sloops to improve their speed to and from the grounds, enabling them to work more pots.

Towards the end of the century the sloops tended to increase in size up to 30 or 35 feet; centreboards became less common and draft increased. A typical sloop was 28ft 6in x 9ft 6in x 5ft draft. The healthy proportions of beam and draft assisted ease of working by one, or at

himself a 19 foot sloop and fished her that summer repeating the process for a few years in his father's boatbuilding shed at Friendship. In 1879 he moved to Bremen and commenced building full time. In 1882 he was back at Friendship building sloops in a shop away from the harbour with brother Jonah as partner and foreman. He designed by model and spent much of his time promoting the interests and name of what became a business specialising in quantity production of small fishing sloops at prices which tempted fishermen to buy, rather than build their own. In 1900 the firm moved to waterside premises at Friendship and their concentrated

output and subtle publicity caused the lobster sloops to become known as Friendship sloops to the rage of builders elsewhere. The firm were reputed to have built between 400 and 500 craft until 1924 when Jonah Morse moved to Damariscotta to start building on his own. Albion Morse, another brother, built about 100 sloops and Charles Morse had a yard at Thomaston. Some sloops were built in the most unlikely places; on the second or even third floors of buildings, and lowered to the ground. Sloops were built in this way at West Waldoboro, and on Morse Island, Warren Morse, son of Olin and third generation of island builders, reputedly built 'up to 45 feet' sloops on the third floor of his shed, but how he got them down is a mystery.

Larger sloops of identical form and rig also fished from Massachusetts. These were fully decked except for a small steering cockpit aft, had wheel steering, and low bulwarks around the sheer. A short topmast set a square-headed topsail and the shrouds were set up with deadeyes and lanyards. The shrouds were made of hemp until the 1890s. These sloops became popular after a disastrous fisheries slump during the 1870s forced fishermen to smaller craft, less expensive to build and maintain than the larger schooners. Between 1880 and 1906 numbers of fishing sloops were built for owners at Gloucester, Boston, Salem, Cape Ann and other places, to work the alongshore and fresh fisheries. Many were built at Essex, Massachusetts, and later some at Gloucester where they were often laid down on speculation in the slack shipyard times of summer. John Bishop, Hugh Bishop and Thomas Innes built many of the Gloucester sloop boats as the type became known, with Innes designing the majority, even for other builders.

The fish hold was amidships and some had wet wells. The fo'c'sle bunked four to twelve hands depending on the sloop's size, and the after cabin had two to four bunks and a stove. Up to eight dories were carried for various sorts of fishing. All the ballast was inside and a few, mostly from Salem, were fitted with centreboards. By 1906 powered craft and auxiliaries were affecting the trade and sloop construction petered out, though one or two yachts were built in imitation.

Sloop rig shared popularity with the schooner for American yachts and the huge sloop *Maria* was among the most fascinatingly original, and was reputedly designed by Robert Stevens, associated with his brothers John and Edwin Stevens, who were also her part-owners. Robert was a noted gentleman engineer and inventor and had previously designed the fast 70-ton Hudson River sloop *Dart* which won the working sloop class in the New York Yacht Club Regatta of 1847. *Maria's* design was evolved by sound reasoning, keen observation and accurate calculation and draughting, producing a shallow draft, beamy and flat-sheered hull with sections like a flat-floored dinghy, having a rounded stem, well cut-away forefoot, hollow forward waterlines and a fine run to a shallow stern. Principal dimensions were 92ft waterline length x 26ft 6in beam x 5ft 2in draft. A 15-ton iron centreboard increased the draft to 15ft when lowered by chains on differential shafting operated by 10 men. Later, a small supplementary centreboard was fitted in the deadwood to aid steering off the wind.

Her mainsail and foresail set 7,890 square feet and the white pine mast was 92ft long by 32in diameter at the deck tapering to 23in at the hounds, and was bored hollow with giant augers to various diameters averaging 9in. The mainsail luff was attached to the mast by slides which were small brass carriages with four wheels, running inside an iron groove on the after side of the mast. Shaped streamlining wedges were later added to the forward and after sides of the mast which had no runners or backstays. Five shrouds were fitted on each side and a single forestay; all of hemp, which stretched badly. The 95 foot long main boom was believed to be the first hollow one used. It was built of staves, dowelled and hooped together like a barrel, 28in diameter at the sheet iron and 13in at the after end. The solid spruce gaff was 61ft long and 9in diameter. The foot of the 2,100 square feet foresail was laced to a hollow boom 70ft long and 8in diameter, which sheeted to a horse across the foredeck. Its luff was 110ft and a greater area was set in this sail than is now carried by a 12-metre yacht's working sail plan. A 40 foot topmast was carried but the small topsail was only set experimentally. Robert Stevens calculated the wind resistance of the sail plan with vertically-cut seams and decided on cross-cut, an early use of this method.

She was built in 1846 by William Capes of Hoboken, being lightly constructed with red cedar and hackmatack, above the bilge. Her owners challenged any vessel in the world to race her off New York for $1,000 and, later that year, the Boston owner of the 66 foot schooner *Coquette* designed by Louis Winde, a clever Danish designer, raced *Maria*. Race day dawned blowing half a gale and Stevens declined to start, offering to pay *Coquette's* owner the stake money if he could sail round the 50 mile course in seven hours 'while I lie fast in my anchorage'. However, after argument *Maria* raced with a double-reefed mainsail. She steered hard and rolled viciously but took the lead off the wind until her huge centreboard broke and the rakishly seaworthy little *Coquette* romped home to luff alongside the buoy which a hand touched with a boathook, then the only formality in finishing an American yacht race. *Maria* reputedly sailed at 16 knots and in 1851 beat the celebrated schooner *America* in her trial races, probably due to the superior windward ability of her rig.

In 1852 the 20 foot 6 inch American sloop *Truant*, designed and built by Bob Fish, was the first American yacht to race on the Thames, beating a fleet of English deep-keel yachts by a considerable distance. She resembled a catboat hull with sloop rig and a small topsail and won races at Liverpool, Kingstown and Queenstown, probably due to well-setting sails and lightness of rig, but she was a poor seaboat. The following year the 205-ton centreboard sloop *Sylvie*, designed by George Steers and owned by Louis Depeau, sailed to the Solent and challenged all comers as her owners regarded her as the world's fastest yacht. She raced five English cutter Yachts and a Swedish schooner, but was soundly beaten by the Ratsey-built cutter *Julia*. *Sylvie* never raced again but cruised successfully in European waters before sailing for America.

The small sloop *Alice* probably did more to demonstrate the practical comfort and seaworthiness of American yachts to English yachtsmen than either *America* or *Sylvie*. She was designed and built by Townsend of Portsmouth, New Hampshire in 1866 for Thomas Appleton of Boston, as a wholesome cruising yacht 54ft overall, 48ft waterline, 17ft 6in beam, 6ft 10in draft aft and 2ft 5in draft forward, where her forefoot was well cut away. Her mainmast was 50ft long, topmast 20ft, bowsprit 18ft outboard and boom 54ft long. Her iron ballast was all inside. She set 1,500 yards of canvas and typified American sloop construction and rig with two hemp shrouds each side and a hemp forestay rove through the bowsprit and set up at the stem. The chain bobstay was set up with a rigging screw and the shrouds, forestay and the throat halliards were shackled to an iron mastband; a great contrast with the eyes of contemporary English practice. All halliards and the mainsheet had their standing parts fast and a single topping lift was fitted. Such refinements as halliard purchases, runner tackles, double topping lifts, mainsheets rove off both ends and wire rigging were not used in the USA until introduced by the English America's Cup challengers *Cambria* and *Livonia*, in 1870-1.

Alice's owner wished to use her in Europe and in July 1866 she was sailed over by Captain Arthur Clark, a Master in sail of American merchant ships with a crew of three hands and a steward; and accompanied by a son of the poet Longfellow and a young yachtsman named Stanfield. With the large cockpit battened down, a storm trysail and squaresail aboard, and the topmast housed, the little sloop made a fast passage of 19 days, 8 hours and 50 minutes from Nahant to the Needles; two days better than the much larger schooner *America*. *Alice* was invariably hailed by shipping as if in distress and needing assistance but arrived in good order. English yachtsmen visiting *Alice* were amazed at her beam and comfortable accommodation but, of course, these lessons went unheeded, probably because *Alice* did not race and, after a pleasant visit to Cowes, the Channel Islands, and France, she laid up in England having proved that a small, well equipped sloop yacht could be safely sailed transatlantic under ordinary cruising canvas. Her voyage did much to promote the celebrated transatlantic race of 1866 between the American schooners *Fleetwing*, *Henrietta* and *Dauntless.*

After 1870 United States yachting began to be influenced by England, particularly by the lean, deep, and overcanvassed double headsail cutters, and this movement of the American cutter cranks as they were called, culminated in the great cutter-sloop controversy of the 1880s. This was largely responsible for the reintroduction of a staysail and jib into the American sloop, besides modifying the sloop's shallow hull form into a deeper drafted and more seaworthy type. Thus the apparent anomaly of the American sloop rig with two headsails was established, and the type reached its extreme peak with the fabulous Herreshoft-built *Reliance* of 1903 which set 16,660 square feet in five working sails on one mast. She defeated the only slightly less extreme *Shamrock III* for the America's Cup.

The largest and most extreme English sloop ever built was *Shamrock IV* of 1914, and her rig had many experimental features which indicate how the gaff rig would have developed for racing yachts had the First World War, and also the bermudian rig, not intervened. Although one of the smallest challengers until that date (110 feet) she set the highest aspect-ratio rig and her hull form resembled a giant rater of the 1890s, having flat sheer, rounded and tumbled-in topsides, full ends, a plumb transom, and a deep, narrow fin keel. Her hull construction featured triple-skin planking on longitudinal frames supported by transverse webs, and she was longitudinally braced with alloy tube struts. The deck was of canvas covered plywood and a vertically-lowering centreplate increased her draft from 13ft 8in to 27ft. Her sloop rig of 9,860 square feet set an enormous staysail of 2,200 square feet area, the foot of which was spread by a hollow boom pivoted on the bowsprit. This worked with double sheets on a pair of travellers across the foredeck, with the sheet tackle led forward.

The 117 foot hollow spruce mast stepped well forward on the keel but was not supported at the deck, with just a rubber collar there to make it watertight. Only eight feet of mast was below deck and, to keep it straight, a jenny stay was strutted out on the foreside to oppose the thrust of the gaff jaws. About $1/3$ of the distance between this and the deck were five radiating struts on the fore side, over which wire stays were set up to check bending. The mainsail luff had mast hoops below these struts but was laced to the mast above them. All halliards belayed below deck and the throat halliard passed through the mast which was 23in maximum diameter but weighed only $2\,3/4$ tons. The short bowsprit was fitted on deck to strut-out the forestay, for it was fitted with a tensioning heel jack. The main shrouds were led to outriggers spreading them out to a secure staying base over the rounded topsides and the spreaders were canted upwards at right angles to the cap and topmast shrouds, one of the earliest instances of this now common practice.

This rig was fast to windward but slower than the cutters off the wind and she was converted to cutter rig. Then another sloop rig was tried with an even larger staysail before she finally reverted to a cutter.

Her Ratsey sails were almost perfectly setting and to increase the mainsail's efficiency, a canvas sleeve or luff casing was also fitted around the mast to fair the entry of wind on to the mainsail by covering the gap between the luff and the mast. Before *Shamrock* raced for the Cup in 1920 (the War having postponed her challenge) a further attempt was made to improve her windward performance by raising the sail plan's height and by rigging a guy from the end of each spreader to the gaff-end, where each lead through a single block and set upon deck forward, with a purchase. Their purpose was to reduce the mean angle of the topsail and mainsail when the sails were sheeted in hard.

With careful tuning and good sailing by her crew of 38 under Captain Albert Turner of Wivenhoe and Sir William Burton, the first amateur to participate in steering a Cup challenger, *Shamrock IV* won the first two races, but lost the series to the American *Resolute*, which ended the last gaff rigged challenge for the America's Cup.

Single headsail gaff sloop rig was little used in England and even in America it had declined in popularity although, theoretically, its large single headsail is faster to windward than the cutter's divided rig. However, this is not always so in practice. Nowadays, in Britain the rig is confined to small cruising yachts, principally sailing coastwise or in estuary waters, the limitations of size being that one man can only sheet home a single headsail of between 80 to 100 square feet in a hard wind, and if it freshens beyond this the single headsail must be changed; a more difficult task in a sloop than a cutter, where one headsail can be backed over to lie the craft to while the other headsail is lowered and changed. Another factor tending to restrict gaff sloop rig to small craft is that the forestay must be set up to the bowsprit end. If the bobstay or bowsprit breaks in a seaway the mast is likely to go too, though as many modern gaff rigged craft have very short bowsprits, and some have none at all, this risk is now minimised, particularly with the higher clewed headsails now used which hold little water when pitching in bad weather.

Gaff sloop rig is simple in having only one pair of headsheets to handle and setting the foresail on a Wykeham-Martin furling gear makes easy work when bringing up, but it cannot compare with the cutter as a rig for sea work.

CHAPTER NINE

THE CUTTER

THOUGH IT HAD DUTCH ORIGINS, the cutter is the most English of all rigs. A modern cutter is next to the sloop in weatherliness and its more divided sail plan is superior to the sloop for handling at sea with a limited crew. There is something noble about the cutter in bad weather. The big, close-reefed mainsail thrusts her to windward in heavy seas and the powerful bows cleave up showers of bursting spray. It is a rig for resolute men.

A true gaff cutter of the last 100 years may be defined as a single masted craft rigged with a staysail, jib, jib topsail, gaff mainsail, usually loose-footed, and a topsail which may be jib-headed or set with a yard or yards. The bowsprit should be reefing and the topmast may be housed in bad weather. The jib and jib topsail are usually set flying; the jib to a bowsprit traveller. The mast is placed between $^2/_5$ and $^1/_3$ of the waterline length aft of the stem.

The exact origin of the cutter is obscure. The term cutter was not used in Blanckly's manuscript *A Naval Expositor*, written in 1732, but a full description of a sloop was given, as quoted in the previous chapter. According to the *Oxford Universal Dictionary*, the first recorded use of the word cutter occurred in 1745, but the vessels existed before then. Cutters do not seem to have originated for naval use but the National Maritime Museum has plans of the British Naval cutter *Royal Escape*, dated 1735. Subsequently many cutters were purchased into the Navy, probably because of their sailing qualities.

The cutter rig has always been associated with speed and the English Act of 1784, for the prevention of smuggling, indicated that, at that date, a fixed bowsprit and jib set on a stay was one of the principal features distinguishing a sloop from a cutter, which had a running bowsprit. Cutters of that period were also noted as being of light construction for speed and, therefore, often clench planked.

In 1815 Falconer's *Marine Dictionary* defined a cutter as 'a small vessel commonly navigated in the Chanell of England, furnished with one mast and a straight-running bowsprit that can be run in on the deck occasionally, except which, and the largeness of the sails, they are rigged much like sloops. Many of these vessels are used on an illicit trade, and others employed by the Government to seize them ...'

A French work of 1783 stated that form of hull was a further distinguishing feature between cutters and sloops; cutters having greater depth of hull than sloops, less freeboard and greater draft, making them stiffer and more weatherly. Cutters were, even then, sometimes carvel but more usually clench built, for lightness, but the fine form of hull, particularly the sharp bottom, meant that the contemporary term cutter built had the same meaning as clipper built half a century later. This was confirmed when Captain Rogers of an Irish Sea packet cutter, questioned in 1819 on the best form of packet for speed, commented, 'a vessel between a cutter and a smack would sail faster than any vessel'.

This is confirmed by the British and European cutter brigs; 100-footers with hulls of cutter form but brig rigged, well armed and extremely fast. Probably the 400-ton *Viper*, which arrived dramatically during the siege of Gibraltar (1779-83), was a similar craft as she was designed as a cutter. Some later Naval 'cutters' were rerigged as ketches but retained their type name, such as the *Arrow* so altered in 1851.

At that period the masts of cutters raked aft, their spars were larger and they set a greater sail area than

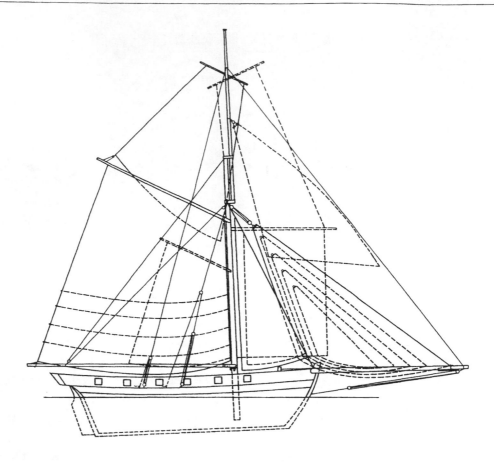

Figure 60. Sail plan of a government cutter, circa 1840. Some rigging omitted; note great variety of sails

sloops. In addition to the stay foresail, mainsail, gaff topsail and jib set flying from a traveller on the running bowsprit, Naval cutters set a greater sail area than the sloop and were the most extreme sail carriers in the Navy (figure 60). A ringtail extended the leech of the mainsail as a vertical strip spread by yards at head and foot, lashed or shipped on to the gaff and boom. Yards were crossed to set a main course, topsail (often with a very arched foot to clear over the jib halliards), a topgallant, and sometimes even royals, moonsails and stargazers, besides studding sails up to the royals, watersails under the boom, and 'below watersails' — an obscure term. Often the gaff topsail head was spread by a further small gaff complete with jaws and hoisted by one halliard. It was probably fitted to evenly distribute the halliard load on the sail head, rather than to gain a small sliver of sail area.

The early cutters may be regarded as derived from the slower sloop by being refined in rig, form and construction for speed in smuggling, smuggler chasing,

Naval use, privateering, and fishing and pleasure sailing, particularly in the North Sea and English Channel.

Development for speed was particularly marked in Revenue cutters, smugglers, yachts and packet boats. Several English builders were prominent among them Philip Sainty of Wivenhoe, Essex, a colourful shipbuilder of great talent and a noteworthy smuggler. During the late-18th century and until 1837, Philip Sainty built fast cutters, sloops, brigs, ships, luggers and smacks, first at Brightlingsea, then upstream on the Colne at Colchester Hythe, and finally at Wivenhoe, his native village. Sainty is reputed to have sometimes had Naval cutters building alongside fishing cutters whose extremely fine form indicated they were to be free traders. This situation was doubtless often repeated elsewhere as smuggling offered great incentive to builders and stimulated design in private shipyards to produce craft which Naval constructors could only imitate.

The principal districts in England noted for building fast cutters were Essex, Sussex, Hampshire and the Isle of Wight. Lynn Ratsey of Cowes and Ransom of Hastings rivalled Sainty's ability. The smugglers' cutters were

Revenue cutter *Badger* stationed at Bradwell, Essex, 1775

often large, well-manned and armed, and frequently faster than the Naval and Revenue cutters attempting to catch them. These latter comparatively short craft made a splendid sight roaring along with decks almost awash, rigging straining and the whole tower of canvas heeling in the puffs as the long bowchasers boomed at the equally fast, fleeing enemy, hoping to bring down a spar and disable him for boarding.

The author does not know the date of origin of Revenue cutters but in 1730 only one was stationed on the Essex coast; the 67 ton *Walpole* built at Ipswich in 1727 and owned and commanded by Captain Martin of Rowhedge. Surprisingly, he hired her to the Government for 30 shillings a ton annually and based her, conveniently, in the River Colne. By 1741 he was hiring

out three large Revenue cutters, the *Princess Mary* 73 tons built at Ipswich, the Dover-built *Good Intent* and the *Cornelius* built at Sandwich, Kent, besides chartering the cutter *Essex*, indicating that cutter construction was widespread in south-east England. The niceties of their rigs are unknown but interesting reference is made to one pursuing a 'smuggling sloop', and they were frequently in action against similar fast craft smuggling wine, spirits, wool, tea, silks, tobacco, lace and other contraband goods.

The description 'smuggling cutter' was commonly used by 1747 to describe a craft purposely built for free trade. One captured at Yarmouth of 30 tons burden was clench built of 1 inch fir and very sharp fore and aft ('the form of her after part is like an Irish wherry'), flush-

West Mersea Smacks, 1935. From Left; *Unity, Kingfisher* (far background), *Gracie* (foreground), two Maldon smacks and the *Boadicea* (extreme right). *Photo Douglas Went*

decked with wide hatches, frames spaced 3 feet apart and the hold unlined; all typical of a cheaply built, light, fast craft intended to recover her cost in a few trips. Some East Anglian smuggling cutters were large, such as Stephen Marsh's 140-tonner manned by 34 hands. Revenue cutters also grew in size and the *Repulse* of Colchester, Essex, built in 1778, was of 210 tons, manned by 50 men and a boy and carried 16 carriage and 12 swivel guns. She was one of six cutters all named *Repulse* owned by Captain Harvey of Wivenhoe, Essex, who, like Captain Martin of Rowhedge across the River Colne, hired them for Revenue service. In 1780 a fleet of four large and heavily armed smuggling cutters, manned by crews of English, Irish and French attempted to capture the *Repulse*, but failed, instead carrying off the Harwich revenue cutter *Swift* to France. Such deeds were typical of the English east and south coast smugglers of

the times and there were some furious battles. Disabled or unwilling smuggling prizes of considerable size were commonly towed in under sail by the powerful revenue cutters which made light of a barge or smack or even a large lugger or cutter.

The interdevelopment and versatility of the cutter type is emphasised by the celebrated racing cutter *Alarm* of 1830 which had the romantic origin of her lines being taken from a condemned French smuggler captured off the Needles and taken to Lymington for breaking. Again, in 1826 the captain of the Cowes-built 100 ton Revenue cutter *Vigilant* challenged three cutter yachts, including *Pearl*, to a race round the Needles and back to Cowes, which the new Government cutter won by such a margin that her cheering crew sailed her back to circle the yachts before they finished. Even the east coast Revenue cutters were not above an occasional race and in 1824 *Scout* beat

Lively, New Charter, Eagle, Fly, Sea Hawk and *Desmond*, over a sea course from the Cork Ledge buoy, off Harwich, round the Kentish Knock to Margate Roads. As further proof, in 1840 this same *Scout* captured the Ipswich smack *Rosabelle* smuggling tobacco, snuff and spices. Built at Portsmouth as a yacht in 1825 for a member of the Royal Yacht Club (forerunner of the Royal Yacht Squadron), the *Rosabelle* had been advertised for sale in 1838 as a pleasure yacht with stowboat gear; for fishing for pleasure was even then reckoned part of the east coast yachtsman's year, though spratting for pleasure would not be reckoned to be everyone's taste.

Naval cutters were maids of all work; scouting and acting as tenders for fleets, carrying despatches, chasing privateers and smugglers, protecting fisheries or charting unknown areas.

Successful cutters were frequently copied. The Revenue cutter *Diligence*, launched in 1799, was so fast that four new cutters were built from her design over 20 years afterwards with typical dimensions of 71ft x 23ft and a tonnage of 161. They mounted eight guns firing through ports in the high bulwarks and their crews of 50 had very cramped quarters.

As smuggling declined during the early 19th century, the Naval cutters had less coastguard work and some were re-rigged as schooners for surveying duties abroad, but fully rigged Revenue cutters were built in the 1840s and the coastguard cutters *Hind* and *Rose* were launched in 1878 and were still sailing in 1900, also being equipped with an alternative yawl rig.

Many early workaday cutters were comparatively short and beamy, with almost level keels and only slight rake of sternpost and round of stem, but the hull sections had great rise of floor and were hollow near the keel, and had fine waterline endings. The small Essex smack *Boadicea*, built at Maldon by James Williamson in 1808, is a survival of a typical little fishing cutter of that period. She still sails Essex waters having been carefully rebuilt, completely and faithfully, by the late Michael Frost of Colchester, who owned her after she ceased fishing in 1938. During reconstruction he removed the last of her original frames which still showed the joggles for the clench planking with which she was built, which was replaced by carvel during replanking at Brightlingsea in 1887. Her form exhibits the general characteristics of the old cutters; the graceful curving stem, chubby bows above water, and fine entry and run. The high-peaked gaff cutter rig, without topsail, was favoured for river and estuary craft working under sail and the transom stern is logical design for so small a craft. She fished from the River Crouch at first and, later, from Bradwell on the broad Blackwater River where being concerned in smuggling, she was noticed by preventative man John Pewter of the nearby fishing village of Tollesbury. He so admired her that after completing Naval service at St Helena Island in 1825 he bought *Boadicea*, which he and his descendants fished until 1871, when she passed to their relatives, the Binks. She continued working in the oyster fishery until 1938, sailing from West Mersea after 1917, owned by E. French. Michael Frost sailed her afterwards and stoutly resisted the fitting of an auxiliary engine, she is probably the oldest active seagoing craft in the world retaining the original rig, form and arrangement. She demonstrates the handiness of the old cutters by steering herself on many points of sailing, and her reasonable speed and general ability commends the old 'cods head and mackerel tail' hull shape for confident comfortable sailing.

After about 1750 the cutter became increasingly used for fishing in England, particularly for trawling, where its powerful rig and efficient hull towed the beam trawl very effectively. Their seaworthiness and speed were desirable qualities when working far from port, as was their handiness in confined waters. Cutters were also favoured for the considerable trade of shellfish dredging, where weatherliness and manoeuvrability were essential. Fishing cutters were in common use in Essex, the Solent, Barking, Brixham and elsewhere well before 1800, and their numbers grew rapidly after 1815. They were of all sizes from the humble 18 foot Solent boats to noble deep-sea trawlers over 70 feet in length, which developed when railways enabled rapid distribution of thousands of tons of fish trawled by fleets of smacks, to feed an ever growing and more concentrated population.

Plymouth owned many of the finest and fastest cutter smacks in England. They used the Cattewater as their home port and most were built on its shores. H. V. Prigg of Plymouth was their best known designer, and his craft were mainly built by Shilston. The cutter *Erycina* was his most noted fast smack and pride of the

Plymouth fleet from her launch in 1882 until laid up in 1934. She was 71ft overall, about 60ft waterline x 17ft 7in beam x 10ft 6in draft aft and 5ft 6in forward. Her fine bow, lean deck plan, and comparatively narrow beam, which was almost a foot less than the majority of slower smacks, were features often found in fast smacks from Essex and Emsworth. For ballast she carried 35 - 40 tons of pig and scrap iron cemented in her bilge. Hull waterlines and sections were beautifully drafted and fair, each section blending with the next. The slightly rounded forefoot matched a well raked sternpost with a deep, narrow rudder blade, and the sweeping sheer ended in a rather square and short counter stern. Construction was the usual elm keel; oak stem, sternpost, frames and beams; pitch pine keelson and bottom planking, with pine and oak above water and pine decks, masts and spars. *Erycina's* hull arrangement was typical of all large trawling smacks: the fo'c'sle had a bulkhead at the mast which formed a sail and gear store, and chain locker; the fish hold was amidships, and the coal bunkers and steam-capstan boiler immediately abaft it, the capstan being fitted just aft of the hold hatch. The crew of between four and six lived aft in a cabin with cupboard-bunks around its sides and a coal stove for heating and cooking, entered by a deck companion just forward of the long tiller. The Plymouth smacks had a gypsy windlass instead of the more common handspike barrel, and a powerful hand-winch was sited in the middle of the foredeck to handle halliards and warps. *Erycina* set the usual cutter rig with the addition of a lower and upper half-squaresail, set when fishing for hake, for which considerable speed was required. These sails were set from yards with jaws which fitted to the mast just below the hounds, and to the upper part of the topmast. They were supported by lifts and sheeted with a single part brace to the upper, and a brace with a purchase to the longer lower yard. When trawling hake the smacks sailed as fast as possible, keeping the trawl just clear of the ground. *Erycina* had the West Country cutters' pronounced characteristic of stepping the mast well into the hull; 28ft 6in from her stem. The boom was 49ft 6in long and the 40 foot bowsprit followed the sheer and had no bobstay. A shorter topmast was sent up in winter. All halliards were usually chain and the mainsheet of cable-laid rope. Some idea of the weight of

gear can be gauged from the deadeyes, which were nine inches in diameter.

In 1872 66 of these big smacks sailed from Plymouth, fishing up to 40 miles from the port; in summer beyond the Eddystone and in autumn and winter inside it, on a favourite ground 21 miles by 9 miles, often working westwards towards Looe and sometimes out of Penzance. Shipping constantly passed through this area and many Plymouth smacks were run down and sunk there. Although they did not venture very far from home waters the Plymouth smacks had to be particularly able seaboats as their grounds were fully exposed to the fierce south-west winter winds. Often, despite their ability, they had to shelter before beating seaward under close-reefed mainsails, with storm jibs bending the long bowsprits .

Erycina followed two other noted Prigg designs; the fast cutters *Vanduara* and *Camellia*, named after contemporary racing yachts which they resembled in many respects. Like a true cutter *Erycina's* best point of sailing was in a strong breeze, close-hauled; but a year after launch she raced and beat *Vanduara*, *Wild Fire* and *Coquette* in light airs. Other builders designed to beat Prigg's cutters, but were unsuccessful. In 22 Plymouth races *Erycina* won 15 firsts, four seconds and two thirds. Plymouth's rivalry with nearby Brixham found full expression in the regattas, particularly at Brixham where Plymouth smacks sometimes beat the native boats. Many Plymouth cutters were converted to ketch rig in the 1890s including *Erycina*, then 12 years old, but their mainmasts remained in the original position, resulting in a comparatively small mizzen stepped well aft but setting a topsail and mizzen staysail.

Before 1900 the Plymouth smacks sometimes mixed coasting with fishing and *Erycina* was one of several carrying early potatoes from the Scilly Isles to South Wales ports in this slack fishing time. Fourteen Plymouth sailing trawlers worked on into the 1920s, and a handful remained 10 years later when a few steam trawlers were owned there.

Fishing cutters worked from many British ports. Roving fishermen from Essex, Devon, Barking and elsewhere used Northumberland and Yorkshire ports and developed Hull and later Grimsby, into the largest trawling and line-fishing ports. Small cutters, often without topmasts, fished the maze of channels, shoals and roaring tides which is the Wash, on the east coast. They hailed from Sutton Bridge, Boston and Wainfleet in Lincolnshire, and Kings Lynn, Norfolk, and trawled shrimps and fish, dredged oysters and mussels, and stow-netted for sprats. An average boat was 30ft waterline x 9ft 6in beam x 4ft 6in draft aft, with long straight keel, square forefoot, some rise of floor and a hard bilge. Designed by half model, they developed extremely fine bows and flat inaccessible counter sterns, which were salted against rot of their red pine planking. They were decked, with long cockpits. In contrast, a dozen 30 ton cutters fished from Kings Lynn, trawling the Wash and North Sea, and oyster dredging on the Dudgeon. Many were built by Harris of Rowhedge, Essex, whence their owners had emigrated to Lynn. They had the local reputation of 'going to windward like a knife' and running, set large squaresails and often also square topsails, in addition to the gaff topsail, as late as 1890.

The famous fleet of Barking smacks removed to Yarmouth, Norfolk, in 1854, but that port and Lowestoft developed for trawling when cutters were being superseded by dandies and ketches.

Large cutters sailed from Aldeburgh, Suffolk; an ancient beach town also served by the tortuous River Alde at its back, which provided haven at Slaughden Quay for a small fleet of smacks fishing for cod. A typical catch numbered 5,000 cod and these were kept alive in a wet well fed with seawater by holes drilled in the smack's bottom, and were sold at Aldeburgh or Harwich, where they were killed before the buyers, with a stick, causing the crews to be nicknamed cod-bangers.

In medieval times fishing craft from Essex, Suffolk and Norfolk had sailed to Iceland and the Faeroes hand lining for cod, and the trade continued for several centuries with the catch salted on board. Well-smacks were introduced at Brightlingsea in 1701, and the cod smacks adopted the idea along with the Harwich fishermen who developed a large fleet of cod cutters. After about 1750 they worked long lines for cod off the Suffolk and Norfolk coasts during December, and on the Dogger Bank from January to April, making 10-day voyages from Grimsby. From May to November they sailed on the year's principal voyage to Iceland and the Faeroe Islands, hand lining for cod.

These Aldeburgh and Harwich cutters were powerful craft up to 80 feet length. Their crew of eight men could lay 20 miles of long line in two hours, its thousands of hooks baited with whelks or small pieces of tallow candle. By 1788 nearly 100 cod cutters were owned at Harwich and nearby Manningtree, up the River Stour, but after 1825 increasing numbers were converted to dandy rig by stepping a small lug mizzen, well aft. Later most were built as dandies and some as ketches. The bulk of this trade died by the mid-19th century but a few cod smacks sailed from Harwich and Aldeburgh until just after 1900. In summer some of them brought cargoes of lobsters and crayfish from Norway and also occasionally carried a few passengers and once, a cargo of Norwegian ponies.

The Essex cutter smacks from the Colne and Blackwater Rivers are described in chapter 11. Small numbers of similar smacks also sailed from the Rivers Crouch and Roach.

Barking, on a creek off the tidal Thames, developed as London's fishing centre and sent a fleet of fishing cutters to sea, trawling and long lining. Much squabbling has gone on as to whether the beam trawl was invented at Barking or Brixham. The argument is void because primitive trawls were recorded in use in medieval times in several places. It is certain that Barking was a well organised and progressive fishing community supplying the London market for several centuries and it developed a breed of cutter smacks around 50 tons, with wet wells to carry catches. In 1852 134 of its cutters trawled the North Sea and 46 long-liners fished for cod and haddock, but were forced to discharge their catches into floating cod chests at Harwich, due to vile pollution of the Thames, which contributed to the migration of most Barking smacks to a new base at Yarmouth in 1854.

The Hewetts of Barking, owners of the famous Short Blue fleet, were prime movers in this change. They also established the use of ice for fish preservation and introduced the fleeting system where large numbers of smacks remained trawling the North Sea in company for about six weeks at a time, before returning to port. Their catches were daily sailed back to Yarmouth or Barking, at first by one of their number who was relieved by a fresh arrival, if possible. Eventually, a system of specially built fast carrier cutters was established. On arrival among the fleet they hoisted a flag and each smack transhipped her catch in fish trunks aboard their boat which was sculled across to the carrier where it was boarded, tallied and stowed. This was extremely dangerous work and many lives were lost in bad weather. The fish carriers set clouds of canvas and their booms swept far outboard over the counter, needing footropes to enable a hand to reach the reef earring. Many of the fast Colne cutters were also used as carriers, but steamers eventually replaced them all by 1870, when the cutter smacks were also being replaced by larger ketches, or lengthened and re-rigged. The Short Blue sailing fleet was laid up in 1901, and sold soon afterwards. Many smacks of Hewett's fleet had suits of sails made by Ratsey of Cowes, in periods when yacht sailmaking was slack and any skipper fortunate enough to be fitted with these reckoned himself a cut above the rest.

Cutter rigged smacks sailed from the Medway where Alston's oyster dredgers brought home Channel oysters before they moved to Leigh-on-Sea. Small oyster dredging cutters worked from Whitstable and the Swale, generally resembling the Colne smacks but were fuller bottomed to take the hard foreshore which served for their harbour. They set a generous rig but could easily be distinguished from the Essexmen by their square, boxy counters and the large ports cut in the bulwarks to shovel the oyster shell and culch overboard.

Cutters fished from Ramsgate before dandies and ketches became fashionable in the 1870s, and Dover fishermen trawled the Channel in cutters alongside the Rye, Sussex fishermen's full-bodied 50 foot cutters. Rye was a noted centre for smack construction for ports such as Lowestoft and Ramsgate until the 1920s. The Emsworth, Hampshire cutter smacks were particularly refined in hull form and typical in rig. The sail plan of *Nautilus* is illustrated (figure 61), taken from the sailmaker's draught by Mr David MacGregor, and is reproduced by his permission. The Emsworth, Hamble, Southampton and Poole cutters are described in other chapters. Brixham fishermen trawled in cutters which they preferred to term sloops until the dandy rig became popular in the mid-19th century. A few cutters worked out of Dartmouth and Penzance but in the Fal a workmanlike type of cutter rigged oyster dredging boat developed, with hull form similar to the Quay Punt

(chapter 15). About 20 of the original wooden craft survive and 10 or more GRP craft have been built for working in the past 25 years; power dredging is still forbidden on local grounds. Each Falmouth Regatta Week they race daily as a class and form the last working sail races in England.

Fishing cutters also sailed from Mumbles and elsewhere in the Bristol Channel, the Welsh coasts, from Liverpool and the north-west, but the rig was rarely used for fishing in Scotland, a stronghold of the lug.

Deep, full bodied trading cutters also sailed the west Scottish coast bringing coal and goods to isolated communities and taking away produce, until they were gradually displaced by the steam 'puffers' after the 1880s. Many had pointed sterns and deep bulwarks. Most set square canvas and all were very strongly built to take the ground when loaded. The 29 ton *Catherine* owned by John Matheson of Kyleakin, Invernesshire built at Clifden in 1849, was typical in being carvel built and setting full cutter rig with running bowsprit, but was described in the Register as a Sloop, illustrating the pitfalls of rig nomenclature.

Cutters were frequently employed in the fruit trade before 1870, making fast passages in ocean conditions. The 265 ton *Margaret* was among the most developed and unusual, and was designed by G. R. Tovell of Mistley, Essex, on his patent principle where every section in her hull was part of a circle and the hull profile a sweeping arc. She was built in 1853 by John Mann at his Colchester shipyard and launched as a speculation for the passenger or fruit trades (figure 62). Her model was exhibited at the 1851 Great Exhibition. The boat's dimensions were 120ft overall x 24ft beam x 9ft 10in depth.

The rig was designed to combine the advantages of the schooner and cutter by stepping the mast amidships and regarding the huge boomed staysail as a schooner foresail. The jib, staysail, mainsail and topsail set 7,200 square feet and she also carried the usual light-weather and running sails of the day. Her mainmast was 88ft long, boom 64ft and gaff 33ft, but the bowsprit was only 17ft 6in outboard. With a flat sheer and open rails she looked rakish and was a very wet craft. John Martin of Rowhedge, the builder's foreman shipwright, recalled her crew claiming that they donned oilskins on getting

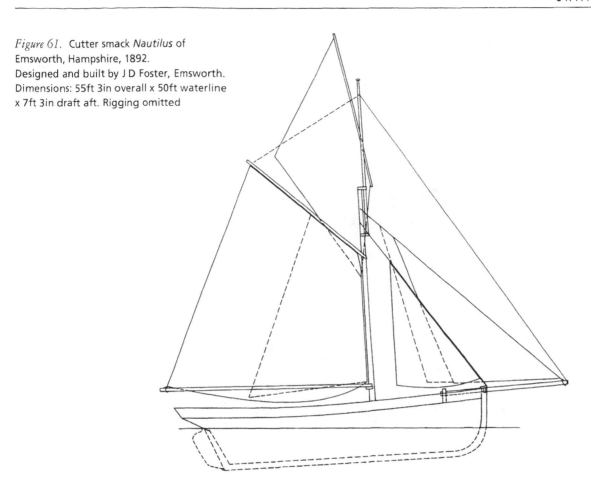

Figure 61. Cutter smack *Nautilus* of
Emsworth, Hampshire, 1892.
Designed and built by J D Foster, Emsworth.
Dimensions: 55ft 3in overall x 50ft waterline
x 7ft 3in draft aft. Rigging omitted

under way and only took them off when brought up. She made at least one record passage from Smyrna to London, in less than 74 days.

Full bodied, cutter-rigged smacks were also widely used in the coasting trades and for fast carriage of passengers, the Leith smacks being most noted, voyaging between Leith and London carrying cargo and about 14 passengers. One is known to have sailed the 460 mile passage in 42 hours, which is remarkable sailing in any age. They were very heavily canvassed and carried all possible sail at all times on passage and Schetky's lithograph of one running 'up Swin' in 1838 shows her setting jib, mainsail, topsail with a small headstick, a large squaresail with a half raffee topsail over, a studding sail, what resembles a spare jib set from the topmast head to the end of the studding sail yard and a watersail under the boom. She is racing along leaving the bulky Geordie brigs bobbing in her wake.

Similar fast smacks operated a service between Leith and Hamburg after 1815. Many were designed by Peter

Hedderwick and were built by Davey of Topsham, Devon, and Good of Bridport, Dorset, including the 196 ton *Eagle*, which in 1819 could sail from Leith to London in 50 hours and logged $10^{1}/_{2}$ knots off the wind. Her closest point of sailing was within six points of the wind and she could beat to windward better than the Harwich packets, in whose company she sometimes sailed.

Fast cutter rigged mail and passenger carrying packets maintained regular services across the North Sea, English Channel and Irish Sea before steamers became reliable. The Harwich to Holland packet service was operating at least as early as 1625 and became officially established in 1661.

For an unaccountable reason, Arundel, in Sussex, was noted for the building of cutters; several of the 18th century Harwich to Holland packet cutters built there included the 77 ton *Eagle* of 1703. She was typical at 53ft x 18ft 6in beam, carvel planked and round sterned, setting what we would now call a cutter rig, with running bowsprit. The packets were manned by a

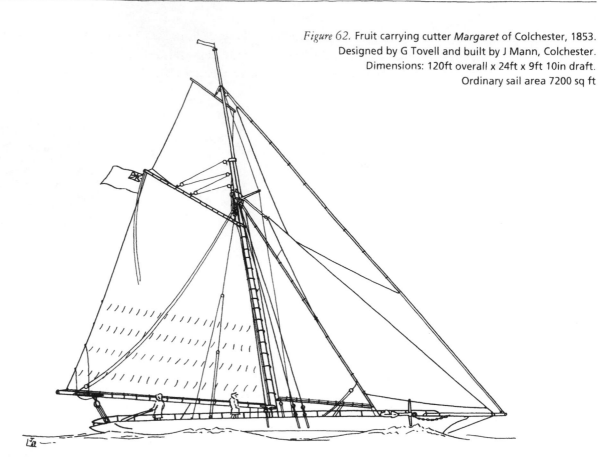

Figure 62. Fruit carrying cutter *Margaret* of Colchester, 1853.
Designed by G Tovell and built by J Mann, Colchester.
Dimensions: 120ft overall x 24ft x 9ft 10in draft.
Ordinary sail area 7200 sq ft

captain, mate and nine hands; and a packet skipper generally became a prosperous man of considerable local standing. Up to 26 passengers were accommodated in a large saloon with double tier bunks around the sides and one can imagine the misery of those often protracted crossings; sometimes the fault of bad weather but, occasionally, of the packet captain who saw the chance of capturing a prize and set off in chase. Four such craft maintained this service which was worked by fast sail craft until 1831. In war, even with Holland, the packet service flourished and usually expanded, the packets being regarded as neutral vessels.

In 1819 an official enquiry on the Holyhead to Howth (Dublin) packet service considered improved designs of cutters submitted by Peter Hedderwick (of Leith smack fame), Philip Sainty of Wivenhoe, and Sir William Symonds, the noted naval constructor. All draughted fast craft but Sainty's lines were by far the most subtle and rakish, and in great contrast to the full-bodied cutters then used, many of which were built at Portland, Dorset. They carried 24 passengers in cabin berths, with four carriages and their eight horses on deck. Despite this clutter they could beat the 54 miles across the Irish Sea in 14 hours or running in six hours.

Cutter-rigged barges, locally called sloops, carried stone and many other cargoes in the extensive Plymouth and Fal estuaries, and for short distances coastwise in Devon and Cornwall. On the Tamar they were also known as Blue Elvan barges as many were owned by the stone quarries of Landrake, Notter and Forder. Many of their skippers lived by the River Plym and some of the type became known as Laira barges. During the 1880s the barges also carried oak bark from woods on the Plym to the Millbrook tannery. Sometimes they traded alongshore to the rivers Yealm and Erme, or to Salcombe. After 1918 several carried stone from quarries at Porthoustock on the south Cornish coast.

Builders included Goss of Calstock and Bankson of the Cattewater. A few were built at Bere Ferrers on the Tavy and Saltash on the Tamar. Many of the fastest were designed (from carved half models) and built by F. Hawke of Stonehouse, Plymouth. They cost £500 complete. *Flora May*, which he designed and built in 1897 for Captain C. Daymond, was typical of the Tamar barges, with dimensions of 50ft 6in overall, 47ft keel x 15ft 6in x 5ft 6in depth in hold. Loaded with 50 tons she drew 6ft 6in aft and 5ft 6in forward and had just 2in freeboard amidships. When working outside Plymouth Sound she carried 5 tons less and had 6 inches freeboard. She had a slightly hollowed stem, flat sheer, long straight keel, raking sternpost and fine, high transom stern, and was remarkably shapely for a barge. Well rounded bilges amidships blended with a fine entry and a very easy run. She sailed well.

They were intended to take the ground loaded and construction was extremely strong with English elm keel, oak keelson, frames, floors and beams; hull planking of elm and oak fastened with galvanised iron bolts and treenails. The pine deck had oak hatch coamings and mast and spars were of pitch pine. Like most of the barges they were cutter rigged, without a topmast, as much of their work was in restricted waters. Topmasts were fitted to some barges later, particularly when working in the Sound or alongshore. *Flora May's* 13 inch diameter mast stepped in a tabernacle, supported by a stemhead stay-fall tackle. Bowsprits followed the sheerline and had no bobstay or shrouds. The loosefooted mainsail had three reefs and the large staysail, one. The long gaff put great strain on the short masthead and the rather square mainsail must have made her gripe. The 40 foot boom went well outboard and fitted to the mast with horns and a collar. The mainsheet led to combined mooring bitts and sheet horse across the transom, above the tiller; an arrangement peculiar to those built at Stonehouse, or for the Tamar. The staysail sheeted to a horse, curved round the mast and sharply curved back, outboard, to tension the sheet correctly on each tack; a feature seldom seen in sheet-horses elsewhere but an obvious improvement.

Jib halliards were chain but other running gear was manilla, the topping lift being unusual in having a tackle at the masthead. Eighteen-inch bulwarks and quarterboards were fitted. The foredeck had a heavy handspike windlass, bowsprit bitts and a large dolly winch for cargo handling. The main hatch stretched from aft of the mast to the cabin bulkhead, leaving only narrow side decks. Below, the fo'c'sle held stores, the

cargo hold occupied $2/3$ of the length, and right aft the crew of two had a snug cabin with a coal stove, entered by a hatch under the tremendous curved tiller which swept the after deck and indicated the amount of helm the barges needed in a breeze, beating up the Hamoaze crowded with the grey might of the British Fleet.

Basically similar, but longer and deeper cutters, sometimes called trading smacks, sailed coastwise with general cargoes in Devon, Cornwall and Somerset, and across the Bristol Channel; serving small rural communities where they were often built and owned in widely dispersed shares. They invariably carried a topmast and set a topsail with a short yard or headstick.

The 41 ton *J.N.R.* of Plymouth was typical, built by William Date and Son, Kingsbridge, Devon, in 1893 for Plymouth owners. With dimensions of 63ft overall, 58ft waterline x 18ft 3in x 6ft 7in depth, she loaded 70 tons on 8ft draft. Her hull form and construction were similar to that of *Flora May* except that her broad stern had a boxy counter, like most of her contemporaries. Sold to Truro in 1930 she relied on sail until 1932, when a tiny 5hp auxiliary was installed to aid her work carrying stone from coastal quarries at Porthoustock to the Upper Fal. Her full rig remained until 1939 and she retained a gaff 'sloop' rig until laid up in 1953.

Scores of these trading cutters carried much of the 19th century West Country trade, lying beached in tiny rock-bound coves with a gin block rigged over the hatch and horse drawn carts backed alongside to load or discharge. A favourite Bristol Channel cargo was limestone from South Wales to North Devon and Cornish limekilns returning with farm produce to Bristol or South Wales ports.

Cutter-rigged trows of the Rivers Severn and Wye were similar but shallower and usually undecked amidships, the hold being covered by tarpaulins from gunwale to gunwale, and protected from spray by canvas weathercloths; a most primitive arrangement, but which facilitated quick loading and discharge as a survival of their origin as lighters. Some larger trows made short voyages in the Bristol Channel, but many of these trows were ketch rigged and often fully decked, with hatches. The trows were poor performers to windward and one suspects that the fierce local tides did much of the work.

Fishing cutters of about 30 tons worked from Southern Ireland 60 years ago, notably with Skibereen registration, and similar craft joined in the coasting trade, predominantly carried in schooners.

Smaller, very deep, round sectioned fishing cutters called hookers sailed from the Kinsale area in the mid-19th century. A typical 19-tonner being 39ft 4in overall, 32ft 4in keel x 11ft x 7ft 6in draft loaded aft and 5ft forward. The bow was very round in profile and section and the counter stern exaggeratedly fine. They were sailed by six men and carvel planked, cost £120.

Until recent times smaller but similar three-quarter decked, cutter rigged boats called hookers were owned in Connemara, Galway and Kinsale. These had the same round bow but a raking transom stern, which had exaggerated tumblehome. No topmast was carried and the mainsail luff laced to the mast while the boom raked upwards considerably and joined the mast well below deck level, in the cockpit. These hookers were originally fishing craft but were also used for passenger and cargo carrying and kelp gathering.

Strangely, a variation of the type migrated to Boston with Irish settlers who built them there for inshore fishing in the 1850s. The rig developed to a laced-footed mainsail and they were known as dundavoes by their owners. They became extinct before 1914.

Cutters were extensively used in Scandinavia, Germany, the eastern Baltic and Holland but need a book to themselves for adequate description. Apart from limited use in the small Boston Irish fishing boats, cutter rig achieved very limited use in America, except in yachts which are described in the following chapter.

Cutter rig became extensively used in France for similar purposes as in England and survived until quite recent times in the west coast trawl and shell fisheries. Early-19th century trawlers from Rochelle, France, set a curious cutter rig, having a pole mast with two shrouds each side and a loose footed mainsail with a long gaff but a very short luff, the gaff jaws coming halfway between the hounds and the deck. A long yard topsail set above this and the rig was of rakish proportions. The beam trawls worked by these craft were rectangular, without 'pockets' or a cod end, and instead of iron the trawl heads had a large circular stone chained to each end of the trawl beam as wheels, on which it moved along the sea bed.

Cutter rig was much used for boarding and landing pilots in England, France and elsewhere, and is described in chapter 13.

CHAPTER TEN

CUTTER YACHTS

THE CUTTER RIG was, and still is, widely used in England, Europe and Scandinavia for yachts of all sizes. Sloop rig and the boomless gaff sail were common in pleasure craft until the late-18th century when cutters were developing rapidly. A yacht club was formed at Starcross, Devon, in 1773 and in 1776 Cowes held its first regatta, sailed by working craft. Southampton followed, and in 1783 a race for Essex fishing cutters was held in the Blackwater and was won by a Burnham smack. A few months later cutters from Rowhedge, Wivenhoe, Brightlingsea and Mersea raced for a silver cup and suit of colours. These events survive in the annual regattas of those places, where smacks still race. Similar racing spread along the east and south coasts.

In all these events yachts were present as a minority of onlookers and few existed before 1800, when there were probably not more than 70 in England, mostly cutter rigged. Such craft were generally built without thought of racing and to no arbitrary limits of size or rig. Most owners required comfortable craft of imposing appearance with ample accommodation for cruising coastwise and to the Continent and Baltic. Their crews were largely drawn from the Colne, Thames and Solent seamen sailing similar smacks and other small craft.

After about 1780, yacht designing and building began to emerge as a distinct branch of naval architecture and craft were built for pleasure sailing and racing. Many were small, but some surprisingly large for those troubled times at sea. At first there was great similarity between the largest and fastest of these yachts and contemporary government fishing, and smuggling cutters. Gradually yachts began to emerge as a type and speed became increasingly important as racing became

fashionable after 1800, often between pairs of yachts for high wagers, and many fine cutters were built in Essex, Sussex, Hampshire and the Isle of Wight.

Notable among them was Philip Sainty's most famous creation, the large cutter yacht *Pearl*; built at Colchester in 1819 for the Marquis of Anglesea, a veteran of Waterloo and confidant of kings and statesmen. Before 1809 Sainty had built, at his Colchester yard, the very fast cutter yacht *Emerald* for the Marquis who, home from the wars, wanted a giant cutter which was to be the fastest in the country. He found Sainty in gaol for smuggling, together with his son and brother Robert, but got them all out by influence, to build the new yacht which Philip designed with characteristic skill, as displayed by her recorded lines. *Pearl* was 113 registered tons, making her a giant among contemporary yachts which averaged between 20 and 70 tons. She displaced 127·5 tons on dimensions of 65ft 4in lwl x 19ft 6in x 11ft 6in draft. Her form was beautifully easy and fair and in an era of bluff bows she was notable for a fine entrance blending almost imperceptibly into a long run terminating in a delicately proportioned counter. Her form was typical of Sainty's work and was to influence other builders for many years. The hull was clench built below the wales and carvel above, as was then fashionable. The construction combined lightness with strength and was of excellent finish, the whole confirming history's high opinion of Sainty as a builder of fast craft.

In contrast to her hull the vast cutter rig would excite little comment from seamen seeking their winter living in fishing smacks and smugglers so rigged. *Pearl's* flax canvas totalled 3,218 square feet in the loose-footed mainsail, staysail and jib, but excluding the gaff topsail. Off the wind she set a square topsail, spread by a barrenyard below it, which was, typically, lowered to the bulwarks when not in use. For racing she set a longyarded gaff topsail and added studding sails to the topsail leech and a ringtail to the mainsail. In light weather the sail luffs were drenched with water to hold a better wind; some yachts being specially fitted with pumps and hoses to aid in skeating, as this practice was called. *Pearl* surpassed all expectations on trials and the Marquis settled £100 a year on Sainty on condition that he did not build a yacht for anyone else, but he broke the contract and quickly resumed shipbuilding at Wivenhoe, continuing to specialise in fast craft.

Cutter racing rapidly became popular but there was no time allowance then for differences in size. In 1825 the 85 ton Lymington built *Arrow*, designed three years previously by her owner, Joseph Weld, challenged *Pearl* to race from Cowes to Swanage and return for £500. The Marquis was so certain of victory that, in accepting, he vowed 'If the Pearl should be beaten I will burn her as soon as we get back.' He must have regretted his words early in the race when *Arrow* led, but *Pearl* eventually won by 10½ minutes and escaped a fiery end.

By then owners had discovered that a good large yacht could beat a good small one and a number of large yachts were built on the south coast to beat the Essex flier, whose bow they copied. In 1826 *Pearl* and *Arrow* met their match racing the 103 ton *Nautilus*, built by Lynn Ratsey of Cowes, which completely outsailed them for a prize of £500, although never again did *Nautilus* distinguish herself or they 'fished her mast and killed her', as the old sailors said.

Instead of the flying start we use today, these early races started with the yachts anchored or moored to starting buoys, about a cable apart, usually with all sails lowered, but sometimes with only headsails down. At the second gun all canvas was hoisted and anchors or springs rattled home in a frenzy of activity.

Cutter racing fever gripped yacht owners and sail areas increased alarmingly, accentuated by the 1826 ban on extra sails, which led to ordinary rig of immense size and consequently larger hulls to carry it. Ballast was increased and a proportion of it was shifted to windward when racing; a practice frowned on by many. All these

early races were sailed by professional crews, hailing mainly from the Colne and Solent areas, and great was their rivalry. They took racing seriously and there were incidents such as the fouling of *Arrow* by *Miranda* when the crews set about each other with windlass handspikes and axes.

In this atmosphere the peak of early rivalry dawned in 1823 with the launch of the south coast cutters *Lulworth*, 127 tons, for Joseph Weld, and the 162 ton *Louisa* for Lord Belfast. During the next two years interest focussed on the tussles between them and the *Meani*; all being stripped out for racing. The rivalry of Belfast and Weld culminated in 1830 with the building of Weld's giant cutter *Alarm*, queen of the south coast and the terror of racing men for generations. This 193-tonner was one of the largest cutters ever built. To match her Philip Sainty's Wivenhoe yard launched *Arundel*, a 188 ton cutter for the Duke of Norfolk, a yacht contemporarily described as 'the finest sea-going cutter ever built', and occasionally she outsailed both *Alarm* and *Louisa*, but inside the Wight the giant *Alarm* was supreme. Competition and experiment was rife and began a craze for lengthening and rebuilding the cutters as they became outclassed by new yachts, and *Arrow* was continually rebuilt and still winning prizes into the 1870s, and was not broken up until 1903.

After a decade of hotly contested racing these big cutters had their day in 1834, when many races were restricted to craft under 75 tons and the most colourful period of yacht racing's early days was closed. Racing and development of cutter yachts lagged for several years while various attempts were made to introduce fair competition by using tonnage rules measuring length on the keel and breadth, but ignoring depth and sail area. Although schooners and even brigs occasionally raced, the cutter rig was recognised as so superior in weatherliness that it was naturally selected for most yachts built for racing.

Space permits no more than an incomplete outline of cutter racing history and, as with all racing yachts until 1939, full credit for success should be given to their professional captains and crews whose considerable skill, judgment and determination often made the reputations of yacht designers and builders. Essex captains and crews from the river Colne were prominent in racing

The Marquis of Anglesey's noted cutter yacht *Pearl* 1818/19. Two views, with single reef in mainsail and squaresail yard stowed.
National Maritime Museum

throughout the period, and the skill of skippers from Rowhedge, Brightlingsea, Wivenhoe, and later Tollesbury on the nearby River Blackwater, was pitted against rivals from Itchen and Hythe, Hampshire, and other Solent ports; and others from Largs, Gourock and Port Bannatyne on the Clyde. The Essexmen found their winter living sailing their smacks; fishing all round the British Isles, North Sea and Channel, breeding fine seamen skilled in helmsmanship at close quarters and the subtle handling of fore and aft craft. Racing captains rose solely by merit in a hard school and were in complete charge of these often huge racers, which might have crews of up to 35 hands. Although the crews were poorly paid, substantial money prizes were then won by yachts finishing first and second, and a proportion of this was divided among the winning crews. Skippers whose

command showed a long string of prize flags at the season's end were well rewarded and many on the Colne built houses and smacks from this source.

Several early racing cutters were built of iron, but it did not become popular as the bottoms could not be kept smooth. In 1844 the iron cutters *Bluebell*, *Belvedire* and *Ariel* from the Thames raced at Cowes but were beaten by a wooden cutter owned and built by the Brothers T. and J. Wanhill of Poole, Dorset, who became noted yachtbuilders. In 1845 the big cutters received a rude shock when the little 35 ton cutter *Heroine*, built by Wanhills to the design of Thomas Musslewhite, their draughtsman, and sailed by John Nicholls of Southampton, beat *Alarm* which had been the champion racing yacht of England for years. The Wanhills' cutters were much finer shaped than any then built and instituted rule cheating by raking the sternpost to reduce keel length, decreasing the beam and increasing the depth, besides having considerable internal ballast to maintain power to carry a large sail area, without increasing tonnage. Unfortunately, many desirable qualities were sacrificed to gain speed. They were uncomfortably wet boats, pitching heavily with lean bows and heavy ballast, and needed a big crew to handle the large sail plans, besides having little space for accommodation.

John Nicholls later sailed the Wanhills' cutter *Cygnet* which was successful whenever she could carry her large yard topsail. The iron hulled cutter *Mosquito* was one of her principal competitors. Built by Mare's shipyard, Blackwall, to designs by Thomas Waterman, she was skippered by the outspoken Tim Walker of Itchen, who, like all yacht captains and crews, resented interference in yacht handling by the owner. In one hard weather race Tim's owner insisted the jib-header be changed for the big yard topsail, which was done after Tim expressed the probable loss of spars, only to lose the topmast in the next squall. Convinced Tim had 'filled on her' purposely to prove his point the furious owner blurted 'Damn you !', to which Tim politely returned: 'And damn you Sir!'

Wanhills' cutters were often winners, particularly *Phantom* and *Vision*, but fast craft were also being designed and built by Thomas Harvey of Wivenhoe, Essex. In 1832 he had bought Philip Sainty's shipyard and continued its tradition of building fast yachts, fishing and merchant vessels. In 1845 Harvey launched the cutter *Prima Donna* which quickly became top of the flourishing 25 ton class. His next big success was the 40-tonner *Amazon*, built at his other shipyard at Ipswich, Suffolk in 1851, and closely followed by the 48 ton cutter *Volante* which, sailed by Captain James Pittuck of Wivenhoe, had a tremendous success and was lying second to the schooner *America* in the historic 1851 race round the Isle of Wight, *Volante* having sailed the correct course, when she broke her bowsprit and gave up. Had the spar held, yachting history could have been radically different.

Thomas Harvey's son John was apprenticed to his father and showed great promise as a designer of very fast craft. When very young he designed and superintended the building of the 40 ton racing cutter *Avalon* in 1850

and next year was running the Wivenhoe Shipyard which during the following 30 years became famous for fast yachts of all sizes besides scores of smacks, and schooners for the fruit trades. In 1852 John Harvey designed and built the noted 10-tonner *Kitten* with very advanced hull shape; her profile being daringly cut away to reduce wetted surface, similarly to that later regarded as revolutionary in the large yacht *Jullanar* of 1875 whose design was actually carried out by John Harvey, though the credit is usually ascribed solely to her owner.

John Harvey evolved and draughted his designs from mathematical principles and was later a founder member of the Royal Institution of Naval Architects. He was famous for his cutters and the best included *Thought*, *Syren*, *Dione*, *Audax*, *Snowdrop*, *Resolute* and *Dagmar*, owned by the Prince of Wales, later King Edward VII.

In 1865 John Harvey had the curious experience of acting as joint umpire with Dan Hatcher, for a series of Solent races between the cutters *Thought*, designed by Harvey in 1852 and the iron hulled *Torpid*, the winner to become owner of both yachts and a £200 stake. *Torpid*, which carried seven racing captains among her crew of 20, won, and both cutters were towed triumphantly up Southampton Water by their victorious owner's steam yacht.

By the 1840s the traditional pattern of British yacht racing evolved; the season commencing on the Thames at the end of May and continuing until the end of August at scores of important regattas from Harwich, round the coast to the Clyde and including Ireland, the west and south coasts and the West Country, the racing yachts sailing several thousand miles during the summer.

Shifting ballast was banned in 1856 but during the 19th century 'the Cutter' became synonymous with a narrow, deep draught, plumb bowed yacht having a counter stern, oversparred and overcanvassed for hard weather, wet in a seaway but able to beat to windward through almost anything, provided she was properly reefed and the crew could stand it and stay aboard. This exaggerated type of cutter yacht resulted from the Thames Measurement tonnage rating rule which was introduced in 1854 and remained in force until 1878. This taxed beam twice but allowed unrestricted draft, forcing English designers to reduce beam and achieve stability by ballasting, at first all inside but later with

Racing yachts running under squaresails circa 1852. The Harvey-built cutters *Thought* and *Avalon* at Lowestoft Regatta. Note broads Lateener and Beach yawl. *Hulton Picture Library*

increasing amount hung on the keel. Length was measured from the stem to the sternpost, on deck, resulting in plumb bows and long counters and, as sail area was disregarded altogether, the powerful and versatile cutter rig flourished and attained excessive area. The legend of this exaggerated type of cutter yacht has persisted and is often, wrongly, regarded as the only true breed. Outclassed racing yachts became useful cruisers with sail plans reduced, and sometimes made seaworthy pilot and fishing craft.

However, the contrast with working cutters, often designed and built alongside cutter yachts, is interesting as in them builders retained to the end the healthy beam and proportions which were denied the racers.

Robert Aldous of Brightlingsea, Essex, built many fast cutters of both types and became noted for the windward ability of his 9 ton racer *Violet* built in 1855.

In 1858 Aldous built another *Violet* which, renamed *Christabel*, became the most famous of his yachts. Like many cutters of the time she was lengthened three years after launch and emerged as a fast boat, winning races all round the coast and the 'ocean' race from Cherbourg to Ryde. In 1866 she was again sawn in two and lengthened to come storming into Harwich winning the Down Swim race. In the boisterous Nore to Dover she led 17 big cutters and schooners, and the enormous 209 ton open lugger yacht *New Moon*, only to be just beaten by the big schooner *Egeria* within yards of Dover Harbour where 10,000 spectators cheered her gallant effort.

Until 1866 Aldous rivalled the five other noted yacht builders of that period: Harvey of Wivenhoe, Wanhill brothers of Poole, Dan Hatcher of Southampton, Inman of Lymington and William Fife II of Fairlie. Until 1880 Dan Hatcher, followed by his son, built many fast

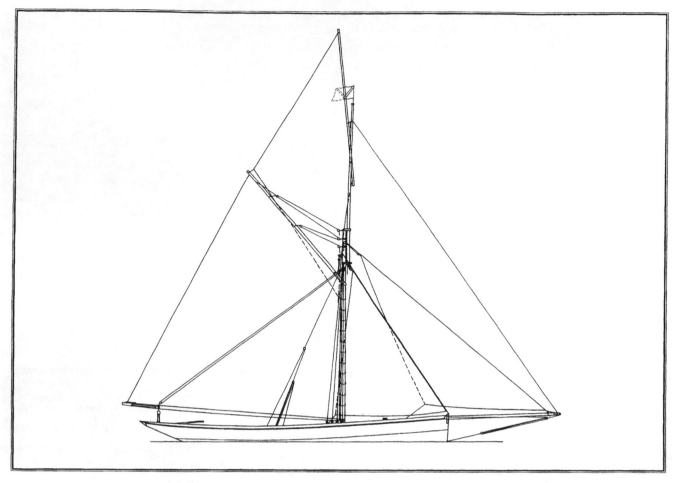

Figure 63. First-class racing cutter *Vanduara*, 1880. Designed by G L Watson. Built by D & W Henderson, Glasgow.
Dimensions: 86ft 1in x 16ft 2in x 14ft 6in draft

cutters up to the 40 ton class, his early success being the *Gleam* of 1855. Most were designed by William Shergold, Hatcher's draughtsman, were of composite construction and had beautiful lines.

When running, early racing cutters set a squaresail with a triangular raffee topsail above. In 1865 Hatcher built the cutter *Niobe* for William Gordon whose skipper was Captain Thomas Diaper (sen) of Itchen. Between them they devised a triangular running sail to be set from the masthead, with its clew boomed out in the manner which headsails had been boomed out for years before, except in racers where their sheets had to be helped by hand. Having tried it outside the Needles *Niobe* first set it racing on the Thames against the iron cutter *Vindex*, beating her by its efficiency when running. The sail became known as the niobe and was known in the racing fleet by this name for four or five years until the yacht clubs wanted a different name for it, when for

no explicable reason, despite its later use in the yacht *Sphinx*, it was called a spinnaker, and has remained so since.

After 1865 the south of England yachts had increasingly to reckon with competitors from Clyde builders. The first William Fife was a Scottish wheelwright who built rowing boats in his spare time at Fairlie, Ayrshire, on the Firth of Clyde. His boatbuilding prospered and eclipsed his wheelwright trade. He established a yard which built small commercial craft and yachts, principally cutters. Fairlie was a most unpromising site for what became a noted yachtyard. For many years yachts were built in the open on the rocky shallow foreshore which made launching of anything but small craft very difficult, and the anchorage was exposed.

Fife's son William (II) joined the business bringing a flair for design. The 55 foot cutter *Cymba*, launched from the yard in 1852, won many prizes but young Fife's masterpiece was *Fiona* of 1865, the most noted cutter of her day when sailed by Captain John Houston of Largs, with a crew of local yacht hands who, surprisingly, turned to weaving rather than seafaring for a winter living. For a time *Fiona* turned the tide of south country racing cutters' success. She was a noted early design of Fife II with dimensions of 75ft 7in x 15ft 8in x 11ft 10in draft.

Houston was a determined racing skipper. Once hard pressed in a breeze by the cutter *Vanguard* sailed by Captain Harry Thompson of Itchen, both yachts were carrying their huge spinnakers too long, running for the finish when they would gybe over the line. To save time Houston ordered *Fiona's* spinnaker to be cut away and with a thunderclap it blew clear as she gybed and won. John Houston carried sail excessively and frequently broke spars but his determination, and *Fiona's* weatherliness, established Fife's name in the racing world. The Fifes' next greatest success came in a roundabout way; in 1872 Boag, also of Fairlie, built the 36 foot cutter *Cloud* for the 10 ton class, and a lanky young Essexman named Lemon Cranfield, from Rowhedge, was appointed captain. He raced her with such success that two years later he was selected to race the 62 ton *Neva* just launched by Fife to challenge the big class. Lemon brought a crew of fellow villagers with him and made her name a legend for speed. At Rothesay in 1876 *Neva* raced *Cuckoo*, sailed by Jack Wyatt of Hythe, with a Hampshire crew, and beat her by one second. Lemon used his wonderful judgment in tacking to fetch the mark exactly, within a foot. During 1877 he won £1,335 in prize money and became the acknowledged genius of yacht racing in which no man, before or since, has had such spectacular success. His five younger brothers were also top racing captains.

Restriction of beam and practical ballasting considerations caused British designers to achieve stiffness by maintaining a relatively low sail plan, long in the base. Booms still extended well outboard and bowsprits were very long. Figure 63 is the typical sail plan of the large class racing cutter *Vanduara* built of steel by D. & W. Henderson of Glasgow in 1880, to designs by young George Watson, who had hitherto only designed small yachts.

The arrangement and proportions of her rig typify

British cutters from 1850 and 1886 as, while designers strove to improve hull form, construction and ballasting, sail plan design remained relatively stagnant. These yachts could be, and were, raced in very hard weather, when topmasts were usually housed and lower sails reefed and shifted. However, they sometimes lost bowsprits through bursting the bobstay and occasionally topmasts were carried away; both usually from pitching in a sea. When these racers were outclassed and converted to cruising the sail plan was reduced from *above*, not from each end. If it had been, probably a quarter century's progress might have been achieved at one step.

By the 1870s, materials were available to heighten rigs and steel wire came into general use for standing rigging, but change came slowly in the cutter rig. Rigging screws were still regarded with suspicion though they had been successfully used in the Plymouth cutter *Ada*, built in 1837, a prizewinner in south coast regattas for 40 years before she was converted to a fishing smack at Brixham.

Sails improved in cut and quality. The leading yacht sailmakers were Messrs Lapthorn of Gosport and Charles Ratsey of Cowes, who advocated use of cotton sailcloth in Britain for many years until its superiority over flax in racing efficiency was at last recognised in the 1880s and became universal until the 1950s.

In 1878, tonnage length was measured on the waterline and plumb stems were generally replaced by graceful clipper bows. Of course there was at first great controversy over these 'ugly' bows, and the proportions of length to breadth became even more freakish after another rule amendment in 1881. The five ton *Oona* built at Wivenhoe in 1886 was the most tragic racing yacht ever built.

Designed by talented 25 year old William Payton, she had only 5ft 6in beam on a 33ft 10in waterline, yet drew 8ft. Her displacement was 12·5 tons, of which her lead keel was 9·6 tons and construction was very light and strong with double-skin planking on steel frames. She set 2,000 square feet of working canvas and sailed from Southampton for the Clyde manned by an experienced professional skipper, two hands, her owner and the designer. Leaving Kingstown, Ireland, she met with a heavy gale, was apparently embayed off Malahide and attempted to beat offshore under trysail, which burst and she drove ashore, where the hull was torn from the keel and all hands drowned. That same year, the Clyde three-tonners carried the rule to its ridiculous conclusion in being 29ft waterline and 4ft 6in beam with enormous lead keels and huge sail area.

Several large racing cutters of the period were designed by Alexander Richardson, a Liverpool naval architect, including the 83ft *Samaena* (1880), *Lorna* (1881), *Irex* 88ft (1884) and *Iverna* (1890); the two last sailed by Captain William O'Neill of Kingstown, Ireland, with crews from Itchen. They were the last large craft built under the flawed tonnage rule before it changed for the length and sail area rule which did not, at first, tax beam or depth, and resulted in a healthier type.

The 1890s brought tremendous upsurge of interest in British yachting in all classes from tiny half-raters to the largest racing cutters, four of which were ordered in the winter months of 1892. At Glasgow George Watson designed, and Hendersons built *Britannia* for the Prince of Wales and *Valkyrie II* for the Earl of Dunraven. William Fife III, at Fairlie, designed and built the unlucky *Calluna*, and at Southampton, J. and G. Fay's designer Joseph Soper worked all through the Christmas holidays draughting the beautiful *Satanita*, the fastest cutter on a reach ever built, sailing at over 16 knots on a timed course off the Isle of Wight. The four yachts were alike in general appearance and size. *Satanita's* dimensions were 131ft 6in overall, 93ft 6in waterline x 24ft 6in beam x 14ft 6in draft. She displaced 126 tons. Her sail plan was the tallest but she typified the quartette. The newly permitted forward overhang shortened the bowsprit and enabled the forestay to be set further forward reducing the jib foot and improving its aspect. However, her total hoist to waterline length was still only 1·2, compared with the 1·14 of the 11 years older *Vanduara*. Total sail area was 10,094 square feet comprising: mainsail 5,264, headsails 3,360 and topsail 1,470. The boom was 91ft long and height from deck to topmast shoulder 114ft. In her second season *Satanita's* sail plan was altered; 3ft was docked from the boom, 5ft from the bowsprit and 5ft added to the gaff; and she improved her speed, averaging 13·7 knots from Gravesend, round the Mouse Lightvessel and back on a broad and close reach, over a much longer course than her previous higher speed.

The increased sail area and speed of these cutters resulted in many broken bowsprits and topmasts, and several sprung masts, for staying was lagging behind the rapid hull development. Their deeper and more concentrated ballast keels and greater beam, allowed by rule, improved power to carry sail and made designers look upwards.

It is an interesting fact that command of all four of these high class racers was offered to racing skippers from Rowhedge; such was then the fame of the Essex village of only 1,000 inhabitants in the yachting world. Thomas Jay sailed *Satanita*, William Cranfield *Valkyrie II* and John Carter was offered command of both *Calluna* and *Britannia*, but chose the Royal cutter. So *Calluna* was sailed by Archie Hogarth from Port Bannatyne on the Clyde, an area which, with the Solent ports, produced the rivals of the Essex racing men from Rowhedge, Brightlingsea and Wivenhoe, on the Colne and, later, Tollesbury on the Blackwater.

It was no mere chance which gained Colnesiders their reputation. Their subtle skill in racing stemmed from sailing the weatherly North Sea cutter smacks – a wonderful nursery for fore and aft seamen. These great Essex sailing masters were above all fine helmsmen, having a natural ability to windward, using excellent judgment, and were more than mere sail trimmers; they possessed the faculty of taking in hand and tuning up the whole fabric of hull and rig, besides concerting the efforts of a crew then numbering anything up to 50 hands. To succeed they had to act more quickly than their opponents, themselves equally experienced men; know the rules thoroughly, and have iron nerves when stretching them to their limits – for a major error of judgment at the helm would not only shatter a reputation but might cost the lives of several men. They were fine seamen, refined personalities in sole charge of these great yachts and their orders were carried out with an unquestioning smartness by crews of selected racing hands from the area.

William Cranfield raced *Valkyrie II* for the America's Cup in 1893 and in 1894, George Watson designed a new super-cutter for Dunraven's final challenge of 1895. The *Valkyrie III* was one of the most extreme racers ever built and her statistics are worth appreciation. Her dimensions were 129ft overall, 88ft 10in lwl x 26ft 2in

King Edward VII's racing cutter *Britannia* 1893.
Note reaching staysail, yankee jib topsail, and the perfect
set of her 10,000 square foot rig

x 20ft draft. She measured 187ft from bowsprit end to boom end and carried a 77 $^3/_4$ ton lead keel. Construction was composite with elm bottom and teak topside planking. Sail area was 13,028 square feet. Mast, deck to head, 96ft x 25in diameter; boom 105ft; gaff 69ft; topmast 62ft 6in x 13$^1/_2$in diameter; bowsprit 34ft outboard x 16in diameter. Five shrouds each side. One set of running backstays to the masthead, and one to the hounds, besides the shifting backstays to the topmast head (figure 64). She was the first large yacht to have a steel mast, but her wooden mast was stepped for the cup races. She was commanded by William Cranfield with a crew of 37 hands, mainly from Rowhedge.

During the next 10 years sail areas continued to increase and steel masts and booms became common in large yachts, but sail plans crept upwards slowly.

The 131ft 6in racing cutter *Satanita* preparing to set her spinnaker 1893.
Note mastheadsman aloft and tiller steering, then typical

Gradually, topmasts were combined into the lower masts, as in *Shamrock III* of 1903 (14,000 square feet), designed by William Fife III, saving weight of topmast housing.

In 1896 Watson designed the splendid racing cutter *Meteor II* for Kaiser Wilhelm of Germany, who had previously owned the ex-British cutter *Thistle* which challenged for the America's Cup in 1887. The 238 ton *Meteor II* was built by Henderson's of Glasgow and commanded by Captain Robert Gomes of Gosport, Hampshire, with an English crew who were gradually

replaced with Germans whom they taught yacht racing methods and tactics, later under Captain Ben Parker of Itchen.

When yacht racing 'went metric' in 1906 and adopted the International Rule whose successor remains in use, gaff rig still dominated the sport. Classes were established from above 23 metres rating down to five metres, of which the 19 and 15 metres were the most handsome. However, there was detailed change in the cutter rig until 1912 when the leading contemporary designer, Charles Nicholson, of Camper and Nicholson,

Gosport, produced the 15 metre cutter *Istria* for Sir Charles Allom. Always a weight-saver and original thinker Nicholson scrapped the topsail yard by increasing the masthead to include topmast and topsail yard height in one spar of then extraordinary height (see figure 50b).

routed her class in 1921. That season, the old 23 metre *Nyria* came out with a spindly bermudian 'cutter' rig and proved its weatherliness in the large class. As usual development had spread upwards from the smaller boats, some 6 metres having adopted bermudian rig in 1910,

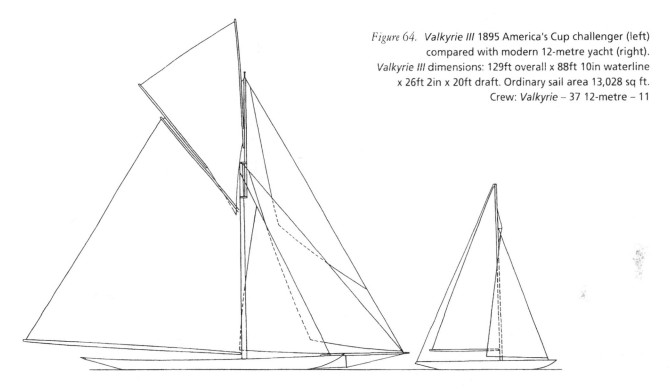

Figure 64. Valkyrie III 1895 America's Cup challenger (left) compared with modern 12-metre yacht (right). *Valkyrie III* dimensions: 129ft overall x 88ft 10in waterline x 26ft 2in x 20ft draft. Ordinary sail area 13,028 sq ft. Crew: *Valkyrie* – 37 12-metre – 11

The topsail luff was set on this by track and slides. The long masthead was supported by two top forestays, opposed by a span leading to the shifter backstay. The peak halliards were set further up the masthead than usual to give a more efficient angle and the running backstay ended in a span to the hounds and opposed the forestay. This rigging and the extreme length of mast caused it to be known as Marconi rig, after the aerials of the radio inventor, then in the news. *Istria* was of large displacement on a short waterline and was unusually planked in Columbia spruce. Under Captain Alf Diaper of Itchen, she dominated the 15 metre class in 1912 and led to the use of Marconi rig by most subsequent gaff rigged racing yachts.

After 1919 there was a vacuum between the British large racing yachts and handicap classes and the much smaller 15 and 12 metres which remained rigged as gaff cutters until the Johan Anker designed bermudian rigged 12 metre *Noreen*, owned by Frederick Last of Burnham,

closely followed by 8 metres, which had been gaff rigged until 1914.

However the big classes clung to gaff rig. Fife built the beautifully proportioned 90 foot cutter *Moonbeam* in 1920 for handicap racing under Captain Tom Skeats of Brightlingsea, Essex, and she won the King's Cup in 1920 and 1923. She came out with a yard topsail rig of 4,261 square feet to drive her 64 ton displacement hull. The continuing upward development of the cutter is illustrated by her bowsprit being only 9ft outboard and the boom 2ft over the counter. She changed to a Marconi mast which was soon fitted in the largest class British racing yachts, which had a grand revival and inspired the building in 1920 of the powerful big class racing cutter *Terpsichore*, designed by Herbert White and built by White Brothers of Itchen. She was the last large gaff rigged racer to be built and became better known as *Lulworth*, sailed by Captain Archie Hogarth from Port Bannatyne, Bute, with a crew from the Clyde, Solent and

Essex, who raced her against the King's old *Britannia*, Sir Thomas Lipton's 23 metre *Shamrock*, and the large schooners *Susanne* and *Westward*, until the bermudian rigged *Cambria* arrived in 1927 to finally prove the superiority of bermudian rig for big class racing. *Shamrock* continued with gaff rig until 1930, *Britannia* until 1931 and *Westward* until she was sunk in 1947.

Cutter rig had limited vogue in the USA, the stronghold of schooners, which however, in many instances such as the Maine sloop, retained the double headsail rig of colonial times to the present day. The American visits of the English schooner yachts *Cambria*

Captains Lemon and William Cranfield at the tiller of *Valkyrie II*, America's Cup challenger 1893

and *Livonia* in 1870 and 1871 led to much interest in English design, rigs and particularly gear and fittings which were then well ahead of American achievement.

In 1866 a New York yachtsman, Robert Center, studied yachting in England and returned to the USA about five years later. He cooperated with Archie Smith, a marine artist who later became noted as a yacht designer under the name of A. Cary Smith, and they designed the iron cutter *Vindex* of 1872. Her performance caused controversy and coincided with a period of severe criticism of the beamy, shallow, centreboard American schooner yachts, several of which had capsized in squalls,

sometimes when at anchor with some sail set. Authorities such as Kunhardt and Stephens declared for the English cutter type with its speed and great range of stability, and a cutter craze broke out in American yachting, resulting, in furious wordy battles between its champions, the 'cutter cranks' and those true to the American sloop. Cutters were imported from England and Scotland, sometimes complete with their crews, and impetus was given to the craze when John Harvey was commissioned to design the extreme cutters *Muriel* (1878) and *Yolande* (1879) built at City Island, New York and sailed by American crews who did not understand them. Harvey closed his Wivenhoe, Essex, shipyard in 1881 and emigrated to New York where he designed many yachts and other sailing craft including the lean deep cutter yachts *Bedouin* (1882) 83ft x 70ft 2in x 15ft 8in x 12ft 2in draft, which set 5,795 square feet, and raced *Puritan*, *Priscilla* and *Gracie* for selection as America's Cup defender in 1885.

In 1881 George Watson of Glasgow designed the cutter *Madge* which was sent out with a Scots crew under Captain John Barr of Gourock and won six of seven races against centreboard sloops whose owners were duly shocked.

Following the success of his cutter *Surf* Harvey designed the 40 ton cutter *Ileen* in 1886 for Mr Paddleford but she was unsuccessful until Captain Tom Diaper (Jnr), of Itchen, went out with an Itchen crew and sailed her to beat the American racing fleet at Marblehead.

Eventually the rival factions were reconciled by emergence of the American compromise sloop which generally adopted the two-headsail rig of the English cutter. The gaff was lengthened slightly in imitation of the powerful English mainsail but the running bowsprit, a prime distinguishing feature of the cutter, was changed to a much shorter fixed spar, which spread back to English cutters 10 years later. The compromise hull form had considerably less draft than English practice but often had a centreboard working through the ballast keel. They were beamier than true cutters and seem to have been as fast or faster in home waters. The compromise sloop *Puritan* beat the typical English racing cutter *Genesta* for the America's Cup in 1885. The following compares the two types at the height of development:

The 42ft 6in waterline length cutter yacht *Lily Maid* was a fine example designed by A.R. Luke and built at Southampton by W. Luke and Co. in 1904. Sail area 2408 square feet

Genesta (1884) British Designer: J.B. Webb	Puritan American Designer: E. Burgess
96ft 5in overall	94ft 0in overall
81ft 0in lwl	87ft 1in lwl
15ft beam	22ft 7in beam
13ft draft aft	8ft 8in draft
141 tons displacement	105 tons displacement
Ballast keel 70 tons	Ballast inside and on keel – 48 tons
Inside ballast 2 tons	
Sail area 7,150 sq ft	Sail area 7,932 sq ft
Mast deck to hounds 52ft	Mast deck to hounds 60ft
Topmast 47ft 6in	Topmast – information not available
Mainboom 70ft	Mainboom 76ft 6in
Gaff 44ft	Gaff 47ft
Bowsprit outboard 36ft 6in	

Afterwards, the two types merged increasingly to form a basic international type in the 1890s, whose influence remains in both countries. The skill of Watson, Fife, Soper, Payne and Richardson was matched against the Burgesses, Herreshoff and A. Cary Smith in America, and some beautiful yachts resulted.

Size continually increased for many years and the American cutter/sloop type found ultimate expression in the huge yacht *Reliance* designed by Nathanael Herreshoff and built at Newport in 1903 to defend the America's Cup against the third *Shamrock*. *Reliance* carried a cutter-sloop rig setting 16,160 square feet in mainsail, staysail, jib and topsail, with a similar amount in a single spinnaker, set by an 84ft spinnaker boom. This was the largest sail area ever set on one mast. Her basic dimensions were 201ft 9in from boom end to bowsprit end, 143ft 8in overall hull, 89ft 8in lwl x 25ft beam x 21ft draft. The topmast was 72ft and gaff 72ft. She was commanded and sailed by Captain Charles Barr, the Scottish-American skipper, with a crew of 64 Scandinavian Americans.

Nathanael Herreshoff designed by carving models from which the lines were taken off by a pantograph for final fairing and calculations. *Reliance's* model is said to have taken him two evenings' carving. He was a most advanced and scientific designer, and his craft were all full of creative thought and light but strong construction of advanced type. They were fitted with ingenious devices for working gear including highly developed sheet winches, which were common in American yachts years before they appeared in British craft.

The close racing characterising the professionally sailed and manned yachts is illustrated by *Reliance* beating *Shamrock III* by only two seconds a mile over the whole course, and like all British challengers before 1958 she had to be sufficiently strong to face Atlantic seas on passage. Designed by Fife and sailed by Captain Robert Wringe of Brightlingsea, Essex, *Shamrock III* was almost as extreme as *Reliance*, and she set 2,000 square feet less canvas.

By 1913 the design of American gaff rigged racers had returned to the single headsails of the old sloops, though *Resolute* defended the America's Cup in 1920 with a double headsail rig, beating the similarly rigged *Shamrock IV*.

Gaff cutter rig has been used by cruising yachtsmen since at least the 1780s, and cutter yachts have sailed to most parts of the oceans. Notable cruising cutters have included McMullen's 29 ton *Orion* which he sailed singlehanded from Cherbourg to the Thames in 1877, and learned the full weight of contemporary heavy gear and the labour involved in keeping it in order; E. G. Martin's converted Harve pilot cutter *Jolie Brise*, was an original amateur offshore racer, which encouraged her owner to found the ORC; and the fleet little *Dolly Varden*, Tom Ratsey's pride, whose story is told in chapter 13.

A gaff rigged cruising cutter should have a well-balanced hull-form with a good length of straight keel and a moderately cutaway forefoot, enabling her to turn readily in stays but to run true on a sea and keep her steady on the helm. The rig is powerful and capable of driving a suitable hull to best advantage in short seas, besides making her quicker on the helm than other rigs, except the sloop. Its principal disadvantage is in handling the large mainsail.

A cutter's mast should step about $2/5$ of her waterline length from the fore end of the waterline and should

Figure 65. "Modern" cruising cutter *Dyarchy* designed by Laurent Giles & Partners. Built 1939. Dimensions: 46ft 6in overall x 38ft waterline x 12ft 4in x 7ft 6in draft

never be further forward than $^1/_3$ of the waterline length. The bowsprit should be as short as possible, sufficient to balance the sail plan, and the boom should only extend to the stern, or very little beyond it. Any cutter should be designed to handle under one headsail and the mainsail, and turn well under this combination. The ability to turn to windward under mainsail and a staysail sheeted to a horse is particularly useful when short tacking with limited crew in a large cutter. Large fishing and pilot cutters were usually tacked by one man when on passage. He lashed the helm a-lee and walked forward to trim jib and staysail while she came round.

The Laurent Giles designed gaff cutter *Dyarchy* exhibits the best of modern cruising yacht design allied to the gaff rig, even though she is now over 50 years old. Her beautifully fair hull has dimensions of 43ft 10in overall x 38ft lwl x 12ft 3in beam x 7ft 6in draft, and displaces 24·2 tons. The total sail area is 1,410 square feet in a graceful rig (figure 65). Her mast is well into the transom-sterned hull and divides the sail plan base. Throat and peak halliards are flexible steel wire, reducing number of parts and windage aloft, and the need for frequent setting up. The mainsail has worm roller reefing gear and is comparatively small and easily handled, while the topsail is proportionally increased and its interesting arrangement is described in detail in chapter 6. Two men could comfortably sail *Dyarchy*, which indicates real progress from the cumbersome rig of *Orion*.

CHAPTER ELEVEN

THE ESSEX SMACKS

THE SAILING SMACKS of the County of Essex evolved through many centuries and their origins are now undiscoverable. They were built, owned and fished from small ports on the rivers Colne, Blackwater and Crouch, with others from Harwich.

Basically the Essex smack was a cutter rigged fishing vessel having various forms and dimensions for different fisheries. There were three principal types: small smacks up to about 12 tons mainly used for estuary dredging and trawling; 12 to 18 tonners which also did this work besides fishing coastwise spratting and oystering; and seagoing smacks over 20 tons which were principally owned on the river Colne, and fished far afield. All were noted for windward ability, seaworthiness and speed.

Several hundred smacks were sailing from Essex by the early 19th century, the majority owned at Brightlingsea, a small ancient town on a creek at the mouth of the river Colne, a limb of the Cinque port of Sandwich. Brightlingsea fishermen dredged oysters in the Colne estuary and coastwise and deep-sea oysters and scallops wherever these could be found around the British Isles and off the French coasts. They spratted with the stow net in winter, carried fish, and 'salvaged' from the large numbers of wrecks off the coast in the days of merchant sail. Brightlingsea shipyards built many hundreds of smacks and small wooden ships. Its seamen took to professional yachting and achieved a great reputation until this noted summer aspect of Essex seafaring ended in 1939.

Six miles up the wooded, winding Colne lies the author's native village of Rowhedge — for centuries the home of an adventurous maritime community of 1,000; fishermen, professional yachtsmen and shipbuilders. Its mellowed buildings looked on a waterside which stirred briskly each high water with a fleet of 60 smacks, mostly large cutters, ranging all round the British Isles, to Ireland, France, Norway and the Baltic in the same fishing trades as the Brightlingsea men. Rowhedge was the home of many noted racing yachtsmen. Its seamen gained international fame as professional yacht skippers and crews from before 1800 to 1939. Rowhedge captains raced four challengers in America's Cup races, besides scores of other large yachts, and hundreds of its sailors formed their crews. The village shipyards built ships, yachts and smacks.

On the opposite riverbank lies Wivenhoe, a larger village which owned fewer smacks but had a fleet of merchant brigs and schooners and shipyards building smacks, yachts and small ships. Wivenhoe men were also yacht captains and crews.

Tollesbury, on the nearby river Blackwater was a fishing community owning about 40 smacks under 20 tons, in the coastal spratting and trawling, and inshore oyster dredging trades. After 1890 Tollesbury men were taken on in the racing yachts; at first as hands, some later to become captains and racing skippers.

West Mersea, on a creek at the west end of Mersea Island, had little interest beyond the Blackwater Estuary after the mid-19th century. Its 55 small smacks worked the extensive estuary oyster fisheries, where sail survived until 1939 and many of the remaining 'smack yachts' lie at Mersea. Bradwell, on the opposite shore of the Blackwater, was a rural parish owning a few small smacks. Maldon, at the head of Blackwater Navigation, 12 miles upstream, owned only a score of little river-dredgers but large numbers of small coasting and seagoing merchant sailing ships and barges were built and owned at the port.

Essex cutter smack *Dove* built by Aldous of Brightlingsea

Further south, on the river Crouch, the ancient town of Burnham owned about a score of big cutter smacks in the same trades as the Brightlingsea and Rowhedge smacks, besides numbers of small smacks dredging the extensive oyster fisheries in the Crouch and Roach rivers, joined by about twenty smacks from the village of Paglesham on the Roach.

Harwich, Essex's premier seaport, had many smacks in the deep-sea cod trade (described in chapter 9) and a score or so big cutter smacks spratting, fish trawling and salvaging. After the mid-19th century Harwich fishermen developed shrimping and adopted the bawley type which was also used from Leigh, at the Thames mouth, and is described in chapter 7.

Essex shipbuilders were producing fine-lined fast cutters well before 1800, but bluff bowed full-bodied clench-built cutter smacks were in common use into the early-19th century. The larger ones had lute sterns (an early form of counter) and many of the smaller had transoms. Indicative of their shape and proportions is the little *Boadicea* built at Maldon in 1809 and still sailing (see chapter 10). Carvel planking was introduced some years later, the earliest recorded being the Wivenhoe smack *Tribune*, built at Ipswich in 1836.

By mid-19th century the Essex cutter smack had been perfected principally by Colne builders, who were much influenced in design by their close association with the building of fast yachts after 1800, which were manned by local seamen who readily adopted, and demanded, unusually fast and weatherly hull forms and sail plans. Thus yachts and smacks had early interacted and Philip Sainty of Wivenhoe, who had shipyards there and, earlier, at Brightlingsea and Colchester, was a noted builder of both until his death in 1837. The Harris family of Rowhedge built over 60 smacks, including many large cutters and some ketches, between 1840 and 1875, and Cheek built there in earlier years. At Brightlingsea the Aldous family built very large numbers of smacks of all sizes and types, and others were launched there by Root and Diaper and later, small smacks by John James, and Stone. Some large smacks were built at Ipswich by Read and Page and later by Thomas Harvey and Son whose Wivenhoe shipyard probably produced the most shapely and well finished smacks of all.

Typical of these was the 15 ton *Beatrice* built in 1848 for Captain Mason, and later owned by the author's grandfather, Captain James Barnard of Rowhedge. Her dimensions were 45ft x 12ft x 6ft 6in draft aft. The plumb stem, slightly rounded forefoot, long raking keel, great forward freeboard and graceful sheer sweeping to a low, well proportioned counter earmarks the type. Hull lines varied in detail with different builders but typically the hull had considerable rise of floor, a moderately firm bilge, hollow bow waterlines and a long and beautifully fair run to the flat-sectioned counter. They had inherent speed to windward and the bold bow kept them going in a sea, but low after freeboard and the 15-inch bulwarks necessary for handling fishing gear, made them wet aft in bad weather when they could stand extremely hard driving. All ballast was inside, of shingle and iron pigs ceiled over. From forward the hull was divided into a fo'c'sle store for paints, sails and ropes entered by a deck hatch; the fish hold amidships was bulkheaded off at the mast and from the cabin aft, and was served by a 5 foot square main hatch with a small spratting hatch adjacent; the crews' quarters for four were in the cabin aft, with cupboard bunks around the sides and a coal stove. The cabin hatch was immediately forward of the tiller and the compass was mounted inside it. On deck a handspike windlass handled the chain cable and an iron geared hand-capstan the trawl warps. Halliards and sheets were brought in by hand.

Like almost all her kind *Beatrice* was rigged as a true cutter, with long housing topmast, and bowsprit housing between deck bitts, one of which was a knee, the other a post from the keelson, also bearing the windlass pawl. *Beatrice* set 1,250 square feet in mainsail, staysail, jib and jib-headed topsail. In light weather a large yard topsail was set besides a balloon staysail or a balloon jib or 'bowsprit spinnaker' from the bowsprit end, sheeting well abaft the shrouds when fetching on the wind or boomed out as a spinnaker when running. A small jib and storm jib completed the sail outfit. No trysail was carried in smacks of this size, but the mainsail had a close reef. Such smacks were built for £10 per registered ton for hull, mast and spars until about 1890. Construction was simple but strong elm keel; pine keelson, oak centreline and sawn frames; pine planking with oak wale and bilge strakes, all iron fastened; oak deck beams; pine shelf; decking, mast and spars. These smacks were well kept, clean and able each representing one man's investment and pride, manned by his sons, relatives or neighbours and fishing on a system of shares of the proceeds of the catch.

Dredging oysters was one of the principal occupations of Essex smacks. The iron framed dredges were shaped like a capital A with a three foot hoeing bar across its feet. A short, twine net spread from the hoe, crosspiece and sides between, and had a chain mesh bottom and a stick to square its after end. The irons' apex had an iron ring forged in it, to which a bass rope towing warp made fast. A smack might tow four to eight dredges across an oyster ground, gathering, besides oysters, much rubbish and pests from which the marketable oysters had to be culled, the rubbish being dumped overboard. All hauling was done by hand and the work required sailing craft which handled well, sometimes under sail reduced to control speed to suit conditions of wind and tide, often in very confined waters and among numbers of other smacks similarly manoeuvring. Such work improved the smartness of the Essex seamen.

Between 30 and 100 smacks up to 15 tons dredged the Colne estuary fishery, administered by Colchester Corporation and the Colne Oyster Fishery Company. Its dredgermen had to be freemen of the river, a privilege much prized in those days of scarce winter employment. The Blackwater and Crouch rivers had large oyster layings worked by companies but without the freemen system.

For several centuries oyster dredging at sea was also carried on by Essex smacks but this trade developed greatly in mid-19th century when railways reached north-east Essex. At Wivenhoe a first effect of the arrival of the railway was to encourage the landing of sprats for distribution by rail as manure. Rumours of an extension to Brightlingsea opened rosier prospects — the solution of the marketing problem by rapid despatch to Billingsgate without having to work tides to Wivenhoe. Grimsby and Lowestoft had been developed by the railway companies during the previous 10 years from insignificant villages into huge fishing ports and who knew what might develop at Brightlingsea? When the railway reached there in 1866 the golden dreams we never realised; the capital was lacking, but not the enterprise. Emboldened by optimism many smacksmen who had money to venture placed orders for new craft, of a size which had never before been owned locally and in such numbers. No companies were created, for the thought of skippering smacks owned by others, common in other ports, was horrifying to the vigorously independent Essexmen, who especially wished to develop their long established dredging of sea-oysters, which then held a high place in the diet of ordinary people.

The result was a fleet of powerful cutters with glorious sheer and rakish rig. Almost all were designed and built in local yards, with the exception of a few constructed at the Channel Island port of Gorey, Jersey, then a noted centre of the shell-fish industry, and much frequented by Essexmen. Aldous of Brightlingsea built 36 of these big smacks between 1857 and 1867, Harris of Rowhedge and Harvey at Wivenhoe built a good number, and some were built on the Blackwater. These 20 to 40 tonners dwarfed the little 10 ton estuary oyster-dredgers and there was as great a difference between them in purpose and voyaging as between the deep-sea motor trawler and an inshore fisherman today. The

Rowhedge *Aquiline* owned by Captain Harry Cook was typical, a bold-sheered cutter of 21 registered tons, she was launched from Harris' Yard in 1865, where her 65ft hull was well built and finely formed, having a beam of 15ft and drawing 8ft 6in when loaded to her hold capacity of 21 tons, and a little less when light. She carried a sail area of about 2,000 square feet in her working canvas, her mainboom was 45ft long and the bowsprit 25ft outboard. When sea-dredging the mainsail was often handed and the mainboom stowed in a crutch, and a large trysail having its own gaff was set in its place, sheeted by tackles to the quarters.

The arrangement of these big vessels differed greatly from the small smacks which still survive. Forward, the usual handspike windlass handled the anchor cable, which ran out clear of the stem on a short davit and, in season, the stowboat gear. A large geared hand-winch having four barrels was fitted immediately forward of the mast for working halliards, running out the bowsprit, working fishing gear or whipping out the cargo. A winch or geared hand-capstan stood amidships and could be worked by two or more hands when dredging or working trawl warps, all of which were hauled by muscle power as, unlike the specialised trawling smacks from the large fishing ports which were equipped with steam capstans, the Essex boats relied solely on 'Armstrongs Patent' to get their gear; a feature they endured to the eclipse of sail. A clench-built boat about 15 feet long was carried on deck, or lashed, capsized over the main hatch in foul weather. In port on Sundays or regatta days most of them sported a long masthead pennant emblazoned with the smack's name or initial. Many had signal letters allotted to them, which is proof of their varied seawork. Few had channels as these interfered with boarding stowboat gear.

Below deck the forepeak was a cable locker and abaft this the mast room, as the space between the peak bulkhead and mast was called, housed a large scuttle butt for drinking water and racks for bread and vegetable stores. Partial bulkheads set it off from the hold which occupied about $1/3$ of the smacks' length. Ballast was clean beach shingle with a proportion of iron pigs, covered by a wood ceiling. All hands berthed in the cabin, aft, entered by a sliding hatch in the deck. In this space of perhaps 14ft x 9ft six men lived for months. Locker seats ran down each side and the four bunks lining the topsides behind them each had sliding panels which could be closed by the occupant to shut himself off for sleep or from his noisy mates. A big double berth across the counter was known as the Yarmouth Roads and could hold two apprentices, who were lulled to sleep by the rudder's groan and kick in the trunk near their heads. On the black coal-stove stood a kettle and a huge teapot which was only emptied when it would no longer hold six mugfuls of water. Knives and forks were stuck in cracks of the deck beams overhead and saucepan lids made good plates in a seaway with the handle gripped between the knees.

These craft formed the finest fleet of fishing vessels ever to sail from Essex. There is no point in attempting to glamourise the fisheries which were always hard, often dangerous, and usually miserably rewarding financially; but they bred men of strong and independent character. Theirs was none of the comparatively idyllic life of the estuary dredgers but a thrusting existence, alive to any and every opportunity which came their way. Most skippers began as apprentices serving aboard their master's smack and living in his house in port. These boys were often at sea at the age of 12, being bound for five, seven or nine years to the owner, who might have two apprentices serving aboard his smack. This system existed in the 17th century but died out by the end of the 19th. The owner found them food but the apprentices supplied their favourite rig of cheesecutter cap, white canvas jumper or smock, over a thick guernsey, and thick duffle trousers tucked into leather seaboots with cobbled soles. It was a life which produced smart seamen and gave fair advancement for those times. There was little of the 'by guess and by God' style of navigation among these smacks' confident skippers, most of whom were progressive enough to obtain fishing masters' certificates when they were introduced in the 1870s.

Their quest for oysters and scallops led them, at various times and seasons, to work the Inner Dowsing and Dudgeon banks, landing at Grimsby or Blakeney; the Ness grounds, stretching from Orfordness to Cromer; the Galloper and Kentish Knock areas of the North Sea; and the Terschelling and Hinder banks off the Dutch coast, landing at Brightlingsea. In the Channel they dredged the Goodwin, Sandette and Varne grounds, with those off the French coast at Caen Bay, Dieppe, St Valery-sur-Somme, Fecamp, Calais and Dunkirk, using Ramsgate, Dover, Shoreham and Newhaven for landing. Down Channel, West Bay provided some work, landing at Weymouth and, in earlier times, the Cornish Fal and Helford rivers were visited — almost raided — by the Essexmen, while the Channel Island of Jersey attracted large numbers of Essex smacks to its fishery for 70 years. Others sailed round Lands End to work on the South Pembrokeshire coast based at Swansea, and Bangor, Western Ireland and the Solway Firth regularly saw the rakish Colne topmasts.

Work aboard was hard when dredging. On the grounds the topmast was often struck and a reef tucked in the mainsail, or the trysail would be set to ease speed and motion in a seaway and keep the boom clear of unwary heads. After this they worked almost continuously, day and night, with only occasional spells for mugs of tea and a bite to eat. It was muscle-cracking work winding up these huge sea-dredges with six foot hoeing edges, from 25 fathoms, by hand. Often six were worked at a time on 65 fathom 3 inch bass warps leading-in through multiple rollers on the rail or through a port in the bulwark. Each weary foot of it was cranked home with one man keeping tension on the slack which jerked back on every sea to etch new scars into hands already torn by the tiny shell picked up by the warp. So it went on hour after hour, without even the respite of sorting between hauls which trawling gives. At about 3am the gear would be laid in and the crew took half-hour watches before recommencing at six, and this might go on for five or six days.

Hard as they were the fisheries flourished and by 1874 there were 132 first-class smacks of 15 to 50 tons registered at Colchester, in addition to 250 second-class of under 15 tons, and 40 third-class vessels. By far the largest part of these were owned at Brightlingsea, with Rowhedge accounting for 29 and Wivenhoe and Tollesbury 12 and eight. Few of these large smacks were owned by West Mersea fishermen, whose interests were chiefly centred in the Blackwater oyster fisheries, and none were owned by them in the latter days of sail.

Rich oyster beds were discovered by the big smacks off Jersey in 1787 and news travelled fast for within a few months over 300 smacks from Essex, Shoreham,

Emsworth and Faversham, manned by 2,000 men, were working there. In a few months the quiet port of Gorey became a boom town. Later, a fleet of 60 Essex smacks sailed there each spring and carried on dredging those waters despite the hazards of the Napoleonic wars. The Jersey fishery declined during the 1840s and became exhausted by 1871.

Wherever there were oysters the far-ranging Essexmen would find them and flourishing local oyster fisheries at Swansea and Cardigan Bays in Wales, Largo Bay, Fife, and off north Norfolk were rapidly worked out by fleets of Colne smacks, to the rage and often despite the resistance of local fishermen. Sea oysters were also dredged on the Terschelling bank, off the Dutch coast, about 112 miles east from Orfordness, from which point the smacks took departure. Trips averaged 12 days during which a haul of 10,000 large oysters could be expected. This was a harsh fishery, a lee shore in the prevailing winds with no handy leeward refuge and many Colne smacks and their crews were lost off Terschelling.

Apart from these activities many Colne smacks were employed on contract for £12 a week as fish carriers for the lumbering Grimsby, Hull, Yarmouth and Lowestoft ketches which trawled the North Sea in fleets. They were ideal for this work, being fast and capable of driving through the foulest weather, and blow high or low, usually got to London's Billingsgate fish market before the iced cargo 'turned'. Colne smacks were also employed in the seasonal carrying of fresh salmon from the Western Irish ports of Sligo and Westport, on the coast of County Mayo, round the northern tip of Ireland, to the Liverpool market. This was, perhaps, the hardest trade of all as the coast is exposed to the full sweep and fury of the North Atlantic, and most smacks took the precaution of reeving chain reef pendants to stand the hard driving of a four-day round voyage to Liverpool; no mean feat in such small ships.

The Rowhedge New Unity, a 39 tonner owned by the author's Great-Grandfather, Captain Thomas Barnard the noted salvager, was one of several Colne smacks engaged at times in the equally arduous voyages to German and other Baltic ports with barrelled herring shipped from Stornoway, in the western Isles of Scotland.

In fact there was nothing these smacksmen wouldn't or couldn't tackle to get a living. Many were in the cattle trade from the Channel Isles to Weymouth at certain times of the year. The Brightlingsea Globe was typical, being licenced to carry 19 swine in the hold and two cows on deck, and the boy had to truss up the boom to clear their backs when going about. The Rowhedge Aquiline and New Blossom were often employed in the French spring potato trade from St Malo and St Michaels to Colchester, and New Blossom frequently shipped Rowhedge's coal from Shields, and often voyaged to the Baltic with coal or barrelled herring before the 1880s.

The big smacks held an annual race with a hogshead of beer as the prize. This event, which died out in the 1880s, was quite distinct from the annual regatta struggles among their smaller sisters. The deep-sea smacksmen scorned an inshore course, though the start and finish were in Colne; racing out round the Galloper Light and back. Perhaps they got a fair wind home, when the great half-squaresails carried before spinnakers were devised, ranted them back into Colne, the winner promptly fixing a metal masthead cock as the finishing gun boomed from a committee boat off East Mersea Stone, announcing the cock of the fleet for that year.

In the latter half of the 19th century yachting rivalled and then eclipsed the Essex deep-sea fisheries as a living, and the smaller 12 to 18 tonners became popular for working the Thames Estuary as they could be more economically laid up all summer while their crews made more colourful pages of sailing history.

The days of the big Colne smacks were numbered, though as late as 1890 Brightlingsea boasted a fleet of 52. Twelve years later several poisoning scares killed the demand for sea oysters and the remaining fleet took to working down Channel from January to March, usually dredging from French ports, notably Boulogne, and sending the catch to London by night steamer. Shoreham was also used when dredging the extensive grounds off Beachy Head, and sometimes Newhaven, but little or no money was made from all this endeavour and hard work. The first world war dealt a great blow to the big smacks and though fish prices were high, several were sold away to Lowestoft and elsewhere, while others worked on Government fishing contracts, mainly stowboating. After the war a few carried on in the traditional ways, supplemented by a few old Lowestoft steam drifters, though with the exception of these and four or five paddle dredgers owned by the Colne, Paglesham and Burnham oyster fisheries for river work, steam found no place in the Essex fishing fleet.

In these days of well-buoyed fairways and elaborate navigational aids it is difficult to picture the appalling conditions which existed for coastwise shipping during the earlier 19th century. Most trade was carried in unhandy and often ill-kept square rigged vessels, and there were few lifeboats or licenced pilots and it is not surprising that ships were driven ashore in dozens during an average winter's gale. The Colne smacks, in common with the Harwich men, were quick to seize salvage opportunities and, until the 1880s, receivers of wreck on the Essex coast had a busy time. Customs' warehouses were crammed with salvaged goods, as were many smacksmen's cellars. The spars, equipment, and even hulls of wrecked and refloated vessels were regularly being brought in, advertised in the Press and auctioned. This work, with the life saving it often entailed, was known in Essex as salvaging.

Captain Thomas Barnard of Rowhedge was among the most noted salvagers and life savers, with his big smacks *Thomas and Mary* and, later, *New Unity*. Throughout his career until he retired from sea in 1881, he saved over 900 lives without the loss of any of his crews. His seamanship was typical of the Swin Rangers, as the old Colne salvagers liked to style themselves. They were bred to sail and learned to handle it from boyhood, learning the hardest way in fishing the treacherous North Sea. Ranging the death trap shoals for whatever luckless ship the sea brought they possessed a courage born from a calculated knowledge of the sea. They were hard, fiercely independent and quick as gulls to seize any chance the sea brought at fishing or salvaging. They frequently risked lives, smacks and livelihoods to take off shipwrecked crews in the foulest weather.

Salvage work was profitable as well as heroic. The fisheries were frequently depressed and salvage presented heaven-sent chances for bolstering the local economy. News of a well-laden wreck spread fast and crowds of mariners thronged village watersides, eager to be taken on as extra hands aboard the salt-rimed smacks already discharging salvaged cargo. To help in this work the salvagers devised special tools, and in winter, most large smacks carried a variety of grapnels, mauls, crowbars and axes. Salvaged cargoes ranged from the usual coal, timber and general goods to the fabulous freight of the 'Knock John Ship' of 1856; a German vessel which stranded on the Knock sand bound for China with barter goods. The wreck was first boarded by a Brightlinsea smack, and from her they took the richest haul ever made on the east coast.

During the early 1800s smack crews might be awarded as much as £80 each for salvaging a small vessel such as a sloop or brig, and for larger ships sums of £500 are recorded. Successful salvaging meant big money in contemporary values. For saving life the salvagers, of course, expected no reward, but many received medals from humane societies and British and foreign Governments, besides tokens such as inscribed telescopes or binoculars. One rescue will illustrate their work. Beating out to the Kentish Knock one wild and bitter night Captain Barnard found a great German barque crowded with emigrants bound for New York, beating her bottom out on the sands in the roaring gale. No time could be lost if all were to be saved and manoeuvring alongside in heavy seas, the terrified passengers were bundled over from the doomed ship to the wildly pitching smack. It was not until *New Unity* made three voyages to Walton, and others to a steamer which hove-to nearby, that the ship's complement of over 200 were saved from the splendid barque which was matchwood 48 hours later.

On the Colne the period 1860 - 1914 was an era of great expansion and growing fame in professional yachting, for the Essex smacks were the finest school for teaching the subtleties of handling fore and aft rigged craft in confined waters and under competitive conditions of workaday sailing. Most of the great yacht racing skippers sprang from this source, which also trained the hundreds of smart hands until the decline of working sail and professional yachting, in the 1930s.

Many fine, fast smacks of 15 to 18 tons were built from the often large sums of prize money won by Colne racing yacht skippers, together with others constructed for masters of cruising yachts. These lean cutters were principally designed for winter spratting and fish trawling, and were a size which could be economically laid up during the summer yachting season. Most of them bore names of the yachts whose smart handling financed their construction. Just as Colne-built smacks and Revenue cutters had influenced yacht design during the early-19th century so, in later decades, local association with the handling and development of the racing yacht affected smack design. Although ability to earn a living in the winter fisheries always remained a prime consideration when building smacks, and few, if any, were built especially for racing in the annual regattas, refinements in hull and rig were continuous and the weatherliness and speed desirable in the fisheries led to the launch of fine cutters from Colne and Blackwater yards.

In September, when the yachts laid up, crowding the rivers' mud berths, the greater number of their crews were soon busy seeking winter employment. Most took on in the stowboaters, as smacks engaged in the winter spratting were called. Some took the train for London or Southampton docks, seeking berths in foreign-bound steamers for six months. But the smacks remained the winter mainstay for there was little else, and 30 or 40 stowboaters annually fitted out from Rowhedge to join the total of scores from Brightlingsea, Wivenhoe and from Tollesbury, the only Blackwater village to participate in the spratting.

Autumn springs saw smacks floated from their summer berths to lay on the hards which resounded all day long with the ring of caulking irons and the chaff of seamen bending canvas, rigging gear and tarring hulls. Along the shore roads men bore on their shoulders the great stow nets and their cumbersome wooden beams. The delicate mesh of the long, brown, funnel shaped nets had been expertly baited by village women before dressing in the bubbling quayside cauldrons. Then they were laid aboard, triced up by fish tackles to air; the half-inch mesh of the sleeve giving a solid look, very different from the coarser trawls. The square spars of the upper and lower beams, each some 30 feet long, were shipped, together with a powerful, long-limbed, stowboat anchor and many fathoms of stout cable. The anchor had to hold both net and vessel against strong tides, wind and sea, the net riding to it, mouth to tide, under the smacks' bottom and away astern by a leg of cable made fast to the lower beam. This beam was ballasted with iron to steady the net's mouth, while the upper was held horizontally above it by the templines; ropes which, made fast to the

ends, passed up to belay on either side of the smack's forward rail. When the net was streamed and fishing its mouth might be 30 foot square. It was closed by the winch chain which started from the lower beam and passed through a ring on the upper one, over a sheave at the end of the stow-davit, fitted in winter on the opposite side of the stem to the gammon iron, and round the windlass barrel. To get the net this chain was hauled in, bringing the two beams together and to the surface, where the net's sleeve was griped-in with ropes, boarded, and its contents emptied into the hold.

It was a fishing gear dating back to the Middle Ages and, because of the Thames approaches' comparatively shoal channels and heavy ship traffic, was used by the Essexmen in preference to the drift net used elsewhere. Traditionally, from about mid-November until mid-February, the stowboat smacks worked their nets by day or night as the tides served. The skippers watched for the paths of the great sprat shoals, their experience guided by the chase of flocks of gulls. Once located, they let go down-tide from them, and waited for the net to do its work. Sometimes it took only three or four hours, sometimes a couple of days, to get a good haul, which was reckoned at 300 to 400 bushels for a 15 ton smack; though the big, Brightlingsea *Masonic* once landed 3,000 bushels in a week. Sometimes days or even weeks passed without a worthwhile landing and during that time the fishermen earned nothing and then there were no unemployment benefits. It was a fickle living and all depended on the movement of the shoals. The wallet and swin were the usual spratting grounds with Shoe Hole on the Maplin's edge, a favourite area. Sometimes the fleet worked the Kentish shore and, at others, the various channels between the Thames Estuary sands, where they were sometimes run down and sunk in fog or darkness. Occasionally hauls were made 'down the Sunk' and in the Whittaker Channel, and exceptional schooling of the shoals would lead them to fish in the rivers themselves, under the Blackwater's Bradwell shore or in Colliers Reach in the Orwell, though this was most unusual. Besides the 12 to 15 tonners, many of the larger first-class smacks, which were becoming fewer by 1900, regularly worked the winter sprat fishery. In rough weather the stowboaters dreaded a full net. This sometimes happened very suddenly and the net would

Rowhedge cutter smack *Wonder* built Wivenhoe 1876

float only as long as there was life in the fish. Then the crew wasted no time in getting the net alongside, for if it sank they might never raise it again without damage. Often a full net had to be parted at the lacings to start the fish out of it, enabling the smacksmen to save the net and most of the catch by bailing out some fish.

In winter Brightlingsea revolved around spratting. The little silver fish occupied the town's merchants and curers, the stowboaters' skippers and hands. Brightlingsea, handling half the east coast landings, was centre of the industry and barometer of its fickle rewards. If prices were low there, smacks from other villages usually sailed their catches home to sell off as manure, though prime sprats were landed and cured at other places. Occasionally the larger smacks landed at Chatham, or, if prices were good, at Billingsgate Market itself. Often three worked in partnership, each acting as runner smack in turn, to land the combined catch. Much

of Brightlingsea's landings was pickled in barrels, with salt, bay leaves and spices, for export to Europe, Russia and Scandinavia. Much prime fish was smoked by the curers and some skippers also smoked their own in tall wooden smokehouses. Big landings glutted the market and prices fell rapidly. Then all the skill was to no avail; large the catch and prime the fish might be, yet the hard-won bushel plunged from shillings to mere pence in a few hours, often to fourpence a bushel for manure, the smacks discharging alongside quays piled high with gleaming heaps of unwanted silver sprats, prime sea food only saleable as manure.

Traditionally fish trawling was a short, seasonal fishery for most Colne and Blackwater smacks, many of which trawled for roker and soles in spring and roker and dabs in autumn, at each end of the spratting season, with occasional winter trawling of offshore codling. Trawling was ancient in Essex; in 1377 fishermen from Colne and

Blackwater villages were using primitive trawls called wonderthons and causing damage by overfishing; and trawl heads are mentioned in Ipswich court rolls of 1566. The beam trawl has long been superseded by the otter type, but suited the slow speed of sail. The beam's length was determined by the distance from the aft shroud to the counter. Most Colne and Tollesbury smacks of 18 tons and upwards had a double-handed iron capstan amidships. Older smacks often had a wooden capstan with deck treads round which the crew walked pushing the bars. Recovery of the trawl was aided by the 'fish tackle' purchase leading from the hounds and also serving as a backstay, when required. Smacks needed two or at most three hands for fish trawling and the catches' proceeds, like most Essex fishing, were divided into shares.

Though most Colne, and later Tollesbury smacks, laid up for the summer while their crews went yachting, a few carried on with shrimping which only needed two men to haul the fine-mesh shrimp beam trawl and riddle, sort and boil the catch on board. Besides the Essex coastal grounds, Colne shrimpers also worked out of Grimsby in summer. Other occasional fisheries were dredging five fingers (starfish) and mussels for field manure; or young oysters for relaying on the grounds. Seines and peter nets were used to catch mullet by the little, transom sterned Maldon smacks which also regularly trawled eels on the grassy Mersea flats in company with a dozen or so Mersea men before 1914. John Howard, the clever Maldon designer and builder of wooden sailing craft, built most of the 10 ton, transom sterned Maldon smacks, which had counters built on after 1919 to add deck space for oyster dredging.

The earliest recorded Essex smack races were sailed on the Blackwater in 1783, and by the 1870s the Colne village regattas became established annual functions each autumn when the yachts had laid up and their crews had fitted out the smacks. Tollesbury held its first regatta in 1900.

Most notable of all the racers of this era was Rowhedge's Captain Lemon Cranfield and his rakish smack *Neva* (figure 66), greyhound of the Colne and Blackwater smack races for over 30 years and built from the fabulous prize money he won thrashing the big class yachts with the Fife-built racing cutter *Neva* in 1876-8, at the height of his fame as the genius of yacht racing. The smack *Neva* was built by Harris of Rowhedge, who draughted a straight-stemmed hull with a hollow entrance filling to a slack mid-section which melted into a long run ending in a graceful counter. Small wetted surface made her a great boat in light weather, when she ghosted in the faintest air. Her cutter rig set 1,400 square feet, excluding the spinnaker which was whisked out to windward by a 38 foot boom – some rig for a 45 foot fishing vessel! The new smack lived up to her owner's reputation between 1877 and 1907 she repeatedly won the smack races of local regattas against stiff competition.

Lemon's skill kept her in front and year after year her mastheadsman proudly jammed the traditional gilt cock at the masthead as she roared across the finishing line and the band struck up, Lemon acknowledging cheers with his cap and her crew cheering second and third, to be

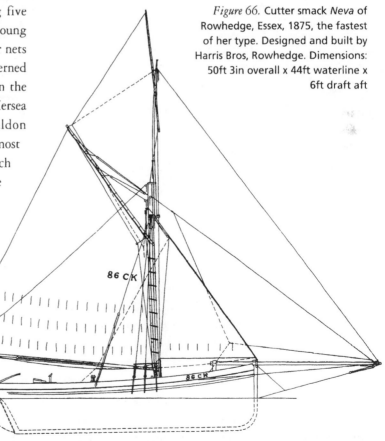

Figure 66. Cutter smack *Neva* of Rowhedge, Essex, 1875, the fastest of her type. Designed and built by Harris Bros, Rowhedge. Dimensions: 50ft 3in overall x 44ft waterline x 6ft draft aft

Rowhedge cutter smacks racing, 1903. Left to right *Sunbeam*, *Neva* and *Xanthe*

cheered in turn. I believe she was beaten on occasions but it was very rare. Not that they were easy victories as the *Neva* was usually pitted against smacks owned by Lemon's five brothers who were all noted yacht racing captains; William of *Valkyrie* and America's Cup fame competed with *Clara* before buying the swift *Sunbeam*, always a close rival to *Neva*; John sailed the *Lily*, Stephen the *Blanche* and Richard the *Ellen*. Other noted Rowhedge racing smacks were Captain Simonds' *Hildegarde*, Captain James Carter's *Wonder*, and the little *Xanthe* of Captain W Cranfield. The *Eudioa* and *Violet* were Wivenhoe's fliers, sailed by the Mason family, and Captain Green's able *Elise* always sailed hard races and once fought it out with *Ellen*, boat for boat over a 22 mile coarse, to finish within six feet of each other. Stephen

Redgewell's *Bertha* was Tollesbury's pride, and Brightlingsea's *Foxhound*, *Varuna* and *Volante* were among the clippers.

Typically Colne starts were off East Mersea Stone and the course was round the Bar, Priory Spit, and Wallet Spitway buoys, then up Colne to finish off the village.

Surmises are often made on the speed of these smacks and the recorded times in Wivenhoe regatta for 1884 show the $20^7/_8$ nautical mile course was won by *Neva* in 2 hours, 23 minutes, 4 seconds; an average speed of $8^3/_4$ knots; *Wonder* finishing second 2 minutes, 56 seconds astern; and the *Eudioa* 1 minute, 30 seconds after her. Sometimes the race was sailed in a gale, when they started with single reefed mainsails and topmasts housed, but set spinnakers on the runs.

These were the crack smacks before 1914. Their crews handled them with an easy grace born of a lifetime's familiarity and their racing was of the highest order; to finish first in that company was prized almost as much as winning a Queen's Cup, for you raced against the same helmsmen.

Such racing revived during the early 1920s but died rapidly with the slump in fishing and general introduction of auxiliary power into even the crack smacks to enable a meagre winter living to be earned. However, there has been a vigorous amateur revival since 1946, with smack yachts racing annually at Mersea Town Regatta, at Rowhedge Regatta and at West Mersea, Tollesbury and Maldon; all keenly contested and attracting tremendous interest as three of the four last sailing fishing craft races in Britain.

Since 1900 Essex smacks have occasionally become converted to 'yachts' by admirers of the type, many since the decline of sail in the early 1930s. They often need considerable and costly rebuilding and, naturally, only the smaller smacks have survived in this way. A few hulls of the old, first-class smacks still lie rotting in Colne and Blackwater creeks, displaying their shapely frames even in decay. But even a small cutter smack is heavy work for amateurs, and many of the smack yachts do not carry topmasts, which is a pity as a topsail is ideal for light weather, and greatly restores their traditional appearance. The surviving estuary dredgers may lack the tallsparred grace of the larger smacks, but all east coasters are indebted to enthusiasts who preserve and rebuild the old smacks to keep them sailing, for the final passing of their tanned canvas from the regattas and the estuary will sever a tradition of fishing and racing under sail reaching deep into the past of the Colne and Blackwater seamen.

CHAPTER TWELVE

THE SOLENT AND POOLE FISHING BOATS

THE WATERS inside the Isle of Wight, with many good harbours and extensive creeks, were the homes of several types of sailing fishing boats, the most numerous being transom sterned cutters between 16 and 23 feet length. For many years yachtsmen have referred to them as Itchen Ferry boats but all old regatta reports referred to them as fishing boats, and the men who sailed them also called them smacks, but that term was usually reserved for the much larger counter sterned cutters and ketches working from the Hamble, Cowes and Portsmouth. They may have developed from the 12 and 14 foot open punts used by fishermen in the Itchen and Southampton Water to work a small trawl or dredge, mainly under oars but sometimes with their rig of sprit mainsail and stay foresail. Some of these had centreplates but most had fixed keels. More probably the small cutters were the result of local seamen being increasingly concerned with yacht sailing in summer, after about 1820, but needing to continue fishing for a winter living, and these boats were economical to lay up for part of the year, being cheap to build and maintain. Small cutters were owned and moored at Southampton, Itchen Ferry, Woolston and Weston on the upper north-east shore of Southampton Water; at Hythe and Ashlett Creek on the west shore, at the edge of the New Forest; at Eling on the Test, above Southampton; on the Hamble River at Hamble, Warsash and Bursledon. Others worked from Cowes, Newtown River, Yarmouth, Lymington, Pitts Deep and Tonners Lake, in the Solent; and from Portsmouth and Gosport, in the east. Similar boats were also used from Sandown, on the Isle of Wight's south shore.

Their numbers were indicated by the 1872 returns for second-class fishing boats, of which 141 were registered at Southampton, 182 at Cowes, and 247 at Portsmouth.

Until about 1850 the boats were commonly clench-built and usually rigged with a gaff mainsail without a boom but fitted with a brail and having an almost vertical leech with the sheet working on a horse across the transom, and a foresail with its tack to the stemhead and sheeted to a horse across the foredeck. The mast was long, the gaff short, and the rig easily handled (figure 67). Occasionally a mizzen was stepped. Reputedly, the rig developed about 1851 when bowsprits and two headsails were introduced, with the staysail tacked to a short iron bumkin from the stemhead. At first, the bowsprit was without shrouds or bobstay. Boom mainsails were used by 1852, causing an increase in sail area, with the boom overhanging the transom, and longer bowsprits. A typical singlehanded boat of 1852 was 18ft 6in by 7ft 8in beam, stem and sternpost plumb and forefoot only slightly rounded. Her sail dimensions were mainsail luff 17ft 6in, boom 13ft, gaff 10ft. The bumkin was 3ft 6in and a bowsprit was also fitted.

By the 1840s hull form was established with basic characteristics which persisted until the last were built 70 years later. The sternpost was almost plumb and the stem above water had very little rake, but the forefoot was moderately rounded. The keel was straight and often had considerable rake. Lengths varied from 16ft to 30ft overall, with an average beam to length ratio of almost 2 to 1. A typical 19ft boat would have 8ft beam and a draft of 3ft 6in. Apart from a small iron keel-shoe, all ballast was carried inside. Despite exaggerated beam, the form was very fair and capable of speed, while the upright sternpost and deep heel helped them run true in a sea.

The bow was fine, with hollow waterlines and the run was easy. The bilge was high and the section buoyant. The lines invariably indicate a 'raking midship section' and when pressed in a breeze these boats carried weather helm due to their short length, full shape, and consequent wave formation. On a 19ft or 20ft boat the reverse turn of the garboards occurred about 2 feet below the waterline, and below this, the lines of the 'fin' and keel were extremely fine. As these boats were never intended to take the ground, other than for maintenance, this was a permissible refinement in search of windward ability, though otherwise it would have been a structural weakness. The bottom of the keel was shod with an iron shoe, which on a 19 foot boat would weigh about 4 cwt, and the considerable remainder of the ballast was inside. Carvel planking appears to have been used in the type at least since the 18th century and, in contrast to many fishing boats built elsewhere, these cutters were usually built of the best materials and were frequently copper fastened. Red pine was commonly used for planking, on oak frames. The centreline structure and deck beams were also of oak and, whether built in a yard or as a one-off job by a contracting shipwright, the workmanship and

materials were good, ensuring a good investment and long life, which in many cases has been fully realised as several of these craft are still sailing a century after their launch. Due to their small size and their work in comparatively sheltered waters, all these little cutters were $3/4$ decked with a large cockpit or well. The foredeck extended about $1/3$ of the boat's length aft and a cuddy cabin was arranged under it, divided from the open cockpit by a wood bulkhead having a door to one side for access. The cuddy had very little headroom, averaging about 4ft to 4ft 6in, and was equipped with two sleeping berths, a cupboard and a coal stove. The mast was stepped through the after end of this space. The cockpit was protected for most of its length by side decks or waterways about 20 inches wide and supported by oak knees from the topside planking. However, the side decks stopped about three feet short of the transom where a wooden 'trawl deck' or 'fish tray' was fitted about 2 feet below the sheer, allowing dredgers or the cod end of the trawl to be emptied by a singlehander, without leaving the tiller, and preventing most of the mud and rubbish from getting into the cockpit and bilge. This was drained overboard on each side, and a lead, plunger bilge pump

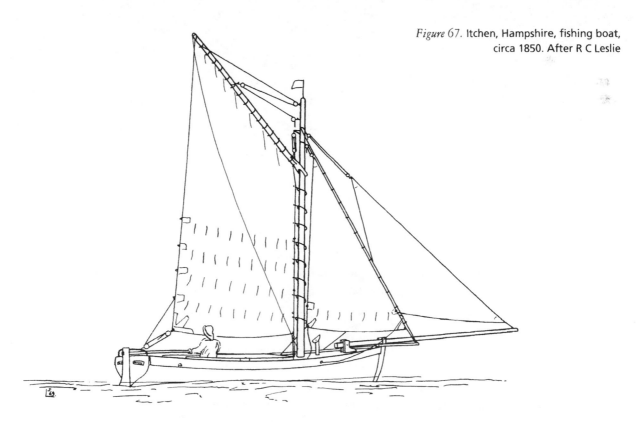

Figure 67. Itchen, Hampshire, fishing boat, circa 1850. After R C Leslie

Solent fishing boats in Portsmouth circa 1900. *National Maritime Museum*

was fitted in its centre. The cockpit was unobstructed by thwarts and its floor, or platform, was of substantial construction. The rudder was of generous area to ease weather helm and to assist in keeping the boat on course when the helm was left unattended for a few moments. Often an additional larger tiller was carried for use in hard breezes, or to concentrate the crew's weight amidships when racing. A footrail was fitted all round the deck edge about 5in high at the stem to $1\frac{1}{2}$ in at the quarter, and the hulls were painted slate grey or black, with oiled or varnished spars and painted mastheads, and some boats affected a varnished bulkhead.

By the 1880s the rig had developed to a powerful pole masted cutter, able to set a yard topsail in light airs. A few of the 27 to 30ft cutters may at one time have been fitted with topmasts but the smaller boats, often worked singlehanded, were happier without them. The mast was stepped well into the hull and the loose-footed mainsail was high peaked and fitted with a boom and four rows of reefs. The gaff was long and spans do not seem to have been used. The foresail (or staysail) was hanked to the forestay which was set up to an iron bumkin fitted to the stemhead. A useful sized staysail was thus achieved in a short boat. It was fitted with one row of reef points and, in bad weather, the bowsprit might be run in and a close reefed foresail set. Three jibs were usually carried: a working jib, a large light-weather jib, and a storm or spitfire jib, also used for correcting balance when trawling. The usual traveller and outhaul arrangements were fitted. None of them appear to have used boom crutches, the boom being let down on the quarter and lashed. As sail area grew, beam increased, and was carried to extremes, one of the fastest boats being 21ft long x 9ft beam, with all inside ballast. In 1880 a similar boat capsized when her ballast shifted in a squall and gradually the desirability of outside ballast, demonstrated by the yachts with which the fishermen were sailing each summer, led to some fishing cutters being built with lead keels of about 1 ton, with some trimming inside. Lead was desirable to avoid trouble with keelbolts and did not waste like iron. Despite this advance, the type continued to be built with great beam. These small cutters were quite fast for their length in smooth water and their fine fore ends enabled them to maintain a good speed in a short sea, though they often dived head and shoulders into it. Leslie wrote of his *Foam II*, a 20ft x 7ft 10in x 3ft 6in draft boat designed and built by Arthur Payne in 1888, 'averaging 7 knots with a fresh beam wind, and sailing from Southampton to Portsmouth and back, about 35 miles, in 5 hours, the wind being fresh NE'.

Among the principal builders of the type during the mid-19th century were Alfred Payne of Northam, Southampton, and Dan Hatcher of nearby Belvedere: both noted designers and builders of racing yachts. Luke of the Oak Bank yard, Itchen Ferry, also built many before he moved to Hamble, and others were launched by Fay's, and Stockham and Pickett, of Southampton. Many of Hatcher's fishing boats and yachts were designed by his foreman, William Shergold, who was also draughtsman and loftsman, designing the 21ft x 8ft *Centipede* which was the fastest fishing cutter launched from the yard. The 18 foot cutter *Wonder* built by Dan Hatcher in 1860 has

fortunately survived unaltered to the present day, and a completely detailed record of her has been made by Mr John Holness of Southampton (figure 68). A few years ago she was acquired by Mr Nicolay, a yachtsman who had known her as a fishing boat 60 years ago, and admired the type. Under his ownership she has been restored to her original trim and still bears her registered number. Her plans are reproduced by kind permission of Mr Holness. The *Wonder* makes interesting comparison with the shallower cutter *Nellie*, also built by Hatcher, whose lines, construction and sail plans were recorded by W. M. Blake in 1931. The 21 foot *Nellie* had been lengthened 20 inches by the bow and otherwise altered by Luke in the 1880s.

By the 1890s the development of these small fishing

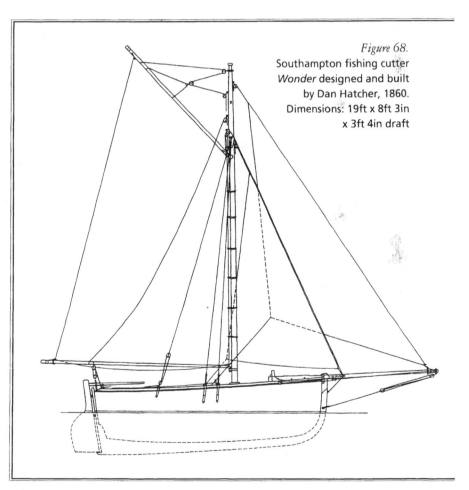

Figure 68.
Southampton fishing cutter
Wonder designed and built
by Dan Hatcher, 1860.
Dimensions: 19ft x 8ft 3in
x 3ft 4in draft

cutters had probably reached its peak and the lines and sail plan of a typical example are recorded by Uffa Fox in his book *Thoughts on Yachts and Yachting*. The *Blue Jacket*, as she was then named, was designed by Arthur Payne and built by Summers and Payne at Northam, in 1894. Her dimensions were 21ft 1in overall, 20ft 9in waterline, 8ft 4in beam and 3ft 11in draft. She displaced $3\frac{1}{4}$ tons and set 417 square feet in the three lowers; 270 square feet of it in the mainsail. Dozens of similar boats were an essential part of Victorian Solent seascapes: fish trawling inside Calshot in a fresh breeze with spitfires aweather; spray and rain glistening the hulls of the little fleet and soaking the tan canvas as they towed across the ebb, the stovepipe smoke blown to leeward, where a big steamer slid up Southampton Water.

The trawl net used by the small cutters was of normal form and proportions, with wings; and the length of the trawl beam was determined by the need for it to stow with one trawl head at the after side of the mast and the other just clear of the quarter. The heads, beam and net were similar to the normal trawl, and the method of shooting and hauling, which was done by hand, was as carried out elsewhere. When trawling under sail, the warp or towrope was made fast to the foot of the mast and was lead out over the quarter between the mainsheet horse and the end of the footrail; the warp could be pinned in to either quarter by rope stoppers spliced to the horse. A single thole pin was also arranged amidships for varying the lead of the warp. When hauling, the beam was run in with the fore head just abaft the mast and the other clear of the quarter, the cod end being emptied into the fish tray and the net re-shot while culling was carried on by the helmsman. Many worked beam trawls for shrimps or prawns in season, and for this work a boomless mainsail was preferred, fitted with four rows of reef points and a peak downhaul to control the gaff when lowering. These were known as sheet mainsails and, as no boom was fitted, their sheeting arrangements were similar to those used in the Essex and Kentish bawleys.

Although boomed mainsails became general in the Solent cutters by the 1880s, at least one, the *Nellie*, carried the old-style sail in the 1920s. These old Solent shrimpers also carried considerably longer bowsprits and larger jibs for the light winds of summer, but no bobstay or bowsprit shrouds were fitted. In fine summer weather a settled cycle brought a northerly breeze lasting most of the morning, giving a run down Southampton Water and after dying to a calm a fresh south or south-easterly set in

and sent everyone bustling about their trawling before hauling round to the west, sending them roaring home from the West Channel with the big jib pulling and the crew criticising the tactics of racing in the Solent. When shrimping, these boats used shrimp-mesh netting, inside an ordinary trawl. These small cutters often worked a beam trawl by driving with the fierce Solent tide without any sail set. The net was shot and the boat, with sails lowered, was brought broadside to the tidal stream by leading one part of the trawl warp to the forward thole and bracing it further forward by another line from the bitts. By varying the angles of these ropes the boat was kept driving roughly down-tide, dragging her net at a good speed. This was normally done in the broader channels but Captain Saunder's *Rosalie* also worked it in the mouth of the Hamble River, bringing up good hauls of soles and barnacle encrusted lobsters which had never been disturbed before.

When very cold weather drove flatfish to the bottom and catches fell off, the small cutters turned increasingly to oyster dredging. Holdsworthy, in 1874, stated that 'oyster dredging in the Solent employed a great many boats from the adjoining shores' and, apart from the extensive layings in the Newtown River, there were also layings at Hamble and dredging was carried on in Southampton Water and Spithead. As late as 1920, 10 or 11 cutters often dredged off Lee-on-Solent, averaging about 200 oysters as a day's catch, then selling at 4s 6d per hundred. Each boat worked two dredges and, if singlehanded, the man worked two dredges as well as handling the boat, which was strenuous. The day's catch was put into sacks labelled with the boat's name and left on the shore to keep until collected by the merchant who called at Hamble on a set day in the week, usually a Thursday. At Warsash, the oysters were usually handled by the proprietors of the extensive crab and lobster business in the village. Wet fish was collected more frequently by merchants or fishmongers with carts. One came from Fareham to buy at Warsash and Norman Cooper was the principal buyer of Hamble and Itchen Ferry fish fifty years ago. Marketing the fish was primitive and individual; at least one boat from Bursledon fished singlehanded for one or two days and nights, usually somewhere by the Needles, and spent the rest of the week disposing of his catch by driving round the district by pony and cart. Typically, he was a yacht hand in summer and a good rigger of small craft. Singlehanded working was quite common in boats up to about 22 feet.

One of the fastest of the small fishing cutters still sailing is *Fanny*, built at Cowes in 1872 for the Paskins family. Her origin is especially interesting as it also involved two notable boats of the type: *Star* and *Dolly Varden*. Before the mid-19th century, a young Itchen Ferry fisherman named Paskins required a new boat. He was approached by a village shipwright who wished to build her for him as his ability as a designer and builder had been ridiculed, and he was eager to refute local criticism by building the best and fastest fishing cutter he could devise. He built for Paskins the *Star*, a 21-footer with red pine planking on oak frames and beams. She entered in the regattas and was, reputedly, never beaten, taking the first prize each year while her owner continued to live at Itchen Ferry. Mr Paskins moved to Cowes in 1853 to found a noted fish merchant business and there *Star* attracted attention by continuing to win the first prize in her class in the town regattas. Her fame caused Mr Grant, then Secretary of the Royal Yacht Squadron, to request permission to have *Star's* lines taken off by J. S. White at Cowes, so that they could enlarge them in direct proportion to build him a yacht of the same form. The yacht was launched in 1872 and was named *Dolly Varden*. Mr Paskins took charge of her when racing and for several years he steered her to victory under Mr Grant's colours. In 1888 she was sold to Mr T. W. Ratsey, of the noted sailmaking firm of Ratsey and Lapthorn, who had admired her performance and was to continue her fame as a racer for the next half century during which she carried beautifully-setting sails and many innovations in sailmaking were tried and perfected on board her. For several seasons she raced in the 10 rating class, her principal opponent being *Nell*, a similar boat built by Halliday of Cowes in 1872 and owned by General Baring. Throughout her long and successful life *Dolly Varden* was regarded with affection as the heroine of the Solent fishermen and yachtsmen.

In 1872, Mr Paskins' son Edmund was old enough to work a fishing boat himself and it was decided to have one built on the lines of *Star* but the length increased to 22 feet and the beam increased in proportion to 8ft 4in.

John Watts of Cowes contracted to build the boat and also the shed in which she was constructed in the owner's back garden at No 1, Pelham Place, Cowes; a site now occupied by Ratsey and Lapthorn's sail lofts. Edmund Paskins assisted John Watts in her building, his job being to coat every piece of timber put into her with two coats of linseed oil. On completion, the end of the shed and the garden wall were removed and *Fanny*, as she was named, was taken on a trolley to be launched at the old town quay. That year both *Fanny* and *Star* were entered in the Town Regatta and *Fanny* won their class by a few lengths from *Star* in half a gale, westerly.

As well as the Cowes fish shop, until 1939, the Paskins family also operated the oyster beds at Newtown, a mud creek on the north shore of the Isle of Wight. The layings were supervised by Edmund's brother Henry, who lived at Newtown and was in charge of the fishing and oyster side of the business. His extensive knowledge of Solent tides and waters made him a useful racing pilot on board the Royal racing cutter *Britannia* during some Cowes Weeks. Although built as a fishing boat in 1872 *Fanny* was not registered until 1902 when the number CS12 was allocated. She was primarily used for prawning, working a 12 foot beam trawl, but her other main work was carrying oysters from the family layings at Newtown, to Cowes. Often several thousand oysters were carried at a time in sacks stowed amidships; *Star* specialised in this work almost every day for over 60 years, until she was sold in 1937. She is still in use as a fishing boat at Lymington. *Fanny* was sold in 1920 and has passed through several owners as a yacht. She was later owned and sailed by D. Cook of Ipswich, Suffolk.

Most of the Portsmouth Harbour boats worked in Spithead and many berthed in the Camber under the grim shadow of the coaling stage. The seven ton *Firefly* was notable in being owned by Captain Robert Gomes of Gosport, a prominent racing skipper in *Petronella* and the German Emperor's *Meteor I* and *Meteor II*. Before 1914 several small cutters were owned at the Isle of Wight resorts of Shanklin and Sandown, working off the beach with pleasure parties in summer and fishing in winter. Hull form and rig were identical to the boats inside the Wight, but some were clench-built for lightness in hauling up, and probably much of their inside ballast was shingle in bags, jettisoned ashore to ease capstan work up the beach skids where they lay strangely out of place among the light pleasure skiffs and gigs. These boats raced annually at Ventnor Town Regatta.

A yard at Fishbourne, on the east bank of Wootton Creek, Isle of Wight, built Solent fishing craft and a photograph exists of two small cutters about 27 feet long, ready to launch, and a counter sterned smack of about 45 feet length in frame, all building in the open. Several small fishing cutters were owned at Wootton Creek and in the 1890s they laid on moorings below the ferry on the west side of the creek. One of them was owned by the Young family of Fishbourne, where their cutter was regularly hauled up on the foreshore in summer. Numbers of these boats were also built at Cowes and about half-a-dozen of them laid on moorings along the East Cowes waterfront, below the chain ferry, until 1914. Counter sterned cutter smacks of about 18 tons and similar to the Essex smacks also worked from Cowes; the last being the Essex-built *Jolie Brise* sailed by Mr Ben Chaplin until 40 years ago. The author has a photograph of one, the *Amorel*, CS5, in what appears to be racing trim in the early 1920s, but cannot discover if these smacks raced as a separate class.

From the late 18th century until 1914, the fastest fishing boats were keenly raced in annual regattas at Itchen Ferry, Southampton, Hythe, Hamble River, Bursledon, Cowes, Wootton and Fishbourne, held in September when the yachts had laid up. Regatta racing directly affected their development as, apart from the striving for improvements in hull form and rig, consistent with fishing requirements, their arbitrary division into classes for racing, measured by overall length, seems to have influenced the owners' choice of dimensions for new boats.

The 1902 Hamble River regatta was typical in being organised by a committee of yacht captains and builders, and was a parochial public holiday with spectator yachts dressed overall and the band of the Hamble training ship *Mercury* trumpeting away aboard the committee barge *Anglo-Saxon*. Traditionally Hamble starts were with anchor down and all sails lowered at gunfire, when rivalry burst into frenzied activity and splendid racing which, like their Essex contemporaries from the Colne, gave opportunity to even up the yacht racing seasons' scores, and cut a dash before the community. All

competitors were fishing boats divided into classes of 23-27ft; 21-23ft; 17-21ft; and below 17ft, with at least eight boats in each class, all setting spinnakers and yard topsails. These were the crack boats of the Solent area, sailed by men who were prominent yacht skippers all summer; Sam Randall's *Freda* of Hythe; William Bevis' 26 foot *Oyster* of Bursledon; George Parker's *Dragon* from Itchen; Bob Edwards' *White Belle* of Hamble; Dan Cozens' *Gazelle* of Weston; and Joe Oatley's smart *Harriet* from Cowes, and many others. The greatest sport was always between Captain Ben Parker's 19 foot *Annasona* and Captain Tom Diaper's *Eileen*, both from Itchen, which for years dominated the 17-21 foot class and were maintained like yachts, of which both men were noted racing skippers; Parker sailed the German Emperor's *Meteor II* and *Meteor III*, and Diaper the 40-tonner *Norman* and other yachts. Often they only beat each other by seconds but their rivalry ended sadly in 1905 when Parker, who lay dying, had his boat sailed to victory by his brother George, who continued to race her subsequently.

Prizes were by public subscription and time allowance was commonly two minutes per foot length, but there were no restrictions on canvas, and very few protests! Hamble gave only money prizes but at the similarly-organised Itchen Ferry regatta, which was reputedly older than the Hamble event, the first class boats raced for a new mainsail and foresail given by Ratsey and Lapthorn, and the second class for a new trawl warp by Tillings, the ship chandlers, in addition to money prizes.

The 1898 Hamble regatta included a race, with nine starters, for 'old class boats'; probably sloop rigged shrimpers with boomless mainsails. A quaint condition added 'No boats over 100 years of age allowed to compete', emphasising the traditional longevity of the Solent type of which the 27 foot *Sorella* of Warsash remained the last rigged example occasionally working. Built as the yacht *Sister* in 1858, by Hatcher, in fishing boat style, she raced successfully in the Solent classes until sold in 1891 to Mr T. Fuger of Warsash, the father of the subsequent owners, before she was rebuilt as a yacht.

Although no new Solent fishing cutters have been built for over 70 years, the type survives with a few working under power and increasing numbers of enthusiasts reviving old hulls and re-rigging them into semblance of the original. These beamy boats offer good accommodation and are comparatively inexpensive to build and maintain, and a joy to all who own and sail them.

In 1872, 17 first class smacks were registered at Southampton; five at Cowes; and 18 at Portsmouth. Many of these were large cutters and ketches belonging to Hamble, Warsash, Titchfield and Bursledon, as the Hamble River fishermen took to yachting later than their neighbours and were more adventurous in using seagoing fishing vessels, some of which mixed fishing with coasting, like the Hamble *Albion* which brought cider from Totnes on the Dart, and potatoes from France. Shellfish dredging was carried on in the Channel by Southampton smacks owned by the Bowyer family at least by 1838, and was a common winter occupation of the Hamble River smacks during the 19th century. Much of this trade was organised by the Scovell's Hamble Fisheries Company, which also owned the Hamble boatyard where smacks were built to high specification, many being copper fastened; including the *Pearl*, *Uzill* and *Jane*. Scovells also owned the old Rowhedge smack *Secret*, and the *Result* from Harwich. Hamble men also dredged scallops in the North Sea, where several of them were drowned, and the smack *Laurel* foundered when her chain plates carried away.

Many Hamble, Cowes and Portsmouth smacks worked stowboat nets between October and February for sprats, in the same manner as the Essex fishermen. Christchurch Bay was their favourite ground, but the Solent and Southampton Water were easier berths. About 1890, 22 smacks and 80 men went spratting from Hamble alone, landing tons at prices varying between 4d and 5s a bushel; but in times of glut, their catches were dumped on the village hards and sold for manure at pitiful prices. The Hamble River fleet included several old Essex smacks such as *Gipsy Queen*, *Jemima*, *Welfare* and *Racer*, a Channel Islands built cutter originally owned by Henry Barnard of Rowhedge and sold to a Bursledon fisherman working her until the 1920s. During World War I the fisheries were under Government control and rapidly declined, so that by 1920 only three spratters were still working from the Hamble, including *Ariel*, built as a Bembridge pilot boat and converted from

clench to carvel planking by Moody at Bursledon in 1870.

Warsash, at the mouth of the Hamble River, was the centre of a thriving crab, lobster and crawfish trade with which Hamble village, opposite, had no connection except that its seamen manned many of the varied craft which had wet wells fitted to them to carry the shellfish to Warsash from Concarneau, the West Country, Ireland and Norway. The crabs were transferred to floating store cages called crab cars, moored in the river, and the carriers laid on Warsash hard to discharge lobsters and crawfish into the vast store ponds on the saltings, now filled in and forming the dinghy park of the Warsash Sailing Club. The old fruit schooner *Peri*, built by John Harvey at Wivenhoe for West Country owners, brought some of the first crabs to Warsash from Concarneau a century ago, and many of her Salcombe crew settled on the Hamble. She was built with two masts, but later stepped a third and her hull was deepened and lengthened, making her the largest carrier, but broke her owners, who spent more on her maintenance than they earned. She was one of the few vessels to get ashore on the Goodwins and get off again, when bringing lobsters from Norway; but she ended her days as a jetty at Hamble. Other carriers included the smacks *Eagle* and *Zebedee*; the old Isle of Wight pilot boat *Cupid*; the schooner *Lotus*, converted from the yacht *Jubilee*. The Concarneau trade ended about 1890, but great quantities were shipped from Ireland, particularly from Newquay, Galway. The Norway voyages were harshest and, on one of them, the Hamble schooner *Imogen* was lost with all hands in the North Sea. A typical West Country voyage was to the Start; anchoring off crabbing villages such as Hallsands, Beesands and Ivycove, with the Red Ensign at the peak as a signal for the locals to launch off with the contents of their store pots which were tallied into the well, where net bags tightly packed with lobsters and crawfish were also hung. A typical cargo was 6,300 female crabs, 650 crabs, 500 lobsters, and 54 crawfish, which was sailed home as quickly as possible. The cutter *Ellen* was the fastest carrier, once running loaded from the Fastnet to Hamble in 49 hours, but rival craft were said to have done the passage in 40 hours. Best remembered of the fleet was the 40 ton ketch *Ceres*, SU2, built at Poole in 1869 for the grain trade then flourishing between Poole and the Channel Islands. On her maiden voyage she got ashore on the Kimmeridge ledge, off St Albans Head, and was badly holed. A crew from Hamble salved her and she was fitted with a crab well, spending the rest of her life in the fishing trade, particularly crab carrying at which she often made twenty trips to the westward each summer, besides voyaging to Ireland and Norway, and carrying oysters from Whitstable. In winter she sometimes dredged oysters in the North Sea, once making two trips in eleven weeks, and occasionally dredging scallops in Caen Bay. For 35 years she was skippered by Captain Bob Williams of Hamble who, when she finished her days on the mud off the village, went along daily to pump her out 'for old times' sake'. After World War I the crab ketches *Mary Leek, Cargo, Macfisher, Gem* and *Ceres* were fitted with small auxiliary motors to ensure regularity of delivery, but their crews still set all sail and motor-sailed, giving a nimble feeling to the tillers of these heavy vessels which, after a few years, were left to rot in the saltings. Most of the large Portsmouth-registered smacks were owned at Emsworth, in Chichester Harbour, and their radical development is described in another chapter.

The Solent and Poole fishing vessels had much in common and until about 1930 a fleet of 30 little cutters sailed out of Poole, fish-trawling, spratting, and oyster dredging; and were kept on moorings off the town. In 1872 most of the 61 second class fishing boats registered there were sprit rigged and averaged 18 feet length. They usually fished in the extensive harbour for shrimps with a net like a wooden oyster dredge, with no scraper, called a keer drag, and in winter dredged oysters from the extensive layings, and trawled fish. Gradually length increased to about 20 feet, producing a beamy, full bowed, clencher built boat, decked to the mast but able to be rowed in calms. The rig changed to yawl with the mast stepped $1/3$ length from the stem, and set a boomless mainsail with a vertical leech and the sheet working on a transom horse. A small sprit mizzen, without a boom, was set on a stumpy mast stepped at the transom and sheeted to an outrigger. Early boats set a large foresail on a long bowsprit, but later changed to a narrow staysail on an iron bumkin, and a moderate length bowsprit with a jib. Ill-fitting yard topsails were set in light weather. The Swanage Bay boats retained this

rig until the 1930s but about 1890 the cutter rig was introduced to the Poole boats which rapidly developed similarities to the Solent type, with some external and much inside ballast, but had more powerful sections and, instead of the long hollow bow of the prototype, evolved a bolder head and higher topsides which were flared throughout, without tumblehome, producing broad quarters and a rather wall-sided appearance. Many were built by Barfoot and until 1900 were mostly between 20 - 25 feet length, often clencher planked. As they ventured further for fish, size increased quickly to about 30ft x 5ft draft in boats like the *Prince of Wales, Sea Gull* and *Jubilee*, which were eclipsed by the 35 foot *Boy Bruce* launched in 1906, and the first fitted with a centreboard which was quickly copied in many subsequent boats; these evolved dinghy sections and were faster and handier, but not more seaworthy, than the keel type. They were heavy work for two men and the limit of size was reached with the 35ft x 12ft *Secret*, designed and built by W. & J. Allen & Co of Poole and a particularly good sea boat which afterwards became the Poole pilot boat, and eventually a yacht. Ashton and Kilner also built many fine boats of the larger type, including *White Heather*, designed by Mr Ashton and launched in 1908 with about 6 foot draft. Their boats had very well proportioned hulls; many were soon sold as yachts, and smaller boats built in replacement.

The favourite type of two-man boat remained about 27ft x 9ft 8in x 4ft draft and were often fitted with a centreplate. They were very stiff, had a plumb stem and transom and fuller bows and bottoms than the Solent boats. Construction was strong and simple and the hull was arranged with a two-berth cuddy under a foredeck carried a foot abaft the mast, with sidedecks reaching to the transom and enclosing a large cockpit or well which had no thwarts, but seats on either side extending from bulkhead to transom. Moveable boards fitted in the front of these seats forming wing pounds for the catch and, when trawling, baskets and boxes were stowed at the forward end of the well, and the beam trawl was worked from the starboard side. The pole masted cutter rig set a generous area with a long luffed, loose footed mainsail. Many had the foresail tacked to an iron bumkin and all had a long bowsprit which carried a big jib in moderate weather, when they were also fond of setting ill-fitting yard topsails. They were powerful boats, often working out by the Needles in a breeze which might freshen to make them reduce to two reefs and the foresail, with their bowler-hatted and side-whiskered crews making splendid weather of it.

Some Poole men went yachting in summer but most of the thirty or so boats fitted-out annually in April for trawling flat fish off Christchurch Ledge, Bournemouth Bay and often at night 25 or 30 of them might be trawling within a square mile, before beating home round the Hook Sand or cheating the tide through the East Looe, close inshore. The Poole spratting season commenced early in November and lasted for three or four, sometimes five months. The crew of two scanned the surface for signs of birds feeding on the sprat shoals which were trawled for with an average 14 foot beam trawl raised about four feet above the top of the heads on iron supports to enlarge the mouth of the net, which was seven fathoms long, had no pockets, and was made of light thread with close mesh. The trawl was shot in the usual way with one end of the warp fast forward and the other fast to a cleat on the transom. The cod end was emptied in the well and the catch stowed in the wing pounds. As everywhere, prices varied with supply, from 3s a bushel to 3d or 4d for manure; a record catch was 240 bushels in a day but they often made more from a small catch, which would be rowed ashore. Large catches were discharged against Poole Quay.

One winter night about 1910 the Poole fleet was working off Hengistbury Head on ground called the Dogger Bank, when a heavy west gale set in with hard squalls. The fleet was about 12 miles to leeward of the harbour entrance, deep loaded with fish, and had a hard beat home; some lost gear and the Life-boat was standing by. Motors were installed in some boats about 1915, usually in the cuddy, and after the war most centreboards were discarded, but sail was retained until the 1930s, and the 20-footer *Sunshine*, built in 1892, was still rigged and working in 1955.

CHAPTER THIRTEEN

PILOT CUTTERS AND SCHOONERS

PILOTS AND THEIR BOATS had great dignity and independence in the days of sail and sailing pilot boats always represented the most seaworthy craft of their area. Unlike other commercial craft they were designed to sail at constant draft and trim, which required no alteration when the best was achieved. Only a few types can be discussed and these varied in requirements: some took or received pilots only a short distance seaward from the port their pilots served; others cruised with numbers of pilots on board 'seeking' ships, sometimes for hundreds of miles in the open ocean. Speed was usually important as pilots were frequently in open competition. When this died, about 90 years ago, with widespread amalgamation of pilots caused by the greater regularity of steamships and the reduction in the numbers of ships at sea, the pilot boats' rigs were reduced.

Because of their numbers the most widely known pilot boats in England were those from the Bristol Channel. These cutter rigged craft were also called skiffs by their crews, or yawls if they hailed from Cardiff; both terms indicated pulling and sailing boat origins. They were usually owned by one pilot, rarely by two or three. The Merchant Venturers of Bristol, and later Bristol Corporation administered the area's pilot services which originated in Cabot's time. For centuries the Bristol Channel pilots worked in competition, and gradually developed cutter-rigged craft averaging 40 feet length, to take pilots from Bristol, or Pill, and from Cardiff, Newport, Barry, Gloucester and Swansea, to meet ships bound for Bristol Channel and Irish Sea ports. They cruised, seeking ships, in those waters, and in the English Channel as far east as the Isle of Wight or Dungeness, sometimes venturing 100 miles west of the Fastnet

seeking large homeward-bounders and racing each other for their pilotage. However, Lundy Island, 80 miles west from Bristol, was the principal seeking ground until 1914 when war conditions ended seeking. Most pilots would put to sea in almost any weather, but had individual limits of distance and preferences for large or certain ships. They could earn an average of £600 annually 85 years ago, but the costs of running the cutter, which amounted to about £300, had to be deducted from this. Even so they lived well by contemporary standards. The élite amongst them were known as western going men, specialising in larger ships whose movements were carefully watched from shipping papers and tips from butchers' runners lists. If several of them were seeking the same ship they might sail to Liverpool, Belfast or Dungeness to make sure of her. Their large cutters carried two men, or one experienced hand and an apprentice; they had to get the utmost from the cutters and were splendid seamen who worked hard and were very proud of their calling. Typical skiffs were manned, apart from the pilot, by one hand known as the pilot's man and an apprentice who 85 years ago was paid five shillings a week in the first year, rising to one pound in the last, with all found, and served five or seven years, before going deep sea in a sailing ship and later, in a steamer. He might become a ship's officer or even captain before a vacancy occurred, but many returned to work in the skiffs as pilot's men until elected a pilot.

The pilot skiffs could ride out extremely bad weather with an easy motion, would lie hove-to quietly and stay reliably in heavy seas, but they were generally very slow to windward. Very few foundered from bad weather alone, but several were run down and sunk, and others

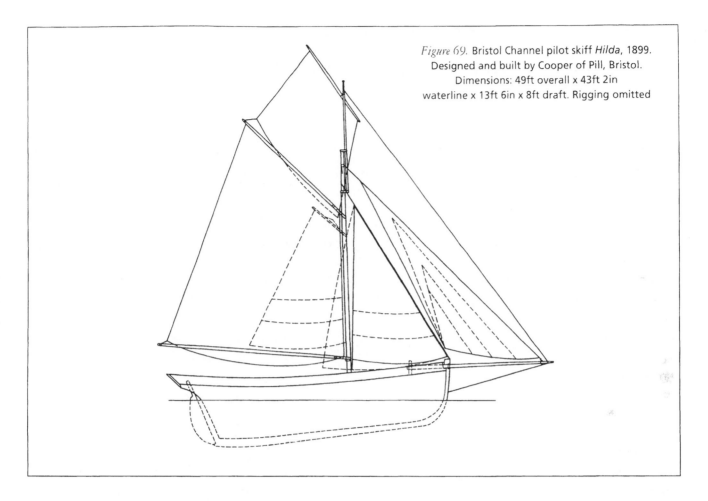

Figure 69. Bristol Channel pilot skiff *Hilda*, 1899. Designed and built by Cooper of Pill, Bristol. Dimensions: 49ft overall x 43ft 2in waterline x 13ft 6in x 8ft draft. Rigging omitted

lost by stranding. However, they occasionally ran for shelter, as in the great gale of October 1880 when Cardiff cutter number nine, pilot W. Morgan, ran for Ilfracombe with topmast housed, mainsail reefed almost to the gaff jaws, and small staysail and spitfire jib; swooping into harbour in tremendous thundering seas, breaking white half way up the coast cliffs. Most accidents in the service arose from 'boating', or boarding ships in bad weather and at night. The skiff's 12ft x 4ft 9in x 2ft 6in deep clencher-built rowboat, locally termed a punt, was carried in chocks on deck and launched and recovered over the bulwarks or through a portable section. On closing a ship the cutter hailed to find if she was bound to her port. If so, and she was a ship the pilot would take, they hailed her to stop. The pilot turned out in shore-going clothes, the man took the helm and the pilot prepared the punt while the cutter sailed under the ship's lee. The apprentice and man lifted the punt overboard and the pilot and apprentice jumped in and towed astern until close to the ship, when the apprentice sculled the

punt over to the jacob's ladder put down for the pilot to climb up the ship's side. The apprentice sculled back to the cutter, head to sea, and the punt was hoisted in. Even in very bad weather lifejackets were only worn occasionally, when many risks were taken to secure profitable service. In fine weather the pilot might scull himself across and board, kicking the punt off from the jacob's ladder, to be picked up by the skiff, which at other times might run alongside the ship.

Many of the 19th-century pilot skiffs were built at Bristol where Hillhouse launched some early in the century, and later William Patterson built in numbers: seven in 1851 alone, six of them sisters. However, Pill, or Crockerne Pill, about a mile from the mouth of the Avon, was the headquarters of the Bristol pilots, whose calling is supposed to have originated in 1497, or earlier. By the 1860s Pill creek sheltered 44 pilot skiffs, and 80 pilots lived in the village, where the majority of the Bristol Channel pilot skiffs were built during the 19th century by William Morgan, Thomas Rowles, Thomas Price and

Charles Cooper, who was succeeded by John Cooper who built there until 1905; and Edwin Rowles, who commenced building skiffs at Pill in 1887 and continued until 1910. In earlier years these yards also built coasting sloops, schooners and brigs; and later fishing vessels and yachts. The best remembered are the Coopers and Edwin Rowles. Skiffs built by the Coopers had a reputation for seaworthiness in bad weather, while those launched by Rowles were generally faster. Rowles also built the 53 foot *Pet*, believed to have been the largest skiff working the Bristol Channel. Both builders designed by carving half models which were approved by the owner (and no doubt half the village) before construction commenced. The elm keels in one piece, and stem, sternpost and frames were of oak; hull planking of oak, elm and pitch pine; and decks of pine. High bulwarks were fitted and cockpits were only introduced in the later skiffs. These yards were small and run in simple fashion. Cooper built the 49 foot skiff *Hilda* in six months during 1899, with the aid of one shipwright and a boy. (Figure 69.) She cost £350; larger skiff's ranged up to £450 at that time.

Hull form varied tremendously. Many skiffs, particularly those built before the 1890s, were transom sterned; others had square counters. Later, elliptical counters were introduced, mainly in skiffs built at Porthleven, Cornwall; and these suffered less damage than other types. Stems were plumb, or raked slightly forward above water, with moderately rounded forefoot. In later years more rounded and cutaway stems became fashionable. Some skiffs were very full forward and fine aft, and had low quarters making them wet in a sea. Others were the opposite shape; but all had high bows. Many had full sections with no hollow, especially the Pill boats which took the ground on legs; but some were very hollow. The centre of buoyancy was generally well forward, especially in the Cardiff-built boats, and the Pill builders favoured a long run starting under the mast. All carried a small rig and many tons of inside ballast in the bilge and side lockers, and only a few boats had a ton or so outside on their keels.

All skiffs carried the cutter rig with the West Country characteristics of the mast stepped well into the hull, resulting in a large staysail and a comparatively small jib set on a moderate bowsprit. Most carried topsails but many had pole masts, including *Pet*. Topmast

shrouds were not always fitted except in the larger skiffs which also had crosstrees. In winter they sent the topmast down and shipped a chock staff in the topmast irons, to fly the red and white pilot flag. The mainsail had four rows of reef points, and large and small foresails and several sizes of jibs were carried: the 'slave' or working jib; the spitfire for bad weather; and the 'spinnaker', set in light airs from the bowsprit end to the masthead. This had to be handed in a breeze as bobstays and backstays were not usually fitted, and although the after shrouds led well aft the headsail luffs sagged badly on the wind. Until the late 1880s the standing rigging was tarred hemp, or even chain, set up with deadeyes and lanyards. In later years wire standing rigging was introduced but the lanyards remained, except in the composite-built Newport cutter *Mascot*, designed and largely built by her owner, pilot Thomas Cox, and at 56 feet the largest cutter in the Channel. Three shrouds were fitted on each side, one a cap shroud. The peak and throat halliards led aft, often to quarter posts in reach of the helmsman. Jib and staysail sheets also led aft. The mainsheet was made fast with a single hitch on the lower block.

Roller reefing worm gear turned by a handle was introduced to the skiffs about 1903 and spread rapidly so that by 1909 70 per cent had them. The booms were docked to end over the taffrail and the mainsheet led to the boom end, throwing great bending strain on the boom when reefed in bad weather. Many booms were broken from this cause and the set of many more mainsails spoiled. In a breeze the jib was often stowed and the bowsprit run in. An experienced hand could do this and roll down a reef in the mainsail within a few minutes, and without calling his mate. The crew of two berthed in the roomy fo'c'sle, where all cooking was done on an iron stove. Aft of this was the main cabin, often with two or four berths at the sides for the pilots, narrow locker seats in front, and a table between, supported by a water tank. In larger skiffs the amidships cabin was sometimes a saloon, with a sleeping cabin and pantry between it and the fo'c'sle. All the accommodation was dark and airless, lit only by decklights as skylights could be smashed in by the winter seas. The 'steerage' aft of the main cabin was a rope fender and oilskin store known as the runs. Watertight cockpits were not fitted until the

Cardiff pilot cutters racing. Photograph taken in 1899. Note pilot flags, cutters' name pennants and large staysails.
Photo courtesy Tom Cunliffe

1890s, and were first drained by lead pipes closed at the hull by leather flaps which stopped water running back up the pipes when sailing. A few later skiffs had patent WCs, in keeping with the dignity of pilots, and most were well rigged and maintained, but some were poorly built. An apprentice once remarked how their cutter kept her decks free of water in bad weather and the man replied that so much water went through her topsides there was none left to reach the deck.

In 1861 pilotage to Cardiff, Newport and Gloucester became independent of Bristol where pilotage became regulated by the City Corporation instead of the Merchant Venturers who had administered it for centuries. At that time Cardiff owned 40 pilot boats; only a few less then Bristol. Many were built there by Hambly, or by Davis and Plain whose *Anita* had the lightest displacement of all the skiffs. Her very hollow section, hard bilge, long easy bow, and full quarters made

her fast, and she won a race round Lundy Island in a westerly gale against the Newport skiff *J.N. Nap*. The 8 ton cutters of the Cardiff 'Hobblers' were another Bristol Channel type engaged in tending on ships lying in the pool, running errands, and ferrying men and stores. When this trade died the craft were sold and worked successfully in the pilot service. They were straight stemmed, transom sterned cutters of similar form to the skiffs and built at the same yards. At Newport, in 1904, 44 pilots worked 36 cutters, mostly built at the port by John and William Williams and Mordy Carney, whose craft were good sea boats but very slow. Charles Cooper of Penarth, related to the Pill family, was another noted builder, while yet another Cooper built at Gloucester, where Jeans and Lemnon also launched skiffs. The Gloucester pilots owned a few skiffs until about 1900, and some fast cutters were built for Barry pilots after they became independent in 1885. Bridgewater pilots

By 1908 about 100 pilot skiffs were working from the Bristol Channel ports and in later years some unusual craft were built: notably William Prosser's Newport skiff *Alpha*, and the Barry skiff *Kindly Light*, both designed by W. Stoba and built by Armour Brothers at Fleetwood in 1911. They resembled large Lancashire shrimping smacks in hull form, with cutaway forefoot, flat sheer and round sterns, and were fast, but could not be pressed in heavy weather. *Kindly Light* is now the yacht *Theodora*. However, the most startling departure was the Barry pilot skiff *Faith*, designed by Harold Clayton of Penarth and built there in 1904. Her form was that of a fast cruising yacht of the period, for which Clayton was noted, and she caused considerable argument among the pilots. She had external ballast and a powerful cutter rig, and was reputed to be the first skiff built from paper plans. She proved fast in light airs, besides being a good sea boat, able to ride out hove-to with the best of the old boats. She was a frequent prize-winner in the races for pilot skiffs held at Cardiff, Barry, Port Talbot, Ilfracombe, Penarth and elsewhere, which sometimes attracted entries of 20 or more until the early 1900s. Special suits of light racing-sails were bent for these occasions, but at Cardiff they once started in a full gale with all canvas for the run, luffing up at the mark and pulling down a reef with wire pendants hauled home by the racing crew of twelve. The Pill-built *Marguerite T* of Cardiff was a noted prizewinner and one of the smartest skiffs in the Channel. Her racing mainsail clew went 6 feet outboard, and in light weather a hand sprayed the leech with a syringe to stop it shaking.

At Newport, and elsewhere, the pilotage board reviewed the cutters annually on two Show days. The cutters assembled at Newport pilots' pill and gear and equipment were inspected. Next day they sailed past in the river, watched by the board on a steamer. Towards the end of sail some skiffs were fitted with auxiliary motors which were reputed not to have been used when racing sailing skiffs for a pilotage job, though this seems hard to believe. Steam pilot boats were introduced at Swansea in 1898, at Cardiff in 1912 and Newport in 1920. The Bristol pilots amalgamated in 1918 but retained their sailing skiffs until a steamer was built for the service in 1922. The fleet of skiffs was gradually sold, some for similar work elsewhere, but the majority to yachtsmen

Falmouth pilot cutter under trysail rig

mainly used pulling boats, and only one skiff, the *Polly*, was owned there. Increase in demand for pilotage after 1860 led to some new skiffs being built outside the Bristol Channel; sometimes for cheapness of delivery; often in attempts to obtain greater speed as, inevitably, the greater numbers of pilots increased competition when trade was depressed. Numbers were built at Porthleven, Cornwall, by William Bowen, and by Kitto who built the elliptical sterned, *Hope of Newport*, reckoned one of the smartest skiffs, and also launched *Jenny* in 1911, among the last of the breed. Slade at Polruan, Trethowan at Falmouth and Angear of Looe, and even Luke at far away Southampton, contributed craft for the service.

who appreciated their seakeeping qualities, if not the weight of their gear.

The small pilot cutters of the Bristol Channel service have been so well documented as to overshadow the larger cutters working from the Scilly Isles, Falmouth, Plymouth and Penzance. The sailing ship trade using Falmouth increased rapidly after the Napoleonic wars, reaching its height during the 1870s and declining quickly thereafter. Competition between pilots and their cutters was keen before amalgamation, and developed craft able to sail in the worst weather, but also capable of speed in all conditions, with light weather ability greatly sought. In sails' heyday they cruised seeking ships between the Bishop Rock and the Start, and 13 cutters were owned at Falmouth, with half a dozen at Plymouth and at least one at Penzance; while each of the Scilly Isles, St Marys, Tresco, Agnes, St Martin and Bryer, owned and manned pilot cutters, many of them of native build. Four of these Scilly cutters were between 65–75 feet length and the remainder 45–50 feet.

Early cutters for this service resembled the West Country fishing smacks but gradually beam was reduced and lines fined in the search for speed. The Plymouth cutter *Allow Me* typified the ultimate development with

dimensions of 67ft length BP x 15ft 5in beam x 9ft depth. She was one of the fastest cutters in the south west, especially to windward, equalled only by the *Leader* also designed by H. Prigg of Plymouth, one of the many brilliant designers now forgotten, who created very fast sailing ships of varied rigs including ships, barques, brigs, schooners and smacks. Trethowan of Falmouth and Gent of Plymouth were other noted builders of pilot boats.

The heavily built hulls compared with contemporary racing yachts in having plumb stems, long straight keels and slightly raked sternposts. The entrance and run were very fine and hollow, and the midship section had great rise of floor with high hard bilges and some tumblehome in the topsides. Counters were short in West Country fashion; well veed to ease slamming in a seaway, and all ballast was in the bilge. The deck was clear except for the bowsprit bitts, staysail horse and small hatches. No windlass was fitted, the cable being got in with a messenger taken round the drum of a winch on the fore side of the mast, which also handled the halliards and lifts. All are reputed to have been salted for preservation, and below deck the fo'c'sle was a store, with the crew's cabin amidships entered from a companion hatch and

Figure 70. Scilly Isles pilot cutter – winter trysail rig

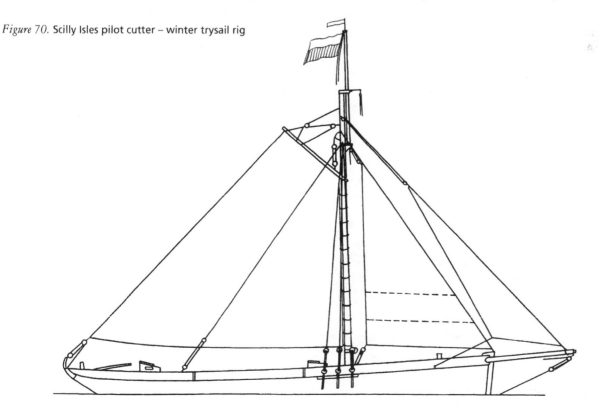

bulkheaded from the large cabin aft, with its tiered bunks for eight pilots and companion opening, just before the tiller, which was often short, but sufficient for the narrow rudders fitted. A deep and beamy 14ft boarding boat stood right way up on deck amidships, ready to be launched and pulled to put aboard or take a pilot from a ship; a hazardous task in bad weather.

In summer these cutters set generous sail areas approaching those of racing yachts, but in winter the tall topmast was laid ashore and replaced by a short pole carrying the pilot flag. The summer mainsail was unbent and a short gaffed trysail set instead, balanced by a smaller staysail and spitfire jib (figure 70). The cutter *Atlantic* from St Marys, Scilly, was 67ft BP x 16ft 4in beam and drew 9ft. She carried a 1,100 square feet mainsail having a boom 48ft long. Her mast was 61ft from deck to cap, and the topmast 31ft long, although Scilly cutters generally set smaller topsails than those from the mainland. Many sizes of jibs were carried and usually two sizes of foresail, to balance reefed conditions. Bowsprits were short and reefing, and powerful topping lifts and backstays were prudent precautions by mariners prepared and confident when working in any weather and at any time in very exposed waters.

As an example of speed, the largest Plymouth cutter, *Verbena*, built at Brixham by Jackman, once ran from the Lizard to Rame Head in four hours 10 minutes; averaging nine knots, ignoring tidal effect. These pilot cutters raced in local regattas and the 1864 Falmouth event was typical: a large number of that port's cutters entered, with the quaintly named *Gorilla* of Penzance the sole outsider. Enthusiasm and local pride led to the *Gorilla's* bobstay and main earring being cut, and she was verbally attacked while in range of her competitors. The *Arrow* won, to the strains of 'See the conquering hero comes', without which no Victorian regatta was complete, while the last boat got 'Cheer up Sambo don't let your spirits go down', another pop tune of the period.

However, the pilot cutters were excelled in competition and speed by the similar sized 'Tailors' cutters of the same period, which were owned by various Falmouth shopkeepers and cruised seeking to put representatives aboard inward-bound ships to secure trade in supplying clothing and other goods. At one time 13 were owned in Falmouth and life aboard them was reckoned more strenuous than in the pilot cutters as, while a homeward-bound ship would heave-to for the pilot, she ignored the tailor's cutter; which usually had to launch a boarding boat ahead of the ship and pull alongside her, to hitch a long boathook in the chains as she swept past, enabling the seafaring salesman to scramble aboard as best he could with his samples and order books. A Falmouth story recalls a whiskered and waistcoated outfitter watching his cutter beating home round Pendennis head and furiously complaining, 'There, they crazed me to buy them a spinnaker and now they have it they will not use it!'

Cutters also developed in the Moray Firth where they hailed from the small villages of Kessoch and Craighton, near Inverness, which until the 20th century had a thriving foreign and coasting trade, most of which passed with steamers. During the 19th century heyday there were two local bodies of pilots, almost all named Patterson; the river pilots seeking ships bound to Inverness, and the Beauly pilots serving the Firth. As trade declined competition among the pilots grew. Originally they owned pulling and sailing boats only, then half-decked boats, then cutters increasing from 10, to 20 and 50 tons, and the race for speed culminated in the river men owning a 60-tonner and the Beauly pilots a 50 ton cutter, both of which lay, always ready for sea, in Kessoch roads, each watched by the other. At the first hint of a ship both were under weigh in minutes, to race for the job.

In calms the pilots rowed off in small four-oared boats, helped by a lugsail, often pulling 40 or 50 miles to seek a ship. Occasionally a small boat raced a cutter; and they were particularly used on calm nights when the noise of getting under weigh might wake a rival crew. There were spirited races between the cutters, sometimes continued to the Pentland Firth or as far east as Fraserburgh, usually seeking Baltic timber barques; but often the cutters cruised for a week without sighting each other.

The Southampton pilots worked in powerful cutters and dandies, similar to the Bristol Channel skiffs but more shapely. They cruised down-Channel and off the Isle of Wight, particularly at the Shingles and off the Nab in company with smaller cutters from Bembridge, which often landed pilots at Seaview. During the age of merchant sail much pilotage was done by the Bembridge and Cowes men for the vast amount of shipping using St Helens roads for shelter, and Cowes roads for convoy assembly during the Napoleonic wars. These early Cowes cutters occasionally sailed races against local fishing vessels and in 1813 the Trinity House commissioners organised a review of local pilot boats which performed evolutions under sail off Cowes, watched by hundreds of wealthy summer visitors who had discovered the little watering place; and this event contributed to the later founding of the Royal Yacht Club.

Sixty years ago some large, pole-masted ketches with auxiliary motors, were built for the Southampton pilots, including the *Solent* and *Totland*, about 80 feet length.

The one or two Poole boats were similar to the fishing cutters of that port: transom sterned, beamy 30-footers with beautifully shaped hulls, rockered keels and well cutaway forefoots, but rather flat sheers and considerable freeboard. The jib set on a long bowsprit and a topmast with crosstrees set off the yacht-like cutter rig. In a breeze, or when cruising on station in the approaches to Poole, they set a spitfire jib and jib-headed trysail laced to the boom and hoisted on the throat halliard; a very snug rig. The cabin had a coach roof and was entered from the steering cockpit. The fo'c'sle was a store.

At Barrow, the pilots owned the old racing cutter *Mosquito*, an iron-hulled flyer built in 1848 which was still weathering winter gales at the turn of the century. Numbers of old, long-keeled, racing yachts were converted to pilot boats in later life during the late-19th century, several of them working from Harwich and Lowestoft, though the Harwich station was earlier served by Trinity House charter of some large Colne first-class smacks, which cruised about the Sunk to land outward-bound pilots. Although this work was greatly desired the smacks were often off station in bad weather as they were owned by fishing captains who were keen salvagers in the Essex tradition and, pilots or no pilots, were ever ready to assist stranded or wrecked vessels, and subsequently to salvage cargo and equipment from them. Rowhedge men were prominent in this work with *Beaulah*, Daniel James; *Deerhound*, Orlando Lay; *Increase*, John Glover; *Scout*, Henry Carder; and *Thomas and Mary*, owned by Thomas Barnard, who was typical in saving the crews of six

wrecks during the four or five years he engaged in pilot work during a lifetime of fishing and salvaging.

Bawleys were also used to board pilots in the Thames mouth, though whether chartered, or built for the work, is uncertain.

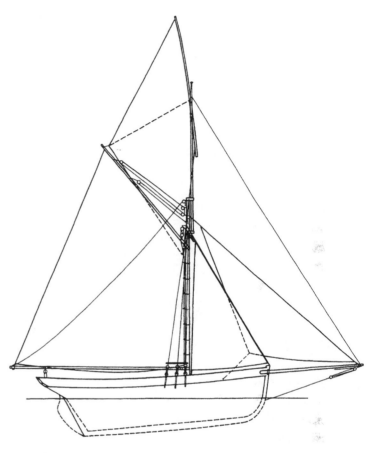

Figure 71. Le Havre pilot cutter, circa 1910. Designed and built by M Paumelle

The French pilot boats of Le Havre and Trouville were broadly similar to the Bristol Channel skiffs, but were generally of superior form and speed. M. Paumelle was a noted builder of the type and seems to have designed by drafting. His fast craft had exceptionally fine runs and elegant counters and general appearance. His most famous creation was the cutter rigged *Jolie Brise* built in 1913. On dimensions of 56ft 2in x 48ft x 15ft 9in x 10ft 2in draft she set 2,400 square feet of working canvas; commonly sailing at 8 knots and making 9–10 knots without appreciable wake. Her mast stepped 18ft 6in aft of the stem and her beautifully formed, heavy displacement hull had a long straight keel, plumb stem,

Figure 72. Swansea pilot boat *Grenfell*
built at Swansea by P Bevan, 1865.
Dimensions 50ft x 13ft x 8ft. Typical deck
arrangement of pilot craft – after J Coates

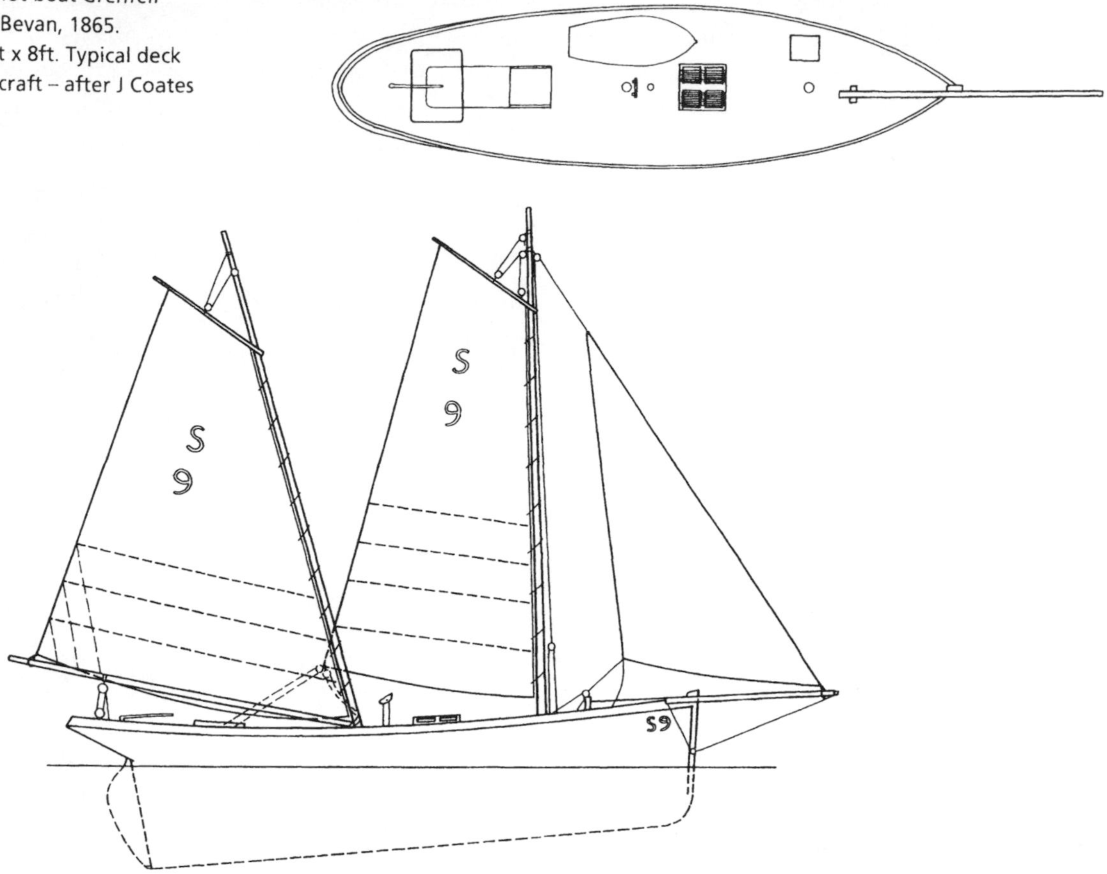

and short counter. When hove-to on station the Havre boats set trysail gear, peculiar in the use of a light boom to extend the clew, which was stretched by an outhaul and secured by a grommet. Sold out of the service she became widely known in the ownership of E. G. Martin founder of the Ocean Racing Club (now the RORC) and probably the finest amateur seaman of his day, who raced and cruised her on both sides of the Atlantic with considerable success. Figure 71 illustrates the final development of the Havre pilot boats before 1914.

In contrast, the yawl rigged pilot boats from Royan at the mouth of the Gironde developed clipper bows, longer counters, and hard bilges, but retained a straight keel for seaworthiness in the Bay of Biscay's sweeping Atlantic seas. The cutters working from St. Nazaire for the River Loire pilots were most extreme in form with well cutaway forefoots, rounded bows, and rakish counter sterns. The keel raked down to a point at the heel of the sternpost, and they were very fast under a cutter rig. The last sailing pilots boats built for Boulogne, in 1911, were

rigged as pole masted ketches, setting 1,954 square feet, including yard topsails on both masts. The hulls were sharp bottomed, had springy sheers and flat counter sterns and dimensions of 59ft x 16ft 5in x 8ft 10in draft. They were designed by G. Soe, a French architect, killed in World War I, who produced many splendid working sailing vessels.

Although the cutter is generally regarded as the principal pilot rig, most others have been used, including brigs; while pilot schooners were fairly common, being used in England at Liverpool and Swansea. The Liverpool service was established in 1766 using bluff-bowed, 50 foot sloops which, when not cruising, frequently laid in Amlwch harbour, Isle of Anglesey, on lookout for inward-bound ships. In 1852 Buckley Jones of Liverpool launched the two masted pilot schooner *Pioneer*. She was slow beating out to shipping waiting to cross the bar, and in 1856 John Harvey of Wivenhoe, Essex, then a leading designer of racing yachts and fast commercial craft, produced the 84ft x 17ft 6in x 10ft draft schooner *Leader*, built at his yard at Ipswich, Suffolk at a cost of £2,500. She became well known for her speed and particularly for beating out to the fleet of ships waiting to enter the Mersey during the great gale of February 1881 when, unable to board her pilots, she mustered the plunging ships and led them in procession into the river, through heavy seas.

The clipper era seems to have infected the Mersey pilots with a desire for fast boats as, the same year, Michael Ratsey of Cowes, a noted contemporary of John Harvey, was commissioned to design and build the schooner *Victoria and Albert*. Others were built by Roydens of Liverpool who launched *Pride of Liverpool* in 1863, and the 79 ton *Mersey* was built in 1875 by William Thomas & Sons at Amlwch, which retains its connections with the Mersey pilots as a boarding station. They were plumb-stemmed and had long keels and short counters. Running bowsprits were fitted and the foremast stepped well aft setting a large staysail and small foresail. The mainsail was moderate in area and a main topsail set. A topmast staysail was set and a foremast chock pole flew the pilot flag. All standing rigging was hemp rope. The most conspicuous feature of these schooners was the two large boarding boats slung in davits amidships and painted yellow for visibility; a recently rediscovered fact. They also carried jump or scramble nets at the main shrouds to aid pilots. Later schooners were more rakish with clipper bows, larger jibs, boomed staysail, and tall main topmast. By the 1870s nine pilot schooners served Mersey shipping but in 1896 they were superseded by steamboats.

The pilots of Fleetwood, Lancashire, also used schooners, including the dainty *Falcon* built by Nicholson and Marsh of Glasson Dock in 1894, with dimensions of 65ft x 17ft x 9ft draft. Her clipper bow hull was a familiar sight in Morecambe Bay until 1919, when she was sold to the Belfast pilots who worked her until 1939. She later sailed as a yacht on the Clyde.

The Swansea pilots sailed uniquely rigged, two masted schooners which had considerable rake of mainmast, and very short gaffs on fore and main. The rig resembled that of the old shallops and was similar to that of American pilot schooners, and some Channel Islands craft. By 1790 most boats in Swansea bay were so rigged and were locally termed luggers. At that time the early pilots used 20 feet, transom sterned, clench-built open boats with the rig, which was handy for beating out and dodging among the crowd of shipping lying weatherbound in Mumbles roads. Growing trade increased the range of seeking, and the size of boat. A bowsprit and staysail were added about 1840, and hull length increased to 30 feet, with full bows and carvel planking, but the rig and undecked hull remained. When the Swansea copper ore trade boomed about 1859 pilots needed range equivalent to those from other Bristol Channel ports and from then until 1870 Bevan and Bowen of Swansea, and Cocks and Perkins of Bideford, Devon, built 11 pilot schooners between 41ft and 52ft waterline length, 10ft to 13ft 6in beam and 6ft to 8ft draft, as they now berthed in a dock. They were sailed by a master and two hands, and generally had three pilots aboard when outward bound. The hull form was deep and slack-bilged, with plumb stem and sternpost. The rig evolved for simplicity as it was often necessary for one man to sail the schooner home and it is worth studying as offering an easily handled and seaworthy gaff rig for larger craft, the disadvantages being the length of mainboom and a most unorthodox, rakish appearance (figure 72). Both masts and mainboom almost equalled the waterline length of about 40 feet, and the mainboom overhung the stern for about half its length, and had to be unshipped in harbour. Its forward end was low and shipped on a mast collar. The foremast was almost upright, but the mainmast had exaggerated rake aft, which counteracted the otherwise forward position of the centre of effort of the sail plan with its short gaffs. The running bowsprit ranged about 18 feet outboard and a single headsail was set on a traveller. Sometimes a standing forestay was rigged with the sheets working round it. In summer, longer masts were shipped and a larger sail area set. The main and fore halliards rove through two sheaves in the masthead, one above the other, and through a single block on the gaffs, $1/3$ of their length out from the throat. No throat halliard was fitted to the mainsail, but sometimes to the foresail where a part of the halliard was taken to the throat. No shrouds were fitted to keep a clear side when running alongside ships, but the foresail halliard was set up by a tackle and belayed at the rail, while the main halliard normally belayed at the mast to clear the foresail sheets, but could be taken to the rail to act as a stay. The jib outhaul was hitched under an iron thumb on the stem at the waterline and this acted as a bobstay.

All sails were made at Swansea, were vertically cut, and had four reefs. The foresail was of heavier cloth than the main as it was usually the last sail handed in bad weather, when a spitfire jib might also be set. The foresail sheets were fitted with tackles and belayed by the helmsman, and the mainsheet shackled to an eyebolt on the counter. The maintack was low and the mainsail was often stowed in hard winds. Both sails were laced to the masts, an unusual feature in England; at first a zig zag lacing was used, later superseded by a spiral; both slacked off before lowering. The deck and hull arrangements were similar to the Bristol Channel skiffs but no windlass was fitted and the cable was hauled in by hand. Early decked boats had a large cockpit aft, and sometimes also round the mainmast, but later craft were fully decked. In calms they were rowed with 20 feet sweeps or towed by the punt. The Swansea schooners raced against the cutter rigged skiffs at Newport and Swansea regattas, and proved their equal except in light airs. In 1890 the

Swansea copper ore trade ended and, with pilotage changes, there was no incentive to seek ships far at sea so the sail plans were reduced. Only two schooners remained when the first steam pilot boat arrived in 1898, and one of them, *Benson*, became a yacht.

AMERICAN PILOTS Although no direct connection can be traced, the Swansea pilot schooner rig was very similar to that of early American pilot schooners serving the approaches to Chesapeake Bay, the Delaware River, New York harbour, and the southern New England ports. By the late-18th century they were 40 to 60 feet length and set a two masted shallop rig with gaffsails, heavily raked. A staysail and short bowsprit were added later, and the many possible sail combinations suited all weathers. In light or moderate winds the three lowers drove the hull fast; in hard winds they worked under mainsail and staysail, or foresail alone, turning to windward under any conditions or combination, aided by the insensibility to sail changes of the long straight keel. No shrouds were fitted and rigging arrangements were similar to the Swansea boats except that the gaffs were proportionately

Figure 73. Virginian pilot schooner *Ann of Norfolk*, circa 1820

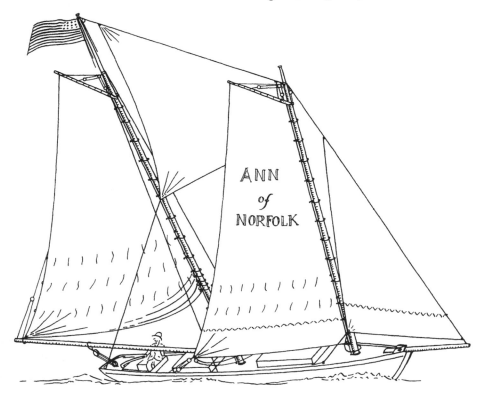

longer and a topmast staysail was set between masts, off the wind.

The light displacement, beamy, shoal draft, sharp bottomed hulls had very fine runs, and inherent speed. These schooners cruised seeking ships a few miles seaward, with lookouts at the foremasthead, and required only elementary accommodation, the hold being fitted as a cabin, and the hatch tops were often removed in summer.

Typical dimensions of a Virginian pilot schooner were 55ft on deck, 49ft 10in x 15ft 3in x 7ft draft aft and 4ft 6in forward. See figure 73. Later Virginia pilot boats were built with very veed sections and an exceptionally easy bilge; while mast rake decreased and masts were stepped further apart, allowing a larger foresail to be set.

Three masted schooners of this type, nicknamed tern schooners, developed around 1780, for pilot work in America and, because of their speed and seaworthiness, numbers of the type were quickly built for naval use, and privateering, under American, British, French and Caribbean ownership. Most carried the long luffed, short gaffed sail plan, with heavily raked fore and main masts and an exceptionally aft-raked mizzen, with its boom well outboard (figure 74). The fore and main gaffs were fitted with vangs to counteract the twist inherent in a boomless sail, and both were furled to the mast with brails. A large staysail set to the bowsprit end and a variety of jibs from the jib boom, extending as far beyond it. Naval and privateer schooners often carried square topsails on fore and main for passagemaking. The hull had the same proportions and features as the smaller pilot boats and construction was light, but they were heavily ballasted with iron concentrated in the bilge amidships. These schooners were among the most extreme and rakish seagoing craft ever built, and were capable of great speed in light or moderate winds. This type of schooner was also adopted by shipbuilders in the West Indies where it became known as the Ballahou rig and was in use for island schooners until comparatively recently. Later some three-masted schooners of the type were built there and rigged with triangular leg of mutton, jib headed or bermudian sails; a rig which lay forgotten in Europe for many years, until its efficiency was rediscovered by racing yacht designers early in the

Figure 74. Three-masted American pilot boat type, naval or privateer schooner in chase, circa 1790. Note vangs to main and fore gaffs and square topsails

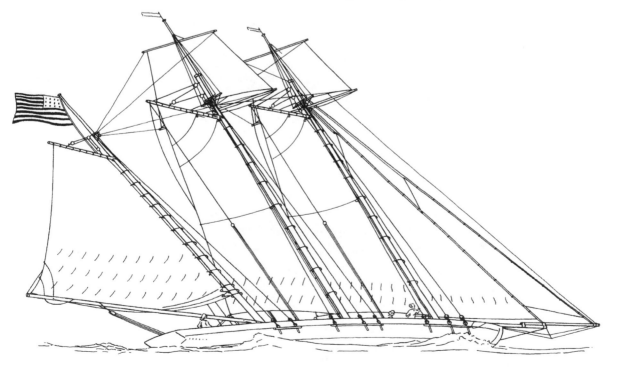

20th century, when it was settled with the name bermudian rig.

New York and New Jersey pilots developed schooners from types similar to those of the Chesapeake. By 1812 five of them served New York and, when war was declared with England that year, the fastest of them was chartered to carry this news through the blockading squadrons to all American ships she could meet at sea, which she did successfully, ending her voyage at Gothenburg, then frequented by American merchantmen, who promptly laid up in neutral ports to escape capture at sea. Three of her sisers became fast privateers, armed with a long tom amidships. By the 1830s they were in competition for pilotage of shipping to the port of New York. This rivalry grew with increasing trade until they were sailing up to 600 miles into the Atlantic for a particularly large vessel, often racing each other for her, if in company. R. C. Leslie's fascinating *Sea Painter's Log* gives his vivid description of a New York pilot boat at work in 1840: 'A low smart-looking fore and aft schooner came rolling over the long Atlantic seas towards the London and New York packet 'Hendrik Hudson' — her mainsail and foresail goose–winged, or spread on both sides before the wind. The schooner's foresail is lowered, and her mainboom swings over as she gybes round under our lee, shooting up in the wind just to leeward of the ship's mainyard; where, as she lies with her staysail aback, we Englishmen look and wonder at her wide, low deck without bulwarks, and just one little round hole in front of her tiller big enough for a man to sit and steer in. Her sides are painted in broad bands of red, green, and white; she is 'No. 8 New York pilot boat'; and as she forges up close under our mainyard for a moment, a small spider-like man is whipped up in the bight of a rope from her deck to ours. This is the pilot, who, except that he wears a tall hat, bears about as much resemblance to the pilot of the old world as his schooner does to the bluff bowed little cutter that we left six weeks ago tumbling in the short sea at the back of the Isle of Wight. The schooner's staysail is let draw; and as she sails away our captain remarks to some of us, 'There! I guess that boat would whip all them yachts of your'n, to Cowes, into fits.' They were prophetic words and George Steers, now best remembered as the designer of

the schooner yacht *America* of 1851, also designed some fast pilot schooners. In 1840, aged 21, he produced *W.G. Hagstaff* for the New Jersey pilots, whom she delighted by beating all their New York rivals. She was followed by *Mary Taylor*, a shallow version of the famous *Moses H. Grinnell* which typified these craft at the height of America's exuberant period of speed under sail. Her dimensions were 73ft 6in x 67ft 4in x 18ft 9in x 7ft 2in draft. The hull had a very fine entry, rounded forefoot, clipper bow, and handsome rounded stern. The rig developed to set a topsail above the main and a larger headsail than the original, but the hull shape became more extreme than any of the Chesapeake craft. The sections had great rise of floor, with a high hard bilge and maximum beam well aft. The yacht *America* was an improvement on this design and was commanded and sailed by Richard Brown, a New York pilot, during her English visit in 1851 when she won the cup which bears her name. Steers sailed in her and on his return to America designed the pilot schooners *A.B. Nielson* and *George Steers*.

These boats were often named for patrons who might be counted on to subscribe towards their construction. They were built for keeping the sea in all weathers, but many were lost in severe gales, or run down. Between 1838 and 1895, 56 pilot schooners were lost at sea, and over 80 pilots drowned. The pilots of this period were dandies, boarding vessels at any hour and in most weather conditions wearing tall silk hats, frock coats and gloves, but they were expert seamen, driving their schooners up to a thousand miles to the eastward and, on the southern 'chance', seeking ships with a lookout at the masthead during daylight. The boarding boat was called a yawl, and sometimes after racing a competitor for many miles under sail for a ship, the contest was prolonged by the schooner's yawls pulling furiously for the ship, the winner ranging alongside the jacob's ladder and the pilot springing upwards only seconds ahead of his rival. Sometimes a schooner would sail for two weeks without boarding a pilot, at others she might board them all in a day and return to the pilot boat anchorage off Tomkinsville, Staten Island, to receive them. One schooner in turn laid hove-to outside Sandy Hook as station vessel to receive pilots leaving outward-bound shipping. She had to stay

Figure 75. New York pilot schooner No. 4, *Alexander M Lawrence*, circa 1885

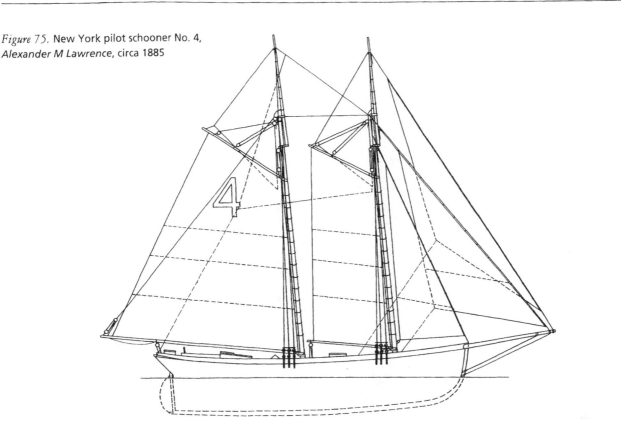

there in almost any weather or, as very rarely happened, pilots would be carried to the ships' destinations, all over the world.

Trade into New York boomed after the Civil War and the increasing numbers of steamships using the port kept 22 New York and eight New Jersey pilot schooners busy. These craft reached the peak of evolution and were among the finest seagoing schooners ever built. Figure 75 illustrates a typical example. Most of the extreme features of the 1850s had been modified, beam reduced and draught increased. The keel was slightly rockered and the stem rounded. The crew berthed in the fo'c'sle abaft which was the galley. The pilot's cabin was amidships with berths for six pilots and a dining table lit by a large octagonal skylight, in defiance of the North Atlantic's storms, when the pilots passed the time playing cards in an atmosphere of choking cigar-smoke. All were well built and finished, and in later years steered by a wheel in a small cockpit, having the main cabin hatch at its forward end. The crew included four sailors, a cook, and the boatkeeper — an apprentice pilot who acted as bosun and was left in charge when no pilots remained on board. He was

also responsible for the condition in which the schooner was kept.

The rig was the ultimate development from the early shallop, and became of almost standard proportions and arrangement. The mainboom was still well outboard but the foresail and staysail became very tall and narrow, and a standing bowsprit followed the sheer, had double bobstays, and carried a large jib, with a long-luffed jib topsail over. The main and fore topsails were small, but a sizeable topmast staysail was set, having a long, low-cut clew sheeted to the mainboom end. This sail was carried whenever possible but the foretopmast was chiefly for show, and the foretopsail was seldom set. In strong winds canvas was reduced to a double reefed mainsail, foresail and staysail, with main trysail and spitfire jib as ocean storm canvas. These pilot schooners had a great reputation for assisting vessels in distress. In 1887 the *William H. Starbuck*, seeking ships to the eastward, came on the disabled schooner *Cordelia Newkirk* and took her in tow, under sail in a rising gale. *Starbuck* was double reefed, but after 36 hours hard towing got the schooner safely inside Sandy Hook. After a storm in 1892 *Starbuck* was given up for lost as nothing was heard of her for

Boston, Massachusetts, pilot schooner *Hesper* 1884.
Designed by Dennison J Lawlor.
Dimensions: 112ft x 22ft 6in x 14ft 3in.
Smithsonian Institution

weeks, but she beat up to the Hook just as her crew's obituary notices appeared in the New York papers. This fine schooner ended her days as a Block Island fisherman.

To settle the inevitable arguments over speed, the pilots occasionally held a regatta, such as that at Cape May in 1873. Some later schooners were converted from yachts, such as *Jessie Caryll*. By the 1880s the increasing size of steamships led to a reduction in their numbers and gradually competition between pilots increased, with their schooners ranging further offshore until 30 of them regularly swept a 600 mile radius from port. Meanwhile costs soared, and a new pilot schooner cost between $18,000 and $20,000 to build. About 1895 the New York and New Jersey pilots amalgamated, all but eight of the schooners were sold, and two steam pilot boats were built to serve the port.

The Virginia pilots' association was among the last in America using sail. In 1915 Cox and Stevens, the New York yacht designers, produced plans for an 80 foot

waterline schooner which closely resembled a large fast cruising yacht of the period, having a very cutaway profile, long bow overhang, no bowsprit, and fine counter stern. She carried 120 tons of outside ballast, with 20 tons of trimming, and construction was heavy. The yacht-like schooner rig set 4,800 square feet in the four lowers, with 2,828 square feet in the mainsail; and the multiplicity of gear, with running and shifting backstays on the mainmast, promised plenty of work for the crew of 11.

Boston pilot schooners had greater draft than their New York contemporaries, and their form and rig were influenced by the fine fishing schooners of New England; which was also reflected in the pilot schooners of Portland, Maine, and Halifax, Nova Scotia. The Halifax boats retained clipper bows and a single headsail into this century, and in winter unbent the mainsail and set a jib-headed main trysail.

CHAPTER FOURTEEN

THE YAWL

THE YAWL RIG evolved for convenience of handling by docking the long boom of a cutter and adding a small mizzen stepped on the counter or transom of the vessel.

For many years the definition of difference between a yawl and a ketch has been that a yawl's mizzen steps aft of the rudder stock and a ketch's forward of it. However, the rig should more practically be defined by the relative size of the mainsail and mizzen rather than the position of the latter.

The date of origin of 'yawl' applied to rig is obscure but appears to originate in the early-19th century, probably contemporary with the many coasting and fishing craft setting the dandy rig, which were in effect short-boomed cutters with a small lug mizzen stepped aft and usually sheeted to an outrigger. The early Clyde yacht *Lamlash*, built by the first William Fife in 1814, was so rigged.

In some cases this mizzen had lugger origins and was found in the Mediterranean, particularly in two and three masted craft of Sicily and Tunisia. Other dandies, including many English drifters, had boomless mainsails with a large lug mizzen, and in France the rig was called a dundis and was carried by various types including tunny fishermen. By the end of the 19th century the dandy rig often referred to a yawl with a triangular mizzen. The yacht *Irene* built by Harris of Rowhedge in 1874 typified this confusion of nomenclature being described as a dandy when launched, but after five years she was always referred to as a yawl, having had no change of rig and setting a gaff mizzen.

Although the yawl is generally closer winded and quicker in stays than the ketch, very few yawls have the ability to sail well under a combination of headsails and mizzen, often quoted as the reason for the rig. Few will

turn to windward thus, even in smooth water, and many will not even fetch properly under short canvas. Yawl rig reduces the size of the mainsail and sometimes handing the mizzen is regarded as the first reef. There is usually great difficulty in staying the mizzen in small yawls, and with its rigging and the usual outrigger or bumkin, the after deck is obstructed and much windage created. It is often difficult to get a yawl's mizzen to set well on the wind and this is frequently due to fitting too light a mast, boom and outrigger; the more rigid these spars are the better chance the mizzen sail has of retaining its shape when sheeted in hard. When sailing off the wind, however, the rig does have the advantage of setting a mizzen staysail.

The shape of yawls' mizzens has changed over the years. Working craft favoured the standing lug, with or without a boom and sheeted to a bumkin. This fashion persisted in early-19th century yachts until gaff mizzens became popular in the 1880s. A bermudian or triangular mizzen is probably the smartest setting, most efficient and lightest sail, which suffers least from the mainsail's backwind but may require a very tall mast, making staying awkward.

It is popular belief that a yawl's mizzen is only a steering sail to make her 'look up into the wind'. In a yacht designed for the rig it should be considered a driving sail otherwise it is worthless.

Old boats are sometimes converted to yawl rig in attempts to improve their steering and some have even had the mizzen boom sheeted to the rudder in attempts to make them quicker in stays, with little effect.

Gaff yawl yachts with moderately cutaway profiles should generally not have the mainmast stepped further forward than $^1/_3$ of the waterline length aft of the stem

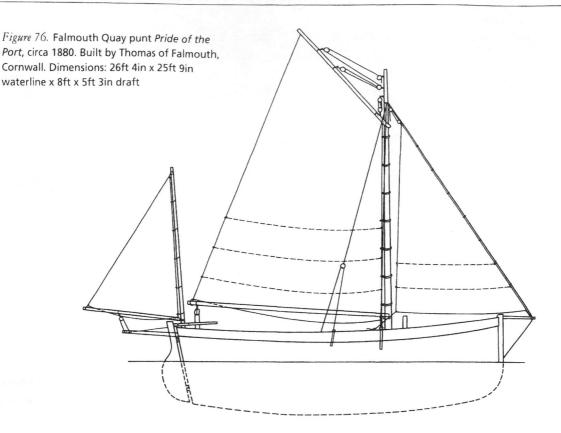

Figure 76. Falmouth Quay punt *Pride of the Port*, circa 1880. Built by Thomas of Falmouth, Cornwall. Dimensions: 26ft 4in x 25ft 9in waterline x 8ft x 5ft 3in draft

and in yawls above about 40 feet length it should be nearer $^2/_5$ of the waterline. Generally yawl rig should not be fitted to a yacht under about 25 feet waterline length and large vessels are better rigged as ketches or schooners, as the yawl's relatively large mainsail takes almost as much handling as the cutter, and probably the advantages of yawl rig are really produced by her resemblance to a reduced cutter, which would have the advantages of a yawl without the disadvantages of a mizzen.

Yawl rig was used by working craft of the Falmouth area, particularly the watermen's boats from Falmouth, often called quay punts. These developed from small lug and mizzen rigged 15ft to 18ft open boats or 'punts', to ballasted and decked craft capable of being sailed by one man and fit to stand bad weather in the approaches to Falmouth, which until 1914 was much used as a landfall and departure port for sailing ships of all nationalities bound up- or down-Channel. Invariably a fleet of ships lay in the anchorage awaiting orders for a few days or repairing damages, all requiring food, water, stores, coal, anchors and cables, sails, rope and communication with the shore. The quay punt type evolved locally to serve

this trade which grew rapidly with 19th century commerce but died gradually in the 1920s.

Speed was not too important as they seldom ranged far beyond the Lizard but seaworthiness, carrying capacity and ease of handling were essential. They were designed by carved half-models and 80 years ago cost about 30s per foot for the hull only; the simple rigging, gear and sails being supplied and fitted by the owner. Until the 1890s they varied from 20ft to 30ft in length, 7ft – 10ft beam and 3ft – 6ft draft. *Pride of the Port*, built by Thomas of Falmouth, was typical of the earlier boats, being 26ft 4in overall, 25ft 9in waterline, 8ft beam, 5ft 3in draft and displacing 6·8 tons. The stem was vertical above water but rounded at the forefoot and the transom stern had slight rake, and the keel considerable rake (figure 76). The bow and stern were very fine and the bilge high. The hull was carvel planked of red pine on sawn oak frames and decked about $^2/_5$ length from the stem, with 12in wide sidedecks and coamings around the large midship well, which could carry up to 2 tons of stores, even in bad weather. An after deck about 3ft long provided warp stowage.

The rig had a gaff mainsail, loose footed with a boom

which was usually slightly longer than the gaff though some boats' gaffs projected beyond the mizzen mast when lowered.

The staysail set to a 15 inch iron bumkin, locally termed a bunkin, from the stemhead, and a jib-headed and loose-footed mizzen was laced to the short mizzen mast and fitted with a boom sheeted to a wood outrigger protruding through the transom, without stays. Many older boats had very long outriggers to which the mizzen sheeted without a boom as in the Cornish luggers. The mainsail had three or four reefs and the working staysail two. When reefing, the mainsail was lashed down by a lacing through the cringle and round the boom, which had no cleats. *Pride of the Port's* spar dimensions were – mast deck to hounds 20ft 6in, main boom 16ft, main gaff 15ft, mizzen luff 13ft, mizzen foot 8ft, and the mainmast stepped 8ft 3in abaft the stemhead. A single wire shroud on each side supported the masts, with manilla pendants and runners to the main hounds which were as high as possible to get a large mainsail. Working boats did not set topsails, but in summer shipped a bowsprit and set a jib, for which purpose a gammon iron was normally fitted on the end of the iron bumkin. The jib was taken in before the staysail was reefed, and storm jibs were seldom carried.

They also often set a large standing lug as summer mizzen, and in fine weather and the regattas, a large reaching staysail. The iron tiller had a bend in it to work around the mizzen which was often offset from the centreline. The staysail and peak halliards, downhaul and mizzen sheet had single blocks and the throat halliard and mainsheet, double and single.

Topping lifts were never fitted, purely from conservatism, and a purchase was often used to peak the sail and hoist the weight of the boom at the same time, aggravated by the long gaff, short masthead and poor lead of the halliards. The mainsail had to be lowered for reefing. In strong winds these craft could beat to windward under staysail and mizzen, if necessary, and sometimes lay-to under this combination but more usually with the foresheet a'weather, when seeking.

The staysail sheets had a single block shackled to the clew and led aft to belay with a half turn under a pin, in the stern benches.

In sail's heyday there might have been four or five

ships arriving on a good day to join the crowded anchorage which made work for the 40 Falmouth punts.

Many left their moorings to go seeking shortly after midnight and sailed westwards, carrying on until daylight without lights which would betray the leading boat, which sought the westernmost berth around the Lizard in prevailing winds. If the second boat were faster she might chase her but usually these, with the third and fourth, jogged about laid-to, awaiting ships bound for Falmouth, while the later boats sailed home to find other work. Competition grew very keen with the decline of sailing ships.

The leading punt would hail the ship with 'Does the Captain want a boat, Pilot?' and, if required, she followed her in and tended her wants throughout her stay, perhaps becoming 'friend' to the ship by obtaining her attendance on future visits to the port. In slack times the punts took out pleasure parties, went long lining, crabbing, drift fishing and mackerel whiffling, or carried the odd cargo of potatoes or wood from the waterside villages on the creeks of the Fal.

Early quay punts had the mast well aft and griped badly, but gradually the mast was stepped further forward, and size and sail area increased until the 1890s. Then a typical quay punt built by the Jacketts or W. E. Thomas would be between 32ft to 28ft 6in x 9ft 4in beam x 6ft draft with ample freeboard, flat sheer, almost vertical stem and transom, slight rocker in the keel which had much of the iron ballast on it, and a well cut away forefoot. Their lines were beautifully easy and fair but the generous displacement and wetted surface, coupled with moderate sail area, made them slow sailers but good sea boats in the long seas down west, although they proved wet and slow in short seas of other waters when taken there as yachts.

Twenty eight feet was the largest size for comfortable singlehanded sailing and old boats carried about $3^1/_2$ tons of scrap iron ballast amidships which, with their fine ends, made them pitch, but ensured stiffness.

Later boats had 10 or 15 cwt straight iron keels but eventually many had considerable rocker in the keel, much outside ballast and considerable round in the forefoot; features introduced in the 32 foot *I.C.U.* of 1898, which made them heavier on their gear and lively alongside a ship, for which they were heavily fendered.

Falmouth Quay punt *I.C.U.* 1900 Racing at Regatta.
Note broken gaff jaw. Osborne Studios

In 1911 W. E. Thomas, a Falmouth builder, used the following proportions for a quay punt's spars: mainmast, deck to hounds 2·5 times the beam, placed 0·32 of the waterline length abaft the fore end of the waterline; mizzen mast deck to hounds 0·6 of the mainmast dimension; bowsprit beyond stem 0·4 lwl; mizzen boom 0·5 length of main boom. Masthead length to suit peak of mainsail and owner's fancy, but kept as short as possible in boats used for tending ships to clear their lower yards when alongside, and usually cut off immediately above the peak halliard block.

At least as early as 1875 Falmouth Regatta saw about a dozen of the fastest punts race a 15 mile course in

Left. Fal oyster dredging boat *Florence* racing 1967. The Fal fishery is still worked under sail by those cutters. *A Pyner*

the 'working sailing boat' classes not exceeding 24ft 6in, 28ft and 32ft; roaring along glorying in big reaching staysails and enormous lug mizzens, *Minnie*, *I.C.U.*, *Maid*, *Ada* and the Thomas's own *Eclipse* were among the champions; all sailed by their owners and their families, eager to beat their relatives and neighbours and urged on by half the Duchy, lining Falmouth waterside.

About 1910 the first waterman's motor launch appeared at Falmouth and tended ships, at first in the harbour and then outside in calms. Then motors were installed in some punts and a few were built as powered craft with shallow draft, little ballast and only an auxiliary sail plan. Many of these and their owners were sent out to the Dardanelles campaign as landing tenders, but were lost when the carrying ship was sunk.

In 1879 Dixon Kemp, the yachting authority, visited

the Falmouth yard of W. E. Thomas who had a 28-footer on the stocks as a speculation and Kemp suggested she should be completed as a yacht. This was done and she was launched as the *Wonderful*, the prototype of many built since, some of which have made long voyages, notably the 28 foot *Twilight* built by Thomas in 1904 which reputedly sailed from Norway to Cowes in $6\frac{1}{2}$ days under short canvas, in bad weather and one day's calm; and the extensive cruises of the 22ft 6in *Joan* which W. E. Sinclair sailed all round Britain, to Spain, Madeira, Sweden and Iceland; finally losing her after dismasting on passage to Greenland and bound for Newfoundland.

Although some British working craft were rigged as yawls the word rarely featured as a type name. The clench built 'Yorkshire yawls' developed about 1840 from the earlier three-masted Yorkshire lugger, and fished the North Sea from Filey, Scarborough and Whitby. They originally set a two-masted lug rig without a jib, but changed to the boomless gaff mainsail about 1870, and eventually to the gaff mizzen; though their rig was never as tall as the drifters of Lowestoft and Great Yarmouth, and its proportions and stepping of the mizzen was that of a ketch.

About 100 were fishing in 1870, varying in size from 48 feet to 65 feet and fitting out in February for coastal long lining before going off to the Dogger Bank in spring with a couple of 20 foot cobles on deck for catching herring bait on the grounds, for the lines. From May to July some trawled and afterwards all took drift nets aboard for the herring season which ended at Great Yarmouth and Lowestoft in November, when they laid up.

The Dutch bomschuit's from Scheveningen and Katwijk used a rig of yawl proportions when herring drifting in the North Sea. These heavily-built craft, almost rectangular in midship section and deck plan, were fitted with leeboards and had a very strong bottom and broad keel, enabling them to be beached in surf on the flat sands of their home ports, and be hauled up or launched off what is usually a lee shore. A fleet of 200 would sail to the Shetland Islands and fish there before drifting their way down the North Sea with the herring shoals, salting the catch on board and transferring it to the jagers, or fast carriers, and eventually stowing their capacious holds full of barrelled herring.

The bomschuits were originally cutter rigged, but later also set a squaresail and a portable mizzen which was overlapped by the main boom and set only on a long sea-tack or when drifting. Later craft stepped the mizzen permanently, right aft, just clear of the tiller and sheeted to each quarter by tackles or to a bumkin. When drifting the mainmast was lowered, bowsprit run in, and the buoyant craft rode to her two miles of nets with the tiny mizzen set, sometimes for many days. How the huge numbers of bomschuits kept clear of each other is as much a mystery as how they worked to windward in a sea.

The Thames hatch boats, described elsewhere, set the yawl rig and were copied by many small yachts racing the London river. The 8 ton clench-built *Don Juan* of 1845, owned by William Cooper, alias 'Vanderdecken' the noted yachting authority, was yawl rigged; but its popularity was established by large racers such as Squire Weld's cutters *Arrow* and *Alarm* which were each provided with an alternative yawl rig in the 1850s.

In 1842 the 80 ton yawl *Corsair* raced the similar sized cutter *Talisman* from Cowes, round the Eddystone and back in an easterly gale which carried away her temporarily-rigged mizzen, but she thrashed on to win the 130 mile race by one minute 30 seconds.

The 112 ton *Ursuline* built by Inman of Lymington in 1858 was the first large racer built as a yawl, and quickly demonstrated the principal attraction of the rig by winning a cross-Channel race on the time allowance she received from the cutters for supposed inferiority of windward performance. Other rule cheaters appeared which were really cutters with short main booms and in 1864 John Harvey of Wivenhoe built the 130 ton yawl *Xantha* for Lord Alfred Paget, which had great success but was challenged next year by the giant 222 ton *Lufra* built at Cowes by Michael Ratsey and the first yawl to win a Queens cup. She sparked a crop of new racing yawls which competed with the great schooners of the 1860s and 1870s under the odd logic that they all had two masts. They sometimes won if there was much windward work, but reaching, the schooners roared past them. The 1870s were the peak of yawl racing with John Harvey's *Dauntless* and *Rose of Devon*; Michael Ratsey's *Corisande*, later owned by King George V; and William Fife's *Latona* and *Neptune* all breaking records but being

The 100ft yawl *Jullanar* of 1875, featured reduced underwater surface and an unusual stern

beaten by the Camper and Nicholson built *Florinda*, a 135 tonner sailed hard by Dan Cozzens of Weston, near Southampton, with a local crew.

She typified large yawls in their heyday; on waterline length of 85ft 8in, beam of 19ft 4in and 12ft draft, her mast was 54ft 6in deck to hounds; the bowsprit was 36ft outboard; the main boom 56ft 6in; gaff 42ft 6in; and topmast 44ft. She set 5,257 square feet, without her topsail and light sails. These yawls seldom, if ever, set mizzen staysails though mizzen topsails were sometimes carried.

The Scottish-built *Neptune* caused consternation in the big class by changing her yawl rig to cutter

alternately, to suit the weather conditions and confuse the handicappers!

In 1875, the noted yawl *Jullanar* was evolved jointly by two Essexmen; E. H. Bentall of Heybridge, her owner, and John Harvey of Wivenhoe, her designer. They attempted to reduce wetted surface, and consequently resistance, more drastically than in any previous English yacht and her profile was radically cut away forward. Harvey's design analysis predicted good performance, yet when her design was exhibited at the Shipwrights Exhibition of 1875, in competition with those of other yachts, the judging committee's decision was that she was fit for neither cruising nor racing. However, she was built by direct labour at Heybridge and completed at Harvey's Wivenhoe shipyard, and after a successful season's cruising to the Mediterranean she fitted out for racing in 1877 under Captain John Downes of Brightlingsea. Her dimensions were 110ft 6in overall x 99ft waterline x 16ft 8in x 13ft draft aft; and her vertical sternpost, comparatively small rudder well under the hull, and immersed canoe stern are reminiscent of present-day trends in racing yachts. She was originally to be rigged as a schooner but was built as a yawl with the mainmast stepped well aft and a comparatively short bowsprit setting off her clipper bow. The concept took advantage of the rating rule and John Downes sailed her to the top of the yawl class in a brilliant season, beating the famous *Florinda* handsomely. *Jullanar's* original form was not copied in such large yachts until Watson designed the America's Cup challenger *Thistle*, thirteen years later.

In the 1880s the 84 foot Harvey-built yawl *Rose of Devon* sailed an unusual race from Lands End to Plymouth against the Naval steam gun-vessel *Cromer*, in a north westerley, *Cromer* being under full sail and steam, but *Rose of Devon* won by four hours, to the jubilation of her Colne crew. Later she won a private match from Weymouth to Ryde against the yawl *Maud* by 6 minutes 80 seconds, indicating the close racing of the period.

Offshore, yawls won several of the North Sea races for the German Emperor's cups, presented in 1897 for an annual race from Dover to Heligoland. The 120 ton *Freda* won the first, skippered by Captain Mathewman and a Rowhedge crew and the following year his neighbour, Captain Charles Simons, won with the 73 ton *Merrythought*, a graceful old straight-stemmer; while in 1899 *Freda* repeated her success. There was no snugging-down in these races, where inshore racing canvas was carried in typical North Sea conditions.

The big racing yawls died out in the 'eighties but revived strongly after 1896 when rating changes caused the splendid large cutters *Satanita, Ailsa, Navahoe, Bona, Kariad, Sybarita* and the German Emperor's *Meteor II*, to dock their main booms and step a mizzen in order to win. The mainsail was much too large in proportion to the mizzen, which was almost useless on the wind causing it to be despised by yacht skippers and crews as being 'only fit to cheat the rule or fly the ensign from'.

Nevertheless some splendid yawls were afterwards built, including Fife's *Valdora, White Heather I* and *Brynhild I*. They had tremendous rivalry with Summers and Payne's 80 foot *Leander* which won the King's Cup in 1902, and these fast yawls expressed the striving of owners for more wholesome yachts than the lightly built and over-sparred craft to which contemporary rule makers had driven the sport. The 110 foot yawl *Glory* (1901) was queen of them all and among the best and last designs of Arthur E. Payne of Summers and Payne, Southampton; Little Arthur, as he was affectionately known. His early death in 1903 robbed British yachting of a great designer of original skill. *Glory's* generous rig raked the wind and with spinnaker set on its 80 foot boom, and her Essex skipper and crew revelling in her power and grace, she sped across a sparkling Solent; a happy ship whose owner, Sir Henry King, gave a dinner each autumn for his crew. One season a hand was lost overboard and the owner vowed *Glory* would never race again, and she was sold.

The racing yawl reached perfection with the 74 foot *Rendezvous* of 1913 and the 79 foot *Sumurun* of 1914, both setting 5,500 square feet and designed and built by Fife. They were the backbone of the large handicap class during the 1920s, skippered by Captain James Barnard of Rowhedge and Captain Gurtin of Tollesbury, respectively. Cruising yachtsmen also look to yawl rig contemporary with the racers, and George Bentinck, MP, was one of its enthusiastic champions who had the 184 ton *Dream* built in 1861, and personally skippered her, living aboard for long periods and despising his fellow squadron members' racing habits. *Dream's* mast was stepped well aft and she carried a large squaresail off the wind, but fast passage making did not appeal to Bentinck who boasted that he once took 42 days to fetch Gibraltar from Cowes! Nevertheless, his discipline was strict; in harbour his bosun always knocked on his door to report eight bells. 'Are the boats up?' Bentinck enquired. 'Yes, sir'. 'Very well, make it so', and after that nobody was allowed ashore.

Yawl rig was used by R. T. McMullen in his 19 ton *Orion*, built as a cutter in 1865 and converted to a yawl in 1873. He made many voyages in her around the British coasts and to the Continent, in all weathers, frequently setting a storm mizzen or reefing his working one. Her cruises are described in his splendid book *Down Channel* whose pages echo to the roar of seas and strain of board-hard canvas.

In 1867 John MacGregor had the 21ft x 7ft x 4ft deep, yawl *Rob Roy* designed by John White of Cowes and built by Forrestt and Son of Limehouse especially for singlehanded sailing; characteristically publishing his experiences sailing from the Thames, down-Channel and up the Seine to Paris in the book *Sailing alone in the Yawl Rob Roy*. She had a single foresail set on a short bowsprit, a moderate mainsail, without topsail, and a standing lug mizzen sheeted to a short bumkin which was removable in harbour.

His example was followed by Empson Middleton, an eccentric who in 1869 had the 23 foot yawl *Kate* built by Forrestt and sailed her around England singlehanded, westabout; visiting the east coast of Ireland and traversing the Forth-Clyde Canal; an adventurous voyage, laborious without auxiliary power due to his desire to lie in harbour overnight whenever possible. It is interesting that he carried a triangular storm mizzen to replace the working standing lug, but it was never used.

Forrestt later moved yard to Wivenhoe and in 1889 built the 22ft 9in yawl *Lady Hermione*, probably the most curious small yacht of all. Her owner, the Marquis of Dufferin and Ava, was a keen yachtsman and singlehanded sailor whose determination to have an easily handled craft anticipated many aids to sailing which are now commonplace. His little yawl was fitted with so many gadgets that she was described as being 'like the inside of a clock', but her sheet and halliard winches, extending tiller and composite construction have become commonplace, even if the remote-controlled windlass and portable sunshade have not!

Joshua Slocum converted his sloop *Spray* to a yawl after sailing across the Atlantic, down to South America, and through the Straits of Magellan, by stepping a standing lug on her transom, sheeted to a bumkin. It reduced the size of her mainsail and improved her steering, on the wind, but he furled the mizzen when running. In later years the 54 foot yawl *Amaryllis* designed and built by Payne in 1882, circumnavigated the world between 1920 and 1923, sailed by Mulhauser with an amateur crew.

At the other end of the scale the tiny 'canoe yawls' provided good sport for active and impecunious owners who made some long voyages in them. In 1865 John MacGregor, an invalided army officer, built a small canoe and paddled and sailed for thousands of miles across Scandinavia, Europe and Egypt. He founded the Royal Canoe Club which quickly established branches on the Thames, Mersey and Humber, where canoe rig developed from a tiny silk lug and staysail to larger craft fitted with centreboards and larger rigs of yawl proportions, to maintain low centre of area. They were the development type of the time with such owners as E. B. Tredwin (later noted for his designs of small barge-yachts) and Baden-Powell as principal innovators of fully battened sails and roller reefing. By the 1880s cruising versions were known as canoe yawls and such owners as Henry Finnes Speed were cruising hundreds of miles each year in yawls like his *Viper* of 1881, a 20ft x 5ft 5in x 2ft 6in depth x 3ft 2in draft yawl having a rockered keel.

The Humber yawl club fostered the type as suitable for the swift tides of the Humber and Trent. Many were designed by George Holmes, a Hull businessman who delighted in sailing and sketching small craft; and others were by Albert Strange, a Gravesend art master who settled at Scarborough and designed and sailed canoe yawls and ultimately larger cruising yachts, many having yawl rig and pointed sterns.

Gaff yawls had limited popularity in America where the schooner rig was preferred in any craft too large to be conveniently rigged as a sloop, but some yawls were built, particularly about 1900 to designs by Laurence Huntington for the newly conceived offshore races with amateur crews; such as the first Bermuda race in 1906 for the Lipton Cup won by the 38 foot yawl *Tamerlane* and established by the efforts and enterprise of Thomas Fleming Day, founder of the *Rudder* magazine and a smart and dedicated sailor who was fond of yawl rig for seagoing.

In 1911 Day and two companions sailed his 26 foot yawl *Sea Bird* from Providence to Gibraltar; making an 18 day passage to the Azores, all of it on one gybe! She often ran under foresail and mizzen (or as Day termed them, 'jib and jigger') and also set a squaresail. She was designed by Charles Mower as a centreboarder but was modified to a keelboat and Day reckoned a chine-boat, such as *Sea Bird*, much drier at sea than round-bilge carvel construction.

A few years afterwards *Rudder* published Charles Mower's design for a 35 foot chine yawl and Harry Pidgeon, a Los Angeles photographer, built himself one between 1917 and 1919, intending to make a world voyage. Naming her *Islander* he taught himself navigation and cruised coastwise to learn seamanship. He made a 2,100 mile voyage to Honolulu and back as prelude to a lone circumnavigation which commenced in 1921 and, after many adventures, the little yawl anchored off Los Angeles four years later with Pidgeon still happy with the rig for seagoing.

The cat yawl was a uniquely American type more correctly termed periauger rig. In essentials, these were large catboats with a big gaff mizzen stepped on the transom and sheeted either to the rudder blade or an outrigger. Both sails were often fully battened but the craft was usually unhandy and overcanvassed.

CHAPTER FIFTEEN

THE KETCH

THE SNUGGEST and most easily handled of all rigs, the ketch was a favourite for seagoing fishing craft, and coastal and short-sea traders. Like the yawl, the ketch has a main and mizzen mast, but the mainsail is smaller and the main boom shorter than a yawl's, and the mizzen, which is much larger, is stepped well inboard, just clear of the main boom and forward of the rudder stock. A bumkin is not usually needed for the sheet. The average gaff ketch sails a point further from the wind than a cutter of equal size as her divided sail plan is far less effective and in a long turn to windward a cutter will beat a ketch by a considerable distance. Ketches are usually steady on the helm but slow in stays, and should have their forefoot cut away to handle in a seaway .

The gaff mizzen of a ketch sets badly on the wind and a bermudian or triangular mizzen is often little better due to the almost vertical leech of the mainsail and relatively tall mizzen. For this reason, a ketch is worse to windward than a yawl. The mizzen is stayed with shrouds and, if large, with runners and shifting forestays set up with tackles or levers. If a gaff mizzen, it is rigged and set as a miniature of the mainsail, and a topsail is often set above in working craft and yachts. The ketch's mizzen topsail was often set in light weather to assist steering, as the rig is often indifferent to steer in those conditions and the mizzen topsail helps to counteract this.

Except in very light airs the ketch should handle under mizzen and staysail, or other headsails, and in a squall she should be able to stow her mizzen and carry on under main and a single headsail, which would be equivalent to double-reefing a cutter. However, in very strong winds her tall, narrow sail plan may affect stability. The short main boom and small mainsail result in easy steering running before a strong wind and sea, but the mizzen then interferes with the mainsail, though gybing is much easier than with other rigs. A mizzen staysail can be set in light weather, but is only effective in large craft.

The word ketch is believed to have originally meant a full-sectioned, tubby ship and was derived from the Italian caicchio – a barrel – and modified into English as the 'catch'. Early ketches were commonly small fishermen or coasters. In 1625 Glanville described them: 'catches being short and round built, be very apt to turn up and down, and useful to go to and fro, and to carry messages between ship and shore almost with any wind.' After about 1650 ketches began to exceed 50 tons and at the end of the century they grew to 90 tons and had developed the curious rig of a square-rigged ship without a foremast. As their length increased they improved in speed and seaworthiness and were often used as fleet tenders and packet boats (figure 77).

In William III's reign larger, heavily-built ketches, called bomb ketches, were built to carry mortars for bombarding shore positions, for which their long foredeck was ideal. One or two big bomb-throwing mortars were mounted there and the ketch aligned to her target by anchoring and warping or backing her sails. The forestays were made of chain to stand the blast of firing, when all the forward rigging was cleared away.

However, the English gaff rigged fishing, and cargo ketches which became popular during, the 19th century owed little to these square rigged craft but had close similarity with the rig of the Dutch galleot and kof types of the 18th century, which in turn may have evolved from the square rigged hoekors.

The ketch rig was a favourite for fishing in England

French privateer *Catch* with a quartering wind. From a lithograph by J Rogers. *National Maritime Museum*

and it is hard to realise that such splendid craft, fully rigged and without auxiliary power were still hopefully being built into the 1920s, for Brixham and Lowestoft.

The trawling ketches from Hull and Grimsby were the largest British vessels working as a fleet but were descended from other ports. About 1840 four or five migrant Brixham smacks habitually landed prime fish at Scarborough during late summer and after a severe gale one found her damaged trawl held many fine soles, indicating the vast quantities she had passed through. In Scarborough, the fishermen, who had no charts, traced the haul's location, aided by masters of colliers lying in port, and returned to fish the rich grounds then named the Silver Pits. The secret leaked and smacks from Brixham, Essex and elsewhere flocked to reap modest fortunes and many owners settled at Hull on the Humber as a convenient base, accelerating its development into a leading fishing port. In 1858 a railway company promoting the new port of Grimsby, facing Hull across

the Humber, offered free port facilities to induce owners to go there and another great migration took place. They worked independently about 100 miles from the Humber, without ice, and the fastest smack in an area acted as voluntary carrier, if required. Catches were so great that only soles, turbot and brill were landed, other species being flung overboard and classed as 'offal'. In 1859 the fleet system began to be adopted and Humber owners built 50 ton cutters as carriers, in imitation of the Barking smacks, and ice was used after 1860.

The Humber smacks grew in size until the 1880s and abandoned cutter rig because of its weight of gear and large crew, in favour of the ketch. The *Othello* of Grimsby was typical; built at Brixham in 1884 for Charles Hellyer of Grimsby, she was 85ft overall x 20ft beam x 11ft draft (figure 78). Her rig was typical of all ketch rigged trawlers except that the jib topsail was rarely set other than when driving for market or acting as carrier, due to the risk of breaking the topmast. She set

166

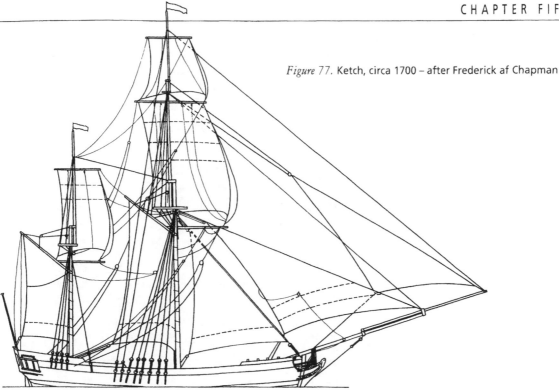

Figure 77. Ketch, circa 1700 – after Frederick af Chapman

the usual loose-footed gaff mainsail and mizzen, working staysail (usually called the foresail in ketches), jib, main and mizzen topsails, and a mizzen staysail when passage-making on a long reach. A large balloon staysail sheeting well aft of the main shrouds was often used when towing the trawl. Four sizes of jib, from very large to storm, were carried, and sheeted through different holes in the bulwarks according to the lead, and were belayed to iron pins on wood kevels on the bulwark stanchions. Yard topsails with spars of extravagant length were once carried, but after steam fish-carriers were introduced the topsails were either reduced in area or effectively replaced by jib headers.

The expansion of British fisheries during the mid-19th century, and increase in the size of fishing vessels, brought the ketch rig into popularity at the busier ports of Hull, Grimsby, Yarmouth, Lowestoft, Brightlingsea, Ramsgate, Rye, Brixham, Liverpool and Fleetwood.

Figure 78. Grimsby ketch smack *Othello*, 1884 built at Brixham, Devon, for Grimsby owners. Dimensions: 85ft x 20ft x 11ft draft aft

The hulls had similar basic characteristics of a plumb stem, usually with small radius-forefoot, long straight keel, raked sternpost and counter stern varying from the boxy, square type, through the graceful Essex counters to the round or elliptical where craft lay close in dock. Hull sections varied from the bluff Humber craft to the graceful shaped Colne and Ramsgate smacks. The hulls were of heavy displacement and very strongly but cheaply built

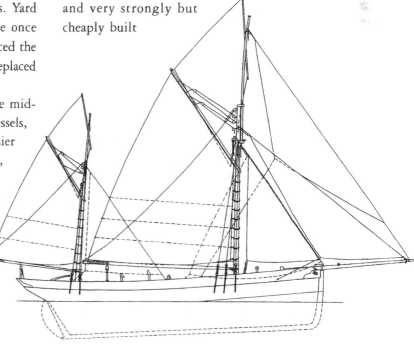

to work hard for about 20 years or so before replacement or sale. Many were lost long before that, overwhelmed by winter weather. Specifications varied but were typically, keel elm, framing oak, keelson pitch pine, bottom planking elm or pitch pine, side planking oak, beams oak, deck planking pine, masts and spars pitch pine or Oregon pine. Ballast was all internal and usually concentrated over about $1/3$ of the length amidships, to keep the ends light and the hull lively in a seaway; it might be shingle cement, scrap iron or iron pigs, or a mixture of these.

The rough work and continual reefing, unreefing and tricing-up of the sails caused much wear, particularly at the seams, so to minimise this, diagonally cut gaff sails were introduced in these craft about 1863. They also stood better than vertical cut, particularly the headsails. The main shrouds were set up with deadeyes and lanyards, on channels, but the mizzen rigging set up inside the bulwarks, clear of the trawl, to a kevel which continued across the quarter timbers and stern. The mizzen mast raked forward and the Grimsby smacks had a jumper stay. The bowsprit was pitch pine, 36 feet overall, 24 feet outboard and completely unstayed. It worked below the rail and through the knightheads, and housed in bitts. Bobstays were unusual in large sailing fishing vessels where the bowsprit was allowed to spring upwards in hard winds and seas, relieving strain on the jib and mast.

The windlass was immediately abaft the bitts and the forehatch, to the store, was between them and a winch used for hoisting sails, heaving up the forward trawl head, hoisting in the cod end, or warping the vessel. The forehold was a store for sails, warps and other gear and its floor covered a dank space known as the dill room where imperishable stores were stowed. The cable stowed in a locker at the after end of the forehold, with a navel pipe over. The main or warp hatch was aft of the mast and the trawl warp as it hove round the capstan, originally worked by hand but in *Othello* and her contemporaries by steam, was led and coiled under deck by the boy, in a compartment on the starboard side. At the after side of this hatch an airtight bulkhead divided the main hold from the fish and ice room which had a very small hatch to the deck and a drain well, pumped out by the slush pump. All hatchways were small as large waves

sometimes swept the deck. The 'dummy' post on deck aft of the mainhatch had the trawl warp attached to it when towing, by a rope stopper, slightly weaker than the warp so, if the smack 'came fast', this parted instead of the warp. The inboard end of the trawl warp passed outside the rigging and made fast to the bitts forward. A dandy winch on the port side, abreast the mizzen, hove up the after trawl head by the dandy bridle when getting the gear alongside. The trawl beam was about 48 feet long and 8 inches diameter. The forward head came up just abaft the after main shroud and the aft one overhung the taffrail. The boiler and bunker room was between the fish room and the crew accommodation, which was entered by a deck companion. The crew of five usually included at least two apprentices and all berthed aft in cupboard bunks, with locker seats and a table with a skylight over, through which the helmsman could shout below. The skylight also housed a compass. The trawler's rowboat was 16ft 6in x 6ft 6in, very deep and heavily built, and carried on the starboard side of the deck.

The smacksmen found their way all over the North Sea and British and foreign coasts with very scant scientific navigational knowledge, but trusted to the lead, the depth of water and nature of the bottom brought up in the lead arming. This, with chart and compass work and keen observation, made them fearless and wonderful sailors, who led harsh lives of heavy toil, usually with little reward. The Hull and Grimsby sailing smacks were rapidly replaced by steam trawlers after about 1890, but sail flourished for many more years at other trawling ports such as Lowestoft, Ramsgate, Brixham, Fleetwood and elsewhere.

The Brixham trawlers caught yachtsmens' fancy more than all the many other ketch-rigged fishermen of equal size and ability. Brixham was a minor fishing port of drift net and line fishermen until it adopted trawling about 1770, when the powerful cutter rig was coming into favour and the trawl had developed to the basic beam trawl as we know it. The Brixham men worked along the Devon coast at first and hauling often took three hours by hand capstan. Local markets were Exeter, Bath, and Bristol, with distribution by road, but growing numbers of trawlers and the adventurous spirit of the Brixham fishermen, spurred by a depressed hinterland, set them voyaging further.

Work aboard a Brixham mule 1930

By 1785 76 decked trawlers sailed from Brixham and fast cutters ran to Portsmouth where catches were rushed to London by horse dray. After the Napoleonic wars the Brixham sloops, as they were called, regularly sailed in season to the Bristol Channel and Irish Sea, and to the spring fishery off Dover and Hastings; later extending their ventures to work from Ramsgate, Lowestoft, Yarmouth and Scarborough, coming into competition with the trawling fleets from Barking, the Colne and elsewhere. As railways spread to the east coast ports, landings could be railed to London, and the industrial north and Midlands. Brightlingsea, Harwich, Yarmouth, Lowestoft, Grimsby, Hull and Scarborough rose rapidly in importance as trawling ports and Brixham smacks joined with scores from Barking and north-east Essex to use and develop them, and concentrated trawling of the North Sea began by large fleets.

The Brixham smacks were still cutter rigged sloops, about 54 foot keel, but east coasters had developed larger ketch-rigged vessels of greater endurance and towing larger trawls, and many cutters were being cut in half and additional midship sections built into them to emerge as big ketches. The Brixham men commenced building ketches in the 1880s and from then until building ceased in 1926, two main types emerged; the larger trawlers over 40 tons gross and those under 40 tons, locally called mules, both being ketches. They were designed by model and the same moulds were often used for many hulls, with midship body increased to suit length.

The large ketches averaged 77ft x 60ft keel x 18ft 6in beam and about 11ft draft aft and 5ft 6in forward. A typical mule was 68ft overall x 14ft 6in x 9ft draft aft and 5ft 6in forward. The sternpost had considerable rake and the raked keel helped them in stays. The sheer was bold, usually ending in a square counter but some later smacks had an elliptical shape. The hull shape had considerable rise of floor and typical West Country, rather slab bow and a long run with a sharp tuck, giving speed and seaworthiness.

The ketch rig's pronounced forward rake of both masts, which was characteristic of West Country trawlers from Brixham and Plymouth, set off the rig's well-proportioned grace and blended with their sweeping sheer. Typical spar dimensions were: mainmast 46ft above deck; mizzen 43ft deck to head; bowsprit 40ft overall; summer topmast 34ft; winter topmast 29ft. A smaller winter mizzen gaff and boom were also shipped for the winter sail.

Winter canvas comprised the loose-footed mainsail of heavy canvas, a moderate sized loose-footed mizzen and triangular main topsail, two working foresails of equal size, a balloon or tow foresail for light weather or hard towing, and sheeting outside the rigging and five jibs ranging from a large fine weather jib to two storm jibs. In summer a similar sized but lighter mainsail and a larger mizzen were bent. The working main topsail was jib-headed but larger topsails with a topsail yard were often set in light weather before the 1920s, and when racing. A yard topsail was also set on the mizzen under the same conditions.

The smacks liked wind and 'a reef and topsail breeze' was most desired for trawling with a single reef in main and mizzen, and a main topsail set above. Brixham craft used reef lacings instead of points to save chafe, but it was a perilous task for the man passing the lacing, crouching in the belly of the loose-footed sails, passing the line through the holes in the sails before his mates drew the lacing taut. In bad weather they shifted to storm jib, single-reefed foresail and double-reefed main and mizzen, or less canvas as required, but always the mainsail remained set. The bowsprit was held with a heel rope instead of a fid, and was reefed with the change of jibs; no shrouds or bobstay were fitted and it flexed enormously in a breeze. Mizzen staysails were set for passage making but after 1914 they were seldom used except in regattas when a temporary jump stay was set up for them.

Before 1914 trawler crews were four hands including one or two boys; usually apprentices but after 1914 few boys took to the fishing. This comparatively small crew was only possible with the aid of the steam capstan which replaced at least two hands. The capstan was mounted just abaft the mainmast and by fairleads or snatch blocks almost every rope in the smack could be led to it and its

power could set the mainsail, get the trawl in and out, hoist the boat aboard, run the bowsprit in and out, and sheet the sails in a breeze.

The crew worked on shares and lived aft in a cabin with built-in bunks. Next forward was the donkey boiler for the steam capstan, coal bunker, galley stores and cabin companion. Further forward was the fish hold and ice stowage with a deck hatch over, abaft the mainmast. Ropes and sails were stowed in the fo'c'sle and the cable stowed abaft the mainmast. Tiller steering was universal and only one man was needed on deck when the trawl was down. They carried a heavy rowboat, locally termed the punt, stowed on deck at sea.

By the 1880s Brixham smacks began to desert the North Sea to fish the Bristol Channel from Godrevy to Lundy, off the Smalls, Cardigan Bay, and the Wexford banks. After the 1890s this 'round the land' fishing became their mainstay, often working confined grounds which steam trawlers would not or could not work due to loss of gear while the Brixham smacks' manilla warps and beam trawls minimised damage from a 'fast'. It was hard fishing with no safe harbour between the Longships and Milford Haven, over 100 miles, and if the winter south-westerly gales flew north-west, they were on a lee shore in the strongest tides of the coasts. At other seasons they fished around the Eddystone for turbot and under Portland for red mullett, towing the trawl fast under a press of sail. At one period great quantities of hake were caught by hook and line; the smacks carrying squaresails. After 1918 they fished mainly between Portland and the Scillies.

In 1914 over 300 trawlers worked from Brixham but many were lost in the War and 10 years later only 80 remained, of which a handful survived into the 1930s and a very few still sail as yachts. In 1925 a large trawler cost £2,500 ready for sea, and few could risk that amount, afford the maintenance or find adequate crew.

As the fleeting system was not used at Brixham, individual speed was unimportant in reaching market quickly, but the fastest smacks displayed their paces at the annual regatta in late August, which became a local holiday. The large and small smacks raced in two classes, sometimes against their Plymouth rivals, usually getting five entries in each class. The lean-headed, fine bottomed *Ibex*, built by Upham in 1896, was the fastest trawler in the West Country for many years, winning 29 firsts in 33 starts, but she was sometimes beaten by the Plymouth ketch *Erycina*. *Ibex* was withdrawn from the regatta by request. After 1918 they raced for the King's Cup and the winner was once timed to average 13·9 miles per hour over the triangular course.

The largest and most experimental sailing fishing vessels built in Britain were ketch-rigged, and sailed from Emsworth, Hampshire, a village at the end of a main channel of shallow and extensive Chichester Harbour, which had considerable shellfish trade and maritime interests during the 19th and early 20th centuries. James Duncan Foster of Emsworth was a timber merchant, owner and builder of wooden ships, and eventually owner of 14 fishing smacks. With typical Victorian vanity he built and named a vessel of each letter of his full name, reading backwards. He built the cutter smacks *Cymba*, *Ostrea*, *Nautilus*, *Aura* and *Una* and the ketch smacks *Evolution*, *Thistle*, *Sybil*, *Nonpariel*, *Echo* and *Echo II*; the ketch barge *Recoil*; the gravel barges *Juno* and *Mab*; the barquentine *Fortuna*; and the steam tug *Dora*, for harbour towing.

During the 1880s he commenced building cutter smacks of conventional form for fishing the Channel and North Sea. They averaged 60ft x 15ft x 7ft draft and all were fitted with a wet well to keep the catch alive in the manner of the Essex smacks which often worked the same grounds down-Channel as the Emsworth craft, which were often manned by Essex fishermen from Brightlingsea and elsewhere in the Colne.

Foster next built some 65 foot ketch-rigged smacks arranged to work more deep-sea dredges and *Evolution* (1888) and *Thistle* (1889) were typical and had well-rounded forefoots. Foster sought speed in passage making to and from the grounds and in 1894 his yard launched the first of a series of large ketches unparalleled in this country. The 97 foot *Nonpariel* had a well cutaway forefoot in imitation of contemporary racing yachts, but to suit her shallow home port she only drew 8 feet. She was well canvassed; her main topmast head was 74 feet above deck and set a yard topsail above that. She was tiller steered and had a donkey boiler to drive three small capstans for hauling warps of the 6 feet mouthed, iron-framed dredges. Her near-sister *Sylvia* was run down and sunk off the French coast in 1927.

Still Foster pondered efficiency and in 1901 launched *Echo*, 110ft x 21ft 6in x 9ft draft, and probably the most extreme sailing fishing vessel in Britain. Her hull was shapely, with forefoot cut well away and rakish clipper stem supporting a short bowsprit. Her elliptical counter was relatively ill-proportioned and long, but increased deck space for working. The shrouds of her ketch rig were set up by rigging screws, her pole mast reached 80 feet above deck and the tall mizzen was stepped well forward. She had an auxiliary steam engine, reputedly of

Figure 79. Steam auxiliary ketch smack *Echo* of Emsworth, Hampshire, 1901.
Designed and built by J D Foster, Emsworth.
Dimensions: 110ft x 21ft 6in x 9ft draft. Note wet wells amidships and steam dredging-capstans

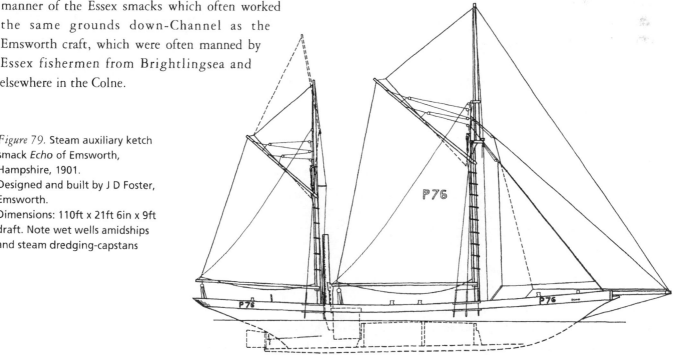

Ketch rigged Ramsgate trawling smack leaving harbour.
Note: bowsprit not yet run-out; main topsail clewed and mizzen
partly set; trawl beam and heads; winch/windlass on foredeck

9 nominal horsepower, driving a single centreline screw protected by an iron cage against damage from the dredges. Thus, her relatively shallow draft, and small rudder, must have affected her sailing qualities. The thin funnel, engine casing and ventilators were immediately forward of the mizzen and she steered by a wheel. Her crew of 11 lived forward and aft of the wet well amidships, which was fitted with valves, enabling it to be pumped dry and reduce draft for entering shallow harbours, principally Emsworth (figure 79).

Echo's success led Foster to build a similar but larger and improved ketch which was launched and fitting out in 1914 when war started, and was never completed and her hull lay at Emsworth alongside a huge wooden floating scallop tank, known locally as Noahs Ark. Foster copied this idea from France but, typically built one four times larger than the prototype. It was intended to receive loads of scallops from the smacks and maintain an even supply for market. It was fitted with sluices to enable the water to be changed but was an unsuccessful project and was mostly used for mooring smacks alongside.

Foster's fleet worked down-Channel and off the French coast into the early 1940s, often landing the catch at Newhaven, particularly when scalloping. Eventually the fisheries paid so poorly that the crews gave it up, but the craft were kept fully rigged to the end and when *Nonpariel* lost her mizzen in the late 1920s it was replaced by an identical spar. On a personal visit to Emsworth in 1959, the author saw J. D. Foster's fleet still laying, just as the crews had walked ashore in the 1930s; the rotting hulks had dredges stacked on deck, pumps rigged and gratings over the wet wells, each distinguished by the large initials J.D.F. painted across their counters. (A quay backed on to a nearby cottage, the occupier of which remarked that the hull of the 50 foot centreboard yacht lying on the quay was commenced in 1914, but was never launched or completed because of the war; wandering through the crumbling hull one could not but wonder at the boldness of design displayed by Emsworth builders, who loved experiment.) Foster and Albert Apps both built small sailing merchant ships, many flat bottomed, and in 1892 Foster launched the 131 ton *Fortuna*, a chine brigantine for Faversham owners which had a huge centreboard dividing her main hold. Shallow draft centreboard sloops with the proportions of American catboats but having heavily cambered decks and large cockpits were built for oyster dredging in the harbour, in contrast to the archaic-looking local luggers with a single black sail set above a shallow, vee-sectioned hull without any centreboard.

Such was the fascination of Emsworth which has now been bulldozed and built over to conform to contemporary living, and little of its old virile seafaring community remains. Space precludes detailed mention of the many ketch smacks owned elsewhere. Large ketches also worked from Scarborough, Hull, Lowestoft, Yarmouth, Aldeburgh, Harwich, the River Colne, Ramsgate, Shoreham, Hamble River, Plymouth, Penzance, Cardigan, Liverpool and Fleetwood; and small ones from Dover and Rye, whose chubby 50-footers were the smallest. Few working craft were so closely linked with the sailing fisherman's hopes and aspirations, misery and hard labour, than the deep-sea fishing ketch.

Large numbers of small ketch-rigged cargo vessels were built for the coastwise and short-sea trades, particularly after about 1860. Later, many topsail schooners were converted to ketch rig to reduce crew and expense. English trading ketches were usually round bilged, but some were flat bottomed with a chine and fitted with leeboards. Size varied from about 50 feet to 90 feet or more and the shape of the hull, bow and stern varied considerably. The deepest draft ketches were owned on the west, south-west and north-east coasts, and in Scotland and Ireland. The shallower chine-built type hailed mainly from the east and south coasts. Rig was conventional ketch, usually with jib-headed topsails on main and mizzen. A staysail and up to three jibs were set in some of them, and many carried a large squaresail on the mainmast when running, and a mizzen staysail was occasionally set. Some set up to three squaresails; course, topsail and topgallant. Large numbers of cargo ketches worked from ports on the south and west coasts, and their history has been well recorded. Many of the tubby, pointed-stern 'billy boys' from the Humber and elsewhere in the north-east were ketch-rigged, and sometimes retained leeboards as a relic of their 'keel' origins.

Small gaff ketches worked from the ports between Keyhaven and Chichester Harbours, inside the Isle of Wight; full-bowed, homely little craft like *Swift*, *Attempt*, *Asteroid*, or the pointed-stern *Bee*, built in 1801 and still sailing in 1914. *Fortis*, built by Albert Apps of Emsworth in 1904 and a regular Cowes trader, had typical dimensions of 60ft overall, 54ft lwl x 16ft x 5ft 6in draft light. The heavy ketch rig set a main topsail. All were tiller steered and sailed by a crew of two who lived aft, the fo'c'sle being used for stores, and a single hatch served the hold amidships. Some had leeboards and many carried goods and passengers to and from the mainland and the Isle of Wight ports until the 1930s. Their form appears to have altered little after E. W. Cooke drew one coming out of Cowes in 1828. Most had rather boxy

Ketch rigged Billy Boy *Star of Hope* of Goole. Painted with the port leeboard raised. Clinker planked, as was common into the late 19th century. Built at Stainforth, Yorkshire, 1862

Leeboard ketch *Leading Light* of Ipswich. Built by Harvey of Littlehampton, Sussex, 1906.
Note roller reefing mainsail and mizzen, and wheelhouse

counters or transoms, but some had pointed sterns, and all had a long life in those comparatively sheltered waters.

English sailing coasting ketches, and schooners, developed rapidly in quality and numbers until about 1890 when increase in trade and demand for speed and regularity of carriage, and cargo-handling, intensified with increasing use of powered craft. This led to great competition in coasting, aggravated by growing numbers of spritsail and boomsail barges entering the trade, whose huge hatches, large box-like holds, shallow draft and small crews of three or four took much work from the less handy ketches, many of which installed auxiliary engines.

Ketch rig was also extensively used by European and Scandinavian seafarers for fishing and short-sea trading. The most curious craft to carry it were the German ewers of the lower Elbe, mainly hailing from Finkenwarder, and fitted with a tank or wet well in the bottom amidships. In the mid-19th century they were built with a straight-line, multi-chine hull and were fitted with a large daggerboard for windward sailing. *Katrina* was typical (figure 80). Constructionally, the tank was a bottom appendage under the main hull, which had a bold bow, fine stern, and sweeping upper chine and low freeboard, resembling modern Japanese fishing sampans. By 1900 the form was modified to shaped sides, a finer bow and well-rockered keel but the narrow stern remained until the more conventional ewer kutter type replaced them, with its broad, shapely counter and smack-like bow and low-powered auxiliary motor. The Danes developed the ewer kutter into larger auxiliary ketches called kutters, which had a shaped form with veed sections, round counter and Scandinavian stem, but retained the fish tank. Fitted with low powered auxiliaries they were forerunners of the many motor kutters built since, which developed to the Danish seine net type of motor fishing vessels.

English yachtsmen were slow to adopt ketch rig as, in the heyday of cutters, yawls and schooners, the ketch rig was regarded as fit only for trading craft with short-

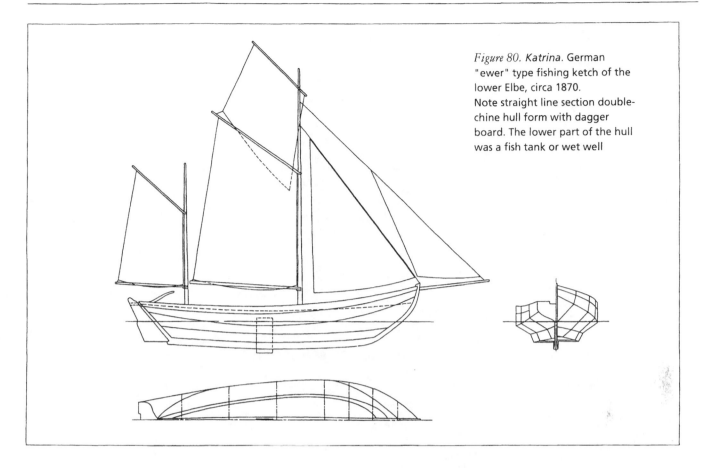

Figure 80. Katrina. German "ewer" type fishing ketch of the lower Elbe, circa 1870. Note straight line section double-chine hull form with dagger board. The lower part of the hull was a fish tank or wet well

handed crews. In 1887, E. F. Knight, that practical cruising sailor, used the rig in his converted lifeboat *Falcon* which he and an Essex barge-hand from Mistley sailed from Hammersmith, on the Thames, to Sweden, via Holland, Germany and Denmark. The *Falcon* was an old teak-built, double-ended P & O ship's boat built by J. S. White of Cowes and converted to a ketch with mainmast in a tabernacle, loose-footed gaff mainsail, and a standing lug mizzen. Like most lifeboats converted for sailing she was slow in stays, despite a false keel and draft of 3ft 6in and, after voyaging to Harlingen in Holland, Knight had oak leeboards fitted and was delighted with her subsequent performance. But there, her story is better told by the book *Falcon on the Baltic*, written in Knight's pleasant style.

The cruising ketch *Iris* of Dublin, built in the 1870s, was an interesting example of original thought in comfortable small cruising yachts. She was 60ft overall x 12ft 6in beam x less than 3ft 6in draft with rounded hull sections and a proportion of outside ballast. Two deep bilge keels 10 feet long were through-bolted to stringers and designed to keep her upright when aground but her

builders also fully understood their aid in sailing to windward as they were tapered in section, shaped outside and flat on the inside faces, and the forward end of each was 1 inch wider apart than the after end. Launched as a schooner she was converted to a ketch for easier handling and was frequently worked to windward in a tideway by one man, and was weatherly, and fast off the wind. Her hull was diagonally planked and a very wide rubbing strake also acted as a breakwater in a seaway, anticipating the knuckled bow of large ships, which has been recently adopted in sailing yachts and in the spray rails of motor boats. *Iris* had 2 feet high bulwarks and an 18-inch high coachroof cabin top, 40 feet long; altogether a comfortable and advanced cruiser, quite uninfluenced by contemporary yachting fashion.

Ketch rig remained generally unused in large yachts until the 1890s when the discerning Lord Dunraven had the splendid 89 feet *Cariad* designed by Arthur Payne and built at Southampton. He jestingly called the *Cariad* his fishing yacht and copied the Lowestoft drifters in staying the mizzen well forward. She was a fine seaboat and sailed well; cruising to Portugal in 1898 and

winning the Vasco de Gama Cup. She was sold to Spain, circumnavigated the world, and is still sailing from Durban, South Africa.

In 1903 she was replaced by the larger and lovelier *Cariad II*, which of all the ketches is the author's favourite. She was another product of Payne's and on dimensions of 95ft x 19ft 7in x 10ft 10in depth, set

6,000 square feet in a graceful ketch rig and was the pride and joy of Captain Bartholomew Smith and his Rowhedge crew, who were justly proud of their yacht's seakeeping powers in bad weather. She usually only raced to make up numbers in the large class for the annual King's Cup, which she won three times and proved the worth of ketch rig in the 1912 race, sailed in wild

Left. Cruising ketch yacht Cariad in the Solent. Designed by Arthur Payne 1896. A beautiful example of the rig. Beken

weather, with the great schooners *Waterwitch*, *Meteor* and *Lamorna*, the cutter *White Heather* the ketches *Corisande* and *Julnar*, *Valdora* and *Cariad*, and other yachts, all throwing spray to their spreaders coming to the Squadron line in the Solent. There was no jockeying, no opportunity for finesse or refinement of racing tactics; the skippers had all they could do to steer those great yachts safely through the squalls. Out by the Warner a tremendous squall swept the racers. Several schooners were in trouble and the giant *Meteor IV* laid over and became unmanageable; down by the head with her rudder almost in the air, and her mainsheet cleats feet under water, the seas pouring in her skylights until at last her German crew got her off the wind and she struggled up the Solent with seven feet of water below, followed by many of her over-rigged contemporaries. The gallant *Cariad*, setting unreefed mainsail, staysail, jib and mizzen, stormed home in clouds of spray to win from the cut-down racing schooner *Lamorna* and the shapely ketch *Valdora*. They were the only yachts to finish.

The 'Cariads' had several contemporaries. The 103ft x 89ft x 19ft x 12ft draft *Lavengro* was designed by Herbert Stow and built by Stow & Son, the now almost-forgotten Shoreham builders who produced beautiful yachts before 1914. She was exceptionally fast for a ketch yacht, yet could be worked by a skipper and 10 hands, while a 60 ton lead keel and 26 inch bulwarks made her a comfortable cruiser and she was representative of the finest type of yacht ever evolved. Fife designed the bold *Julnar* in 1909 and his shapely yawl *Valdora* was altered to a beautifully proportioned ketch rig in 1912, but seldom left the Solent. That year saw a vogue for ketch rig, the old yawl *Corisande* being converted and the outclassed 23 metre racer *Nyria* became a cruising ketch with auxiliary power and was renamed *Lady Camilla*. She was later reconverted for racing in 1921, when she became the first large yacht setting bermudian rig.

Ketch rig was commonly used in North America for nearly two centuries from the earliest European settlement, then it lapsed for almost a century.

Open centreboard fishing boats on the Great Lakes, called Mackinaw boats, set the rig. The pointed stern,

Lake Huron Collingwood skiff type averaged 30 feet and set an unstayed ketch rig with a staysail on a short bowsprit. Some had a cuddy forward and many were clench planked. Transom sterned variants also sailed Lake Huron, without a staysail, and a third type had greater sheer and more beam to suit the Western Lakes, and usually set a boomless lug-sheet mainsail. All worked the freshwater gill net and other fisheries were sometimes transported to other lakes and usually laid up in winter. Rather similar ketch rigged open boats, without a centreboard and staysail, fished along the exposed Newfoundland coast until the 1930s.

The Lake Erie centreboard fishing boats were round bilged with short counters and flat bottomed, with a transom. Rig was a curious ketch having sails like a main and topsail in one, spread by a gaff but hoisted by a single halliard at the top of the tall raking mast. No staysail was set, and with the mainmast stepped in the bow, it closely resembled a triangular-sailed sharpie rig, from which it probably originated, and had much in common with the English Bembridge rig devised there for racing dayboats in the 1890s.

Many other types of North American boats set a spritsail ketch rig, such as the no-man's land boat, but do not come within the scope of this book.

Until the mid-19th century, cargo carriers sailing the Great Lakes were square-rigged brigs and barquentines, but gradually, the two-masted topsail schooner rig became fashionable. Later, topsails were discarded and a mizzen was added to produce three-masted schooners, usually with four headsails. This allowed handling by a small crew and kept down wages to compete with the steamers. This rig was inefficient close hauled, when the gaffs of the three narrow sails sagged to leeward, and off the wind they steered wildly. They were very slow in stays and a triangular raffee topsail was set to a long fore yard, helping to box-haul these cumbersome craft round, and sometimes a square sail was set below it for running. By the 1890s steamers had taken most Lake trade and the schooners were reduced to meagre freights and became poorly kept.

In 1896, one skipper from Grand Haven, Michigan, found his main mast rotten and, unable to afford another, sawed it off above deck and led the triatic stay from the mizzen to the mainmast stump, setting an old jib on it as a mizzen staysail. With this picturesque rig, he traded to the jeers of contemporaries, who soon copied it, naming it the Grand Haven rig, though it was actually a ketch. The new rig proved better to windward, easier on the helm, less strain on the hull and easier to handle than the three-masted schooners it replaced.

The ketch *Resolute* of 1891 was the first American beam trawler, ordered by progressive Gloucester owners from designs by her builders, Arthur D. Storey of Essex, Massachusetts. She was a strange hybrid, rigged like a North Sea trawler and sporting a loose-footed mainsail, but her flat-sheered, full bowed hull and long standing bowsprit with double bobstays, were pure schooner. Her dimensions were, 85ft x 22ft x 9ft 6in depth and on deck and below she was arranged as an English trawler, complete with an imported steam capstan and boiler and sailed by an English skipper and a crew of eight. Mishaps with the trawl gear in deep waters and Gloucester's distrust of these new-fangled English ideas led her owners to convert her to schooner rig after two years.

A few small auxiliary ketches fished from Provincetown about 1920.

Ketch rigged yachts were not popular in the USA except for those produced by Ralph Munroe, an amateur designer of the best examples of American shoal draught cruisers; practical craft entirely free from measurement rule influence, and good seaboats. At first, in 1877, he designed sharpies for use at his Biscayne Bay, Florida home, then a newly-settled swamp region where now is the city of Miami.

American sharpies were generally ketch-rigged to obtain a low centre of area and centre of gravity of rig for these tender craft and Munroe carried the same reasoning into his gaff rigged sharpies. After producing six sharpies the type began to be used for cargo carrying on the east Florida coast and from Lake Worth, where 50 schooner rigged sharpies such as the *Bessie B* sailed cargoes to Jacksonville for years before the railroad.

Munroe improved the sharpie form for sea work by designing the *Presto* in 1885, a compromise having 10 per cent greater depth than a sharpie, with round bilges, flaring sides, more draft and rise of floor. She was 41ft overall x 35ft 6in x 10ft 6in beam at deck to 9ft at waterline x 4ft 3in depth x 2ft 6in draft amidships, and a large centreboard was fitted. Sail area comprised: mainsail 412 sq ft; mizzen 266 sq ft; foresail 200 sq ft; topsail 274 sq ft. She proved an excellent seaboat but was slow to windward. This was partly rectified in her many successors by slight increase of beam. One of them, *Micco*, sailed 105 nautical miles in 10 hours against a $1\frac{1}{2}$ knot current, and for four hours she was under reefed trysail.

Initial stability came from hull form and ultimate stability from internal ballast, low in the bottom and heavy enough to give good righting moment when hove right down. The US Coast Survey copied the Presto type for two survey ketches built by Browns', City Island, for Florida service.

Munroe designed many more shoal draft ketch yachts up to 80 feet, developed from the *Presto* of which Henry Howard's *Alice* is best remembered from his book describing her cruises. As a commodore of the Cruising Club of America his enthusiasm for such shoal draft yachts caused many to think, and maintained interest in the still-developing shoal draft ketch type. Munroe's rigging plans were sparse. Many of his small craft had no shrouds or forestay; few had backstays and most had single halliards to the gaff. He was fond of full-length ash sail battens in the lower part of the main and mizzen to obtain better set to windward and extend the leech of the narrow main, besides making point reefing easier.

His work attracted the notice of Nathaniel Herreshoff, America's leading yacht designer, and he became a close friend of Ralph Munroe, who died in 1933. His type of shoal ketch has been revived in cruisers designed by S. S. Crocker of Manchester, Massachusetts, particularly fine examples being *Mackaw* and *Jingo*.

CHAPTER SIXTEEN

THE SCHOONER

THE BASIC two-masted schooner rig may be described as a purely fore and aft rig having a single headsail, gaff foresail (usually with a boom), and a gaff and boom mainsail wide in the foot and generally taller than the foresail. Other sails may be set: a jib or jibs; jib topsail; gaff topsails; a topmast staysail; and square fore topsails, without altering the type name of schooner. More masts may be added, three being common in Europe and between four and five in America, where six or seven masters were also built. Schooners were built for cargo carrying, fishing, pilot services, as minor warships, privateersmen, for surveying, smuggling and slave carrying.

Before his death in 1707, Van de Velde the younger in *A Brisk Gale*, one of his splendidly accurate drawings, shows a fleet at anchor attended by small craft one of which, under the English flag, has a schooner rig; setting a staysail to an unstayed bowsprit, a short gaffed, boomless foresail, a larger mainsail with a boom but loose footed. A similar craft is shown in the background and a similarly rigged English open boat is shown in a plate by Kip of the Thames at Lambeth, dated 1697. Neither artist was prone to drawing unusually-rigged craft and as one example shows decked seaworthy vessels and the other an open boat, the rig was evidently widely used.

The commonly quoted reference to the origin of the schooner is that it was devised at Gloucester, Massachusetts, about 1713 by Andrew Robinson in a vessel at whose launch a spectator cried 'Oh, how she schoons' and of course Captain Robinson instantly replied 'A schooner let her be!'. There is no word 'schoon' in English and as the account was written on oral evidence in 1790 it seems quite improbable. It may be true, but extremely doubtful, that Robinson built the first

schooner-rigged craft in North America but she was certainly antedated by vast numbers of English and Dutch craft which were not then named schooners but had previously developed the rig, which was probably taken to America by colonists from those countries.

A drawing of Boston, Massachusetts dated 1725 shows two schooners. One, a square-sterned, full bowed craft with raking masts setting a staysail on a short bowsprit, foresail and main, both loosefooted but with booms and short gaffs. Three shrouds were fitted each side and this was the typical American schooner rig of the 1700s. Another in the same drawing sets a foresail whose head is spread by a very short sprit, reminiscent of the later Barbados schooners. Similar gaff rigged craft appear in drawings of Newport, Rhode Island, 1730; Savannah, Georgia, 1734 and 1741; Charleston, South Carolina, 1735 and 1759, indicating spread of the schooner rig.

The world's oldest schooner is preserved at Castletown, Isle of Man, England; a 26-footer clearly linking the rig with the two-masted shallop. The *Peggy* was built in 1791 as a general service boat for the Quayle family who were Island bankers and businessmen, owning several small craft for pleasure sailing and trips to the mainland. She is an open boat 26ft 5in overall, 24ft 5in waterline x 7ft 8in beam. Her hull form is typical of contemporary small craft, with steep rise of floor, slightly flared sides and a lute stern. She has a full forward deckline but a fine entry and run. The clench planking is stiffened by sawn frames notched over the lands. The foremast is stepped in the extreme bow, shallop fashion, and is clasped to a thwart. The mainmast is just aft of amidships and both are stayed with two shrouds each side. The foresail is boomless and the main

Topsail schooner and cutter. Engraved by E W Cooke 1828

boom overhangs the stern considerably. A long bowsprit spreads a large staysail which turns her from a two-masted shallop to a small schooner.

In 1796 *Peggy* sailed to the mainland and was trailed behind horses to Lake Windermere where she raced against local craft before being brought back. Before *Peggy* was launched George Quayle, elder son of the family, appropriated her and, despite her good sailing form, had three experimental daggerboards fitted; a device then under experiment in the warship *Trial* and much discussed in maritime circles since the British Naval officer Captain Shank had first tried a 'sliding keel' at Boston, in 1774.

For well over a century *Peggy* has lain preserved in a Castletown boathouse which has become a museum. She has marked similarity with the form and rig of contemporary 'Chebacco' boats built in New England for fishing.

Working craft rigged similarly to *Peggy*, were common in the Irish Sea until the mid-19th century, and were known as wherries. Often their foresail was slightly larger than the mainsail and they were fast craft. The 40 ton wherry *Bull* of Howth, Ireland, was recorded in 1848 as sometimes carrying the mails across the Irish Sea to Holyhead in weather which the 70 ton packet cutters could not beat through. The rig also survived in the contemporary Mumbles fishing skiffs and, later, in the Swansea pilot boats.

Larger English craft of the late-18th century were also rigged as schooners, but many would now, technically, be regarded as ketches. Their distinguishing feature was that the mizzen mast was stepped well forward, almost amidships, was heavily raked aft and set a triangular mainsail which was loose-footed with a long boom, its luff hooped or laced to the mast and hoisted by a single halliard. The taller foremast set a gaff foresail and topsail, often a square topsail, and sometimes a main course and topgallant as well. A staysail and jib completed the sail plan. Such schooners were common in the early-19th century and the triangular sail was probably a survival of Dutch influence. The gaff mainsail predominated however, and the term schooner referred always to rig and not to hull form or other characteristics, which differed widely internationally .

Two and three masted gaff schooners, with and without square topsails, were commonly used in the British coasting and short-seas trades until World War 1, and a few lingered on under sail until the 1930s. They became very popular after about 1820 because of their speed and weatherliness compared with the square-rigged brigs, which they began to displace.

Generally British cargo schooners were deep hulled, with short counters and a clipper bow. It is impossible to attempt description of the scores of type varieties in this chapter and much of the English south and west coast schooner story is told in detail by Basil Greenhill's *The Merchant Schooners*.

However, the fruit schooners have been little recorded. An Ipswich ship landed a part-cargo of lemons from the Mediterranean in 1727, but it was not until 1740 that lemons were sent to Britain from the Azores. The Mediterranean, Azores and Canaries fresh-fruit trades developed rapidly and special fast schooners were built, about 100 feet long with fine lines and by the 1840s they had clipper bows and square counter sterns, averaging 100 to 160 net registered tons. The two-masted topsail schooner rig, with square topsail and topgallant, had heavily raked masts and the press of canvas carried, the smart crews and clean trade made them yacht-like ships revelling in names like *Tally-ho, Dart, March Hare, Cock o' the Walk, Little Wonder, Forest Fairy, Fruit Girl* and so on.

Many fruiters were owned at Liverpool, Falmouth, Salcombe, Brixham, and a few at Dartmouth and Plymouth, but the centre of the trade was with London merchants of whom Messrs Adams owned many schooners in the orange and lemon trades in the 1840s. This was the height of the fruit schooners' fame; a dozen or more would be discharging at New Fresh Wharf, near London Bridge, pouring six million oranges from their holds in a deliciously scented stream. The record passage home from the Azores by one of these clippers was five days to the Lizard and seven to London; about 1,100 and 1,475 miles, respectively. In very bad weather the fruiterers often beat out to the westward under reefed fore and aft sails, sometimes thrashing out and loading to return, passing through the same fleet of vessels windbound in the Downs, which they had passed bound out. Life aboard when homeward-bound was rather spartan as only occasional fires were allowed for restricted cookery, in case the heat affected the precious cargo which was most carefully stowed and ventilated at sea.

Many of these extremely fast schooners were built in Essex and Suffolk; Bayley's of Ipswich built several and James Howard of Mistley, Essex, launched the *Countess of Wilton* in 1834, fully rigged with yards manned, and she sailed on her maiden voyage the day after her launching. Thomas and John Harvey, the noted Wivenhoe yacht and shipbuilders, designed and built many fruiters and also equally fast schooners for the foreign butter trade where speed was also important. The *Peri* was one of the most noted 'Buttermen' schooners designed by John Harvey and built at Wivenhoe in 1858 with typical dimensions of 95ft overall hull, 85ft waterline x 19ft 3in beam x 11ft draft. The two-masted schooner rig set a boomless gaff foresail, a loose-footed, boom and gaff mainsail, staysail, four jibs, a large square topsail with a topgallant and royal above it, two main topmast staysails and a very large main gaff topsail with luff laced to the topmast and a topsail yard setting its head vertically, above the main topmast head (figure 81). Fleets of these heavily rigged schooners loaded with milk, cheese, butter and eggs at Guernsey, in the Channel Islands, and sailed as fast as possible for London, racing all the way, the first schooner to berth getting highest price for her cargo.

Small fore and aft craft had carried cargo between ports along the Atlantic coast of North America since its early settlement by Europeans. This trade was

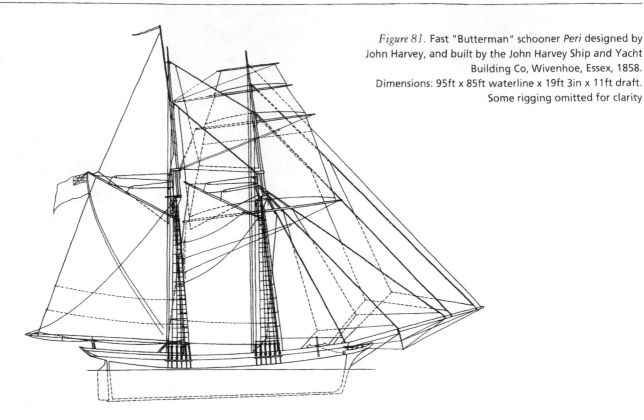

Figure 81. Fast "Butterman" schooner *Peri* designed by John Harvey, and built by the John Harvey Ship and Yacht Building Co, Wivenhoe, Essex, 1858. Dimensions: 95ft x 85ft waterline x 19ft 3in x 11ft draft. Some rigging omitted for clarity

consolidated by a 1789 Congressional Act excluding foreign craft from the coastal trade, which was mainly carried by schooners, at first having two, then three, and eventually more masts as size increased, culminating in six masted giants and one enormous seven-master.

Until about 1850 most of these schooners were two-masted and few were over 100 tons. They were rigged with gaff and boom foresail and mainsail, with gaff topsails and topmast staysails over, a staysail, and one or more headsails. Sometimes a squaresail or square topsail was set on the foremast. Hull form was shapely and generally they sailed well, with a small crew, mostly being skipper-owned. These schooners carried the bulk of cargo alongshore between Nova Scotia and the Caribbean. Larger, three-masted schooners had been built in the Caribbean and the Chesapeake around 1800, in the search for speed under sail in war. In 1827 the 380 ton, three-masted cargo schooner *Pocohantas* was built in Matthews County, Virginia, and was fitted with iron rod rigging to minimise stretch of the usual hemp shrouds. Four years later similar schooners, but with orthodox rigging, began to be built in Maine, which became the principal cargo schooner building State. The largest were about 200 tons and voyaged to South America, Europe and Africa,

besides the usual coastwise carrying which flourished tremendously after the Civil War when great urban and suburban expansion occurred along the Atlantic coastline and in reach of its tidewater, needing tremendous quantities of building timber shipped in fleets of schooners which often returned with cargoes of southern grown shipbuilding timber. Southern cities needed ice, which was cut naturally, in winter, principally on the Penobscot and Kennebec rivers, and stored in ice houses for shipment by schooners. This profitable trade reached its peak in 1890 when three million tons were shipped, indicating the numbers of craft trading. Stone and lime for building were also shipped in considerable quantities. The lime schooners were sound little two-masters; but the stone carriers were old craft ending their days in this heavy trade.

Southern coal became an increasingly important return cargo for New England schooners, after the War, to supply railways, factories, power stations and an increasing domestic consumption in coalless New England. It was loaded at Hampton Roads ports, and on Chesapeake and Delaware Bays. This trade gradually increased the size of the coasting schooners.

Many of the two and three-masted schooners were

Five-masted centreboard schooner *Governor Ames* 1764 tons. Ready to launch at Waldoboro, Maine 1888. Built by Seavett Storer. Wrecked off Cape Hatteras 1909. *Smithsonian Institution*

deep draft, shapely vessels while others were shallow draft and fitted with huge wooden centreboards working in heavy wooden cases stretching from keel to deck and dividing the midship hold. Boards 28ft long and 6in thick were common, raised and lowered by an iron rod connected by a tackle to a masthead. Even larger centreboards were fitted later, to schooners of greater size, but always they were a source of leaks and eventual rot, and in some schooners were ultimately removed. The board was usually placed alongside the keel and any mast in the way was stepped slightly offset, on the opposite side. During the 1870s a compromise type developed, having moderate draft and a centreboard, but by the 1890s few schooners were built with boards.

Construction was extremely heavy with sawn, closely-spaced frames and heavy keel and multiple keelsons to minimise the inevitable hogging of those relatively long wooden hulls which frequently took the ground in loading and discharging berths.

Expansion of trade and search for economical carriage rapidly increased the size of schooners after 1875 when three-masters of 800 tons were being launched. In them, masting and rigging had almost reached the limits of three masts and a small crew's strength and in 1879 the 758 ton *Charles A. Boylan* was launched at Bath, Maine, with a steam windlass also capable of handling sheets and halliards, pumping the bilge, and working cargo, and similar installations quickly became universal in the schooner fleet.

In 1880 the *William L. White* was built at Bath as the first four-masted coastal schooner. She was 205ft x 40ft x 17ft depth of hold and carried 1,450 tons of coal. Her masts were named fore, main, mizzen and spanker, and this large vessel was sailed by a captain, two mates and five seamen, indicating the type's economy and the tremendous labour on board. Inevitably the four-masters multiplied as schooner size increased during the 1880s until many were over 1,000 tons. Forty four-masters were built in 1890, but by then the type had again increased in size.

In 1888 Captain Davis of Somerset, Massachusetts ordered a new large schooner to be built at Waldboro, Maine, and designer Albert Winslow of Taunton, Massachusetts designed a five-masted schooner to reduce the size of individual sails to manageable limits. This 1,764 tonner, named *Governor Ames*, was fitted with a centreboard and was totally dismasted on her maiden voyage. After a few years she sailed round Cape Horn to trade on the Pacific Coast for four years, making one voyage to Australia, eventually returning to the East Coast coal trade, and was lost in 1909.

Taunton was a manufacturing centre and the home of many notable schooner owners. Mr Winslow was a pattern-maker by trade and how he gained the knowledge to design many large schooners by half model is uncertain. Later noted designers of these schooners included B. B. Crowninshield and John Frisbee, of Boston; John Wardwell of Rockland, and Frederick Rideout.

After 1889 the schooners' freight rates were increasingly threatened by economic conditions and the ever growing numbers of large wooden barges towed coastwise by steam tugs, almost independently of tides and winds. Eventually these too were supplanted by steam colliers. As a result, by 1890, many schooners were built with fuller ends and flatter bottoms, to increase capacity. This degenerated them into little better than huge sailing barges, heavy on the helm, unweatherly and poor seaboats; unable to go to windward in bad weather, when many had to anchor or run back for shelter, losing many days lying in roadsteads where they frequently had to wait for the anchorage to clear before they dare get under weigh. In winter they were sometimes driven offshore and forced to lie-to for days, unable to carry enough sail to work back inshore to find a lee and sail to their ports. Light or loaded they commonly refused to tack in a seaway and some would not wear unless after canvas was stowed.

Despite these failings even larger schooners were built, though it was ten years before another five-master was launched. Meanwhile, four-masters of 2,000 tons were sailing, the largest being the 2,015 ton *Frank A. Palmer* built at Bath in 1897 and altogether 56 of the type were built, almost all in Maine and the last in 1920. At that stage of development a 2,800 ton schooner cost $175,000, could carry 4,500 tons of coal, an averaged 11 complete annual voyages between Maine and Hampton Roads, returning with coal. Such a schooner was manned by a captain, two mates, a steward, eight seamen and an engineer; all well housed in two large deckhouses. The steam boiler worked sail- and cargo-handling machinery, the windlass, provided electric light, steam heating and telephones to aid communications from forward to aft.

The *George W. Wells* of 3,200 tons was launched in 1900 at Camden, Maine, as the first six-master of the ten built until 1909, mostly by Percy and Small, at Bath. Largest was *Wyoming*, a 3,730-tonner which carried 6,000 tons of coal. She was the ultimate development of the large wooden schooner, but suffered severely from hogging. She was one of the largest wooden ships ever built and was lost off Chatham, Massachusetts in 1924, anchored in a gale. Such schooners could sail at 13 knots

Six-masted schooner *Wyoming* 1909, 3730 tons. Built by Percy and Small, Bath, Maine.
Note typical open rails and steam from winches. *Smithsonian Institution*

off the wind, but generally all these large schooners were poor to windward and many took a tug in approaches to ports

A few later schooners were of steel construction, including the unique *Thomas W. Lawson* of 1902, designed by B. B. Crowninshield. This giant's dimensions were 395ft overall x 50ft beam x 32ft depth and she loaded 10,000 tons of coal on 28ft draft. She was rigged as a seven-masted schooner and sailed well off the wind but was difficult, and often impossible, to tack. She proved unsuccessful and was converted to a tanker, being lost on the Scilly Isles in 1907 on her first transatlantic passage.

Steam colliers were introduced in 1910 and few new schooners were built between 1909 to 1915 when, with wartime freights soaring, schooner construction feverishly recommenced until 1921, when the boom subsided (figure 82). Most were launched as purely sailing ships but some had auxiliary engines and were usually rigged without topmasts. Many made transatlantic voyages; some foundered and others were torpedoed, some scared their crews stiff; but owners were happy with pre-paid war freight.

American post-war depression caused numbers of schooners to be laid up or sold. The 1925 Miami building boom brought the last profitable fling for the coasting schooners, which carried building materials there for many months; their rows of bowsprits jutting over the waterfront in scenes which were reminiscent of the mid-19th century. Afterwards a few schooners lingered on with odd uneconomic freights until 1939.

Large multi-masted schooners, often having centreboards, were built and traded on the Great Lakes. On the American West Coast similar coasting schooners

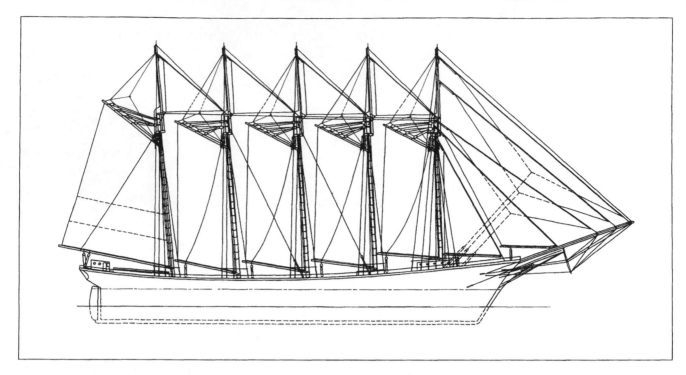

Figure 82. Five-masted American cargo schooner, 1916. Dimensions: 291ft x 255ft waterline x 48ft x 24ft draft. 2,100 tons gross. Note topsail clewlines. Bulwark details and some rigging omitted

developed, but were of more weatherly and shapely form. The 172ft x 39ft 6in x 14ft depth of hold *Resolute*, designed and built by Hitchings and Joyce at Hoguian, Washington, in 1901, represented a better type of coasting schooner with rig improved by setting a jib-headed spanker having a ringtail set over it; a feature common in Pacific coast craft. She could load 800,000 feet of timber and was a successful ship.

Schooners worked the most profitable American fisheries, in the Gulf of Mexico. Fleets of them fished oysters and shrimp from Biloxi, Gulfport and Pascagoula, Mississippi; Pensacola, Florida and Louisiana ports; supplying the extensive canneries processing the abundant landings.

The local oyster industry probably existed since French colonial times, and shrimps were fished after the mid-19th century, when rapid rail transport became available. At first, shallow hulled centreboard luggers were used, the only working luggers built in the USA, but by 1870, shallow draft, centreboard schooners were used. Many, usually over 50 tons, were built and owned by companies; other equally large schooners were owned privately, often in shares, and sold their catch to the canneries. Biloxi was a centre of the trade with 17

canneries packing oyster and shrimp. In 1918 the port had 300 craft over 5 tons fishing oysters alone, and was the third largest oyster port in the world.

The Biloxi schooners were built on the shores of Biloxi Back Bay by shipwrights whose yards were a foreshore plot and a shed for tools. They were designed by half models made by builders including Henry Brasher, Martin Fountain, F. B. Walker, and later, J. Covacovich. Hull length varied between 50 to 74 feet and by 1900 there were two principal types of hull: one had a clipper bow and short after-overhang ending in a raked transom; the other had a slightly rounded stem with little overhang and the stern was often almost plumb, carried to the waterline in a wide transom. In both, the bottom had little rise of floor, bilges were of small radius and sides were almost vertical. Sudden violent storms sweep the Gulf of Mexico, and seaworthiness was essential. The Mississippi and Florida coasts are shallow and draft was restricted to around 4 feet, increased by a wood centreboard working in a case extending between the masts, from keel to deck. This was raised and lowered by an iron rod leading up through the deck forward of the mainmast and hoisted by a tackle to the masthead. Sailing over shoals one man worked the tackle while another kept a hand on the rod, awaiting the tremor telling she touched. Then the board was hoisted a few inches. Construction was strong but not too heavy to spoil sailing performance, for speed was desirable in the climate and trade. All the main structure was of cypress, fastened with galvanised bolts and dumps, and the schooners were long lived. Later a few were built with pitch pine planking. The shallow hulls were prone to hogging and many were built with considerable sheer to counteract this, but many had a flat sheer. Much iron ballast was carried and they were all rigged as two-masted schooners with gaff and boom foresail and mainsail, boomed staysail, narrow jib, main gaff topsail and main topmast staysail. The foremast was well sprung forward and the long bowsprits were lightly stayed, bending in a breeze. Crew accommodation was forward and under a low, square house aft. They were wheel steered. Few of them had auxiliary engines and schooners were prohibited from fishing under power at least until 1930, when fully rigged examples were being launched at Biloxi. Before World War I these schooners occasionally carried goods coastwise and similar fishing schooners were built at Pascagoula by Sideon Krebs, and at Apalachicola, Florida by Samuel Johnson. Smaller sloops and catboats were also occasionally built for these fisheries.

The schooners fished a coastal ground varying from 10 to 50 miles width and stretching from Mississippi Sound across the Mississippi River, westward to the Sabine River, Texas, and including tidewater oyster beds and shrimp grounds in coastal shallows, bays and creeks, all teeming with edible seafood. Shrimps were fished in August, September and October, up to 100 miles from port. The iced catch was sailed to market or sent by carrier. Oysters were dredged from November until the end of April. Another shrimp season lasted from mid-February to mid-April, and from May to August the schooners refitted. Originally they stayed at sea a week and could hold 300 barrels of shrimp. Later, they stayed out a month and the catch was sent in by motor craft. Small motor shrimp-trawlers were introduced by 1900 and gradually ousted the schooners from shrimping, but they continued dredging oysters under sail into the 1930s.

Until about 1930 Biloxi Regatta featured a fishing schooner race, usually with six or ten entries. The Championship of the Gulf Cup was often won by *Julia Delacruz* owned by the Foster Packing Company and sailed by Skipper Willie Green. Other fast Biloxi schooners were *Curtis Fountain*, *Rebecca Fountain*, *Mary Foster*, *H.E. Gumbel* and *I. Heidenheim*. They started with sails hoisted at gunfire by the straw-hatted racing crews of 35 hands, mainly to replace the ballast which was removed for racing! They sailed several times round a 12 mile course with a grand finish under the verandahs of the yacht club.

Similarly rigged but more gracefully proportioned two-masted centreboard schooners dredged oysters in winter and carried cargo in summer, on Chesapeake Bay. They evolved in the tradition of the fast 18th century Bay built sloops and schooners nicknamed Baltimore clippers, through little 50 foot shoal-draft, sharp bottom keel schooners called pungies, with saucer-like, fine-ended and fast hulls used for oyster dredging and cargo carrying. Centreboard schooners appeared on the Bay about 1825, with flatter bottoms and increased cargo capacity. They too were fast hulls and in about 1850 adopted the 'clipper' bow, which gradually became a feature of all Chesapeake sailing craft. Typical dimensions were 73ft x 22ft x 6ft 6in draft aft and 4ft 6in forward, without the large centreboard which was always fitted alongside the keel. The entrance was convex, the greatest beam well forward and the run long and flat. These handsome schooners had flaring bows, well proportioned short counters with elliptical transoms, spacious decks and low bulwarks. A yawl boat was slung in pipe davits across the stern and the crew accommodation was aft, the fo'c'sle a store, and the hold amidships. The gaff sails were laced at the foot to boom jackstays and the staysail had a large clew club.

Many were built by William Skinner at Baltimore, Maryland for the usual trades, or for carrying oysters to New England. Other, particularly fast, schooners were built as 'buy' boats cruising the Bay purchasing oysters from fishermen and sailing them to market or a packing house. Smaller, 50 foot schooners were built for oyster police duty, armed with rifles or small machine-gun.

Some schooners had round sterns as did a few bugeyes, which were occasionally also rigged in imitation of the centreboard schooners and were then for some reason called square-rigged, on the Bay. Both pungies and schooners were gradually ousted from Chesapeake oyster dredging by the jib-headed, ketch rigged bugeyes, originally hewn from great logs, by the little Brogans, and, later, by the chine-built, jib-headed skipjacks or bateaus.

The two-masted centreboard schooner type also dredged oysters in the Delaware under sail until 1945, when power dredging was allowed and the fleet unrigged. Delaware Bay oystermen hailed from Leesburg, Bivalve and Port Norris, on the Maurice River; Newport on Nantuxet Creek, and Greenwich, in New Jersey; and Little Creek in Delaware. Their schooners were smart sailers among the swift tides, muddy foreshores and winding tributary rivers and creeks of the Delaware. Their rig, arrangement and dimensions were typically as for the Chesapeake Bay schooners, but after World War I topmasts were generally discarded and auxiliary motors installed, though dredging was done under sail. Until about 1905 they worked many small dredges, hauled by hand, but later a pair of large dredges, one each side, were hauled by a motor winch amidships.

Many of these schooners were built by Barton at Marcus Hook, Delaware, and others by William Parsons at Greenwich, New Jersey. About 1900 the Delaware schooners began to be built with round stems and this type culminated in the Parsons' built *Nordic*, a 101-footer launched in 1926. She was 81ft 6in waterline x 23ft 7in x 8ft draft with centreboard raised. She was built without topmasts.

The average dredgers remained like the Chesapeake centreboarders and the rougher conditions of wind against tide in the Delaware limited the use of the low freeboard bugeyes and kept a little fleet of sailing schooners dredging the various grounds under canvas, winches puffing and decks piled high with shell, fore and aft.

Many fishing schooners hailed from the Long Island Sound port of Noank, Connecticut, where the Latham family were noted builders and captains of fishing sloops and schooners. Since colonial times their yards launched craft for cargo-carrying, whaling, sealing and general fishing. *John Feeney* typified Noank fishing schooners which developed from the sloops during the 1850s. Built in 1885 by John Latham and Son she had a long keel, full-bodied hull, 65ft on deck, 58ft 6in waterline x

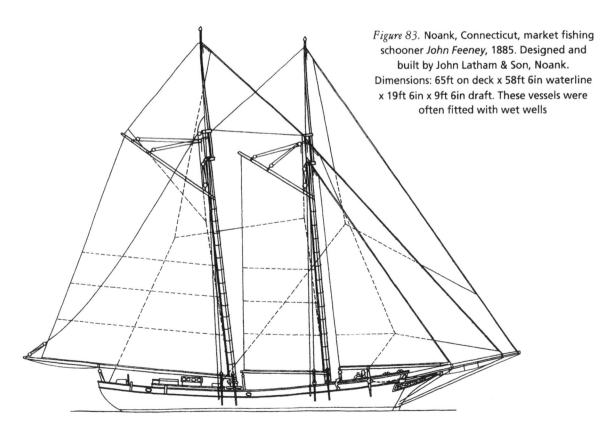

Figure 83. Noank, Connecticut, market fishing schooner *John Feeney*, 1885. Designed and built by John Latham & Son, Noank. Dimensions: 65ft on deck x 58ft 6in waterline x 19ft 6in x 9ft 6in draft. These vessels were often fitted with wet wells

19ft 6in beam x 9ft 6in draft. Her moderate clipper bow was balanced by a round stern, which seems to have been peculiar to later Noank craft. The bowsprit followed the sheer and a jib net and double bobstays were fitted. Her sail plan was typical of the American fishing schooner with tall, narrow foresail, moderate mainsail, and long-luffed headsails. She was a particularly able schooner and the three rows of reef points in main and foresail told of the need to claw off a lee shore. A live fish well was fitted amidships and these craft were much used in the New York market fresh fishery (figure 83).

Manned by a captain and eleven hands *John Feeney* cruised between Fire Island Light and Long Island, to the Delaware from April to October, for bluefish; and from November until the end of March was codfishing between Block Island and the coast of Maryland.

The fishing schooners of Barbados set an unconventional rig with masts of almost equal height. The foremast had moderate rake aft and the mainmast considerably more. The bowsprit was steeved down towards the water an carried a large jib with a heavily-roached foot which hung almost to the waterline in light airs. The gaff foresail had reasonable proportions except that its boomless foot sheeted well abaft the mainmast, abreast the rudder stock. The mainsail had a long boom and a gunter type yard at the head, locally called a sprit, which set almost vertically and was half the length from peak to tack. The foresail and mainsail luffs were laced to the masts. When bringing up, the staysail and foresail were lowered and as she luffed-up the main boom was topped to its fullest extent, gathering the mainsail loosely abaft the mast.

They were very fast and hull lines were very fine, the sheer flat, stem plumb and the counter short and square, but the low freeboard and lofty rig gave a rakish appearance to craft which probably descended from the similarly built and rigged fast schooners which developed during Napoleonic times in the West Indies for naval blockade running and piratical use, known as Ballahou schooners or Bermuda schooners. These differed from the Barbados schooner in having very short gaffs on each sail.

CHAPTER SEVENTEEN

THE FISHING SCHOONERS OF NEW ENGLAND AND NOVA SCOTIA

FEW SAILING CRAFT have had so much written about them as the north-east American fishing schooners, particularly those hailing from Gloucester and Boston, Massachusetts; yet their true development as a type remains obscure and seems not to follow the pattern of evolution previously accepted as their history.

Coastal and inshore fishing had always been part of the settlers activities in New England, while the prolific cod fisheries on offshore banks had led to very early local settlement in north east America by English and European fishermen. Colonial New England fishing vessels included sloops, shallops and the schooners which developed from them, many of the largest being owned at Marblehead, Massachusetts, which was the principal fisheries centre; but small two-masted schooners predominated.

The colonial shallop was increased in size and refined in model and rig for coastal fishing. Reputedly, development first took place at Chebacco, a parish of Ipswich, Massachusetts, later renamed Essex. The type was built in two forms; the pointed-stern chebacco boat and the slightly smaller square-sterned dogbody. The rig in each was two-masted cat schooner, with the foremast in the eyes of the hull but without headsail, bowsprit or topmasts. The foremast was only slightly longer than the main and, at first, both sails were loose footed, but after about 1812 were laced to the booms. Two shrouds were fitted each side in some; none in others. They were decked, with a cuddy cabin forward, and usually had two transverse cockpits or standing rooms for the crew of four to stand in when hand-line fishing over a bank. One

cockpit was just aft of the cuddy hatch and the other immediately abaft the mainmast. Right aft was a similar steering cockpit for the helmsman at the tiller. Two small hatches were arranged aft of the standing rooms to stow the catch below. A low rail surrounded the deck and a windlass occupied the foredeck which was afterwards raised to the rail to increase headroom. Average dimensions varied between 36ft x 11ft x 5ft depth to 42ft x 12ft 6in x 6ft depth, but after 1812 larger chebaccos were built and were a fast type, seaworthy enough to trade to the West Indies, or to fish the Gulf of St Lawrence. They were also built all along the north east seaboard, from Massachusetts to Nova Scotia and Quebec.

After the Revolution of 1776, many fishermen could not afford to replace large, expensive schooners lost during the war, and in 1799 a Salem parson recorded 300 chebacco boats fishing out of Cape Ann's coves and harbours, about half of them sailing from Sandy Bay, now known as Rockport. Fishermen and chebacco boatbuilders were busy after the Revolution and the 1812-14 war, encouraged by fish bounties and grants to vessels. By 1817 the same parson, visiting Essex, noted 'Jebacco boats building for the bay fishery not only at every landing place but in the yards of the farmers some distance from the shore ...'

The dogbodys were rather fuller-bowed and had square lute sterns. They were known as jakes in Canadian waters, when built for cargo carrying. The village of Chebacco, Massachusetts, had a rare place in maritime history. Sited on the narrow, winding and tidal Essex river, five miles from the coast in Ipswich Bay, its

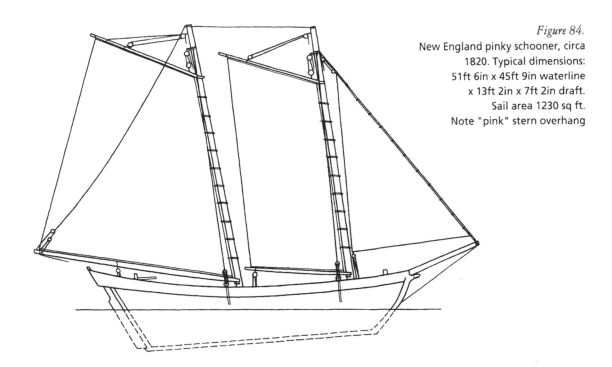

Figure 84.
New England pinky schooner, circa
1820. Typical dimensions:
51ft 6in x 45ft 9in waterline
x 13ft 2in x 7ft 2in draft.
Sail area 1230 sq ft.
Note "pink" stern overhang

saltings, wooded banks and clustering waterside houses are reminiscent of the villages of north-east Essex, England, by whose emigrés it may have been renamed. Having vast stands of good oak timber all around in colonial times, it naturally became the home of skilled shipwrights who built at first for themselves, and then for ports elsewhere.

Chebacco boats and dogbodys were followed by pinkys and fishing schooners by hundred, and the village became the most prolific fishing craft building centre in North America. Long after its natural timber resources were exhausted, Essexmen were launching the last of the great American fishing schooners in the 1920s, vessels which dominated the village when on the stocks, and filled its tiny river when launched.

After 1812 larger schooners of the pinky type began to supplant the chebacco boat in the fisheries. The pink had been brought to America in colonial times by settlers from England and Europe, and was distinctive for its pointed stern, which often had the bulwarks carried out over it in a short overhang. Ketch rigged pinks sailed the American coast in the 1700s and some, schooner rigged, fished from New England before 1790. After 1812 the pinky type fishing schooner became widely used in all the New England offshore fisheries, and was very popular between 1818 and 1840. They were larger but similar in hull and rig to the chebacco boats, but were fully decked and the after sheer swept up abruptly, the deep bulwarks overhanging the pointed stern to meet at a 'tombstone' transom which had a score in its top forming a crutch for the mainboom, when stowed. It also sheltered the helmsman and a hole in the overhang made a lavatory for the crew, who lived forward in the cuddy, which had a brick fireplace in those days before iron stoves. The foremast passed through the cuddy, and had a circular table round it which, when not in use, was stowed up to the deck. Pinkys stowed their salt in the extreme stern, bulkheaded off from the hold.

The peaceful post-war years free from risk of privateers and warships, favoured the building of the heavy displacement, safe and seaworthy pinky, and they fished in hundreds. A typical pinky would be 51ft 6in overall, 45ft 9in lwl x 13ft 2in x 7ft 2in draft, setting 1,230 square feet in staysail, foresail and main (figure 84). A sharply-steeved bowsprit and the staysail were added to old pinkys and fitted in new, set off by a small knee-head on the stem. Many set gaff topsails and a main topmast staysail and all had a great reputation for seaworthiness and easy motion in bad weather, being well ballasted. These little schooners, manned by four men and a boy cook, voyaged 800 miles up to St Lawrence Gulf and fished for cod or mackerel, a compass and leadline their only navigational aids. Pinkys were long-lived: *Julia Ann* built at Essex in 1819 was still fishing in 1908; and they varied greatly in speed and hull form. The 53 foot Gloucester pinky *Tiger*, built at Essex in 1830, became notorious in Canadian waters for her speed when poaching and many similar fast pinkys were built for the mackerel fishing where, as the great shoals moved to windward at about four knots, weatherliness was necessary. Some pinkys, particularly in Maine, had wet wells for carrying live fish or lobsters, and were known as smacks. By 1815 the type had spread to the Canadian Maritime Provinces. Many were built at and fished from Cape Breton Island, where numbers of small, clench-built pinkys were still working in the 1890s.

Good seaboat though the pinky was, her cod's head and mackerel tail form did not allow extreme speed and the fisheries were demanding larger and faster schooners. The 45 ton *Accumulator* launched at Essex in 1835 had a square stern for improved deck room and was soon joined, at Gloucester, by many others which gradually became finer lined, more heavily rigged, and faster. By 1845 the last pinky had been launched at Essex and the square-stern schooners became fashionable at larger ports. Many were owned at Marblehead which was a great fishing centre in colonial times and about 1730 began to fish offshore on George's Bank. Between 1830 and 1850 Marblehead sent 80 schooners of 50 to 70 tons hand-line fishing for salt cod, making the two trips over spring, summer and autumn and drying the catch on extensive flake tables around the harbour where the salt fish matured in the crisp New England sunshine.

Until about 1820 New England fishermen caught mackerel by trailing lines like the whiffelers of the West of England. About that time the mackerel jig was introduced and enabled a whole schooner's crew to take quantities of fish while hove-to over a school. The jig was a hand line with a lead sinker having a baited hook protruding from its side. With this, skilful fishermen caught enormous loads of mackerel, often bringing the school to the surface by 'tolling' with ground fish-bait known as chum. Hand lining flourished greatly between about 1820 and 1865, when it was replaced by the mackerel purse seine.

Cohasset, Massachusetts, was a principal mackerel port at that period. Before 1754 Samuel Bates built a wharf for his few vessels, at the Cove, as Cohasset Harbour is called. His enterprise changed it from a farming community into a leading port in the mackerel fishery, rivalling Hingham and Scituate. By 1840 Cohasset ranked fourth in Massachusetts fishing ports behind Gloucester, Boston and Wellfleet, and by 1848 was the principal mackerel port in the State. The Bates' fleet grew and prospered, being rivalled by those of H. T. Snow and A. H. Tower, other Cohasset owners. In 1851 44 mackerel schooners sailed from Cohasset jigging mackerel, manned by 561 hands. Typical was the 85 ton clipper-bowed schooner *Georgiana*. Most were built at Cohasset in the yards of Bela Bates and his son Jonathan, but at times they were so busy that some Cohasset schooners were built at Essex. Cohasset's mackerel fishery suffered severely in 1863 through the emancipation proclamation. The slave owners of the South found mackerel the cheapest food available and the freeing of the slaves wiped out a vast market. The Cohasset fisheries declined until many schooners were broken up in the 1870s. The last Cohasset fishing schooner was launched in 1883 and John Bates sold his last four, two years later.

In 1821 Gloucester schooners started fishing George's Bank and by 1835 a large fleet fished there, mainly pinkys, for cod and halibut, each man working a hand line around the bulwarks, heavily weighted against the strong tides sluicing around the anchored schooners, rolling and plunging in the sea over the Bank. The New England fishermen extended their regular voyages to Canadian waters, leading to their prohibition from many areas by patrol vessels. This, and the distances sailed to fish, increased striving for speed; accentuated by the building of railways to Boston in 1836 and Gloucester in 1846, created opportunity for vast markets and expanding the fisheries. Many new schooners were built, fitted with wet wells to bring home live fish, but development of ice-keeping at those ports led many schooners to ice the catch at sea and sail hard for market and higher price than salt fish, resulting in short trips and fast schooners. Some fished for mackerel in summer and in winter carried oysters from the Chesapeake to relay in Massachusetts waters. Many were built on the Chesapeake and others borrowed features from these shoal-draft fast centreboarders which had, in turn, descended from the Baltimore clipper type schooners which New England builders studied together with the speedy local pilot schooners to evolve new designs.

Typically, in 1846, Andrew Story of Essex designed and built the schooner *Romp*, on speculation, and attempted to reproduce the speed of the Southern schooners, with increased seaworthiness. She had a raking keel, flat sheer, much rise of floor and hard bilges, with a broad, round-bowed deckline. She was long for sale as fishermen distrusted her form but, at last, one who had lost his schooner bought her but had difficulty in finding a crew as she was thought 'too sharp' for safety. She proved her speed and seaworthiness on her first voyage and her design was copied in new schooners, appropriately named file bottoms or sharpshooters. They were rakishly rigged with masts angled well aft and a narrow foresail. The mainsail was broad footed and set a thimble-headed topsail over a short gaff. A main-topmast staysail set between mastheads and a single large staysail or 'jib' set to the bowsprit end and was fitted with a bonnet and sheeted from a club which spread the load on the clew. Trailboards and a clipper bow set off the long bowsprit with its double chain bobstays, footropes and shrouds. All standing rigging was hemp and the shrouds were set up with deadeyes and lanyards, while the staysail hanks were of wood. A low, raised quarterdeck extended aft from just forward of the mainmast. The boat was carried in wood davits across the stern which was a variant of the English lute form with a large rectangular transom which caught the gybing mainsheet with its corners and resulted in later well-rounded quarters. The foredeck had a two handled pump windlass and a wooden staysail horse, between bulwarks, with a bolt standing proud at each end as traveller stops. Abaft the foremast was a sliding hatch to the fo'c'sle and a small hatch to the hold. The 5 foot square hatch was amidships and just abaft this the deck raised 10 inches forming the quarterdeck, traversed by the iron foresheet horse. The mainmast was abaft this, with a wood bilge pump on either side and the square after hatch behind it. A large coachroof or trunk top covered the after accommodation, entered by a sliding hatch, and the wheelbox was fitted at the stern with the mainsheet working on a horse abaft it.

At this period there were two main types of fishing for the schooners; bankers, which stayed fishing on banks until they had a full hold or fare of salted fish; and the market boats making quick trips to closer grounds bringing home the catch fresh on ice or in wet wells. Bankers fished principally on George's or the Grand Bank. The mainsail was usually stowed or unbent on the banks and a triangular, loose-footed main trysail or riding sail set in its place. The schooner usually anchored, lying to an old-fashioned wooden stocked anchor and a 9 inch coir cable. Fishermen used hand lines from the schooner's deck until about 1855, when dory fishing was taken up as it cut the time of a trip by about half compared with hand lining from the schooner. The sharpshooters carried between 8 and 15 flat-bottomed double-ended dories on deck. They varied between 13 feet and 18 feet, had removable thwarts, and could be nested inside each other for close stowage.

These narrow, chine rowing-boats sometimes carried a small spritsail and were remarkably seaworthy in skilled hands. Arrived on a bank the skipper caught a few cod for critical examination before letting go anchor and setting the riding sail. At sunrise the 12 to 16 hands hoisted out dories and rowed off widespread, anchored, and each threw out two lines, one in each hand, standing up and moving them up and down a few feet from the bottom. Having got a bite one was made fast and the other hove in with probably an 80lb cod. The rebaited line was flung over and by then the other probably had a bite. When the dories were finally recalled about mid-afternoon each often held 1800lb of cod which were pitched on deck, cleaned, salted, and stowed in the hold before the schooner was pumped out and washed down, ending the long day. There was always great loss of life in dory fishing, capsizes, fog, and bad weather annually taking many fishermen.

Although sharp vessels, the bankers were of greater displacement than the faster market schooners which had larger sail areas and also set a fore topsail. Before 1857 market schooners developed beamier, shallow and very fine-ended hulls, until the extreme clipper type was common and replaced more moderate schooners. Clipper schooners had taller masts, fore and mainsails, and sported long topmasts each setting a topsail. A long jib-boom extended the bowsprit and a large jib and jib-topsail were set. Their beamy, shallow hulls were very fast, racing fresh cargoes home under clouds of canvas, but they lacked range of stability, and many were overwhelmed at sea, several being capsized. In 1862 15 Gloucester schooners with 120 men, were lost in one night. In winter the schooners discarded the topmasts and jib-boom and the staysail could be reduced by removing the bonnet, a feature persisting until the 1880s. Trail boards, clipper bows and wooden-stocked anchors all gave the schooners an old-fashioned, clumsy appearance. The extremely large mainsail was their principal weakness, difficult to take in and frequently replaced by the main try-sail in bad weather.

The increase in fresh fishing, with ice, after about 1870, led to great improvement in speed and sail carrying by the schooners. Many carried topmasts through winter, even on the long voyage to the Grand Bank. After 1865 halibut, and also later, haddock, were caught regularly, dory trawling or long lining. These had ground lines up to two miles long with short lines, or ganglings, at fathom intervals, with baited hooks. One end of the ground line was anchored and the other buoyed. The lines were baited and coiled in wood tubs and set from a two-man dory. After a time the line was under-run and the catch removed before re-setting or hauling. Cod and mackerel gill nets were also worked in a similar manner. Great attention was paid to sailing trim and haddock trawlers carried more ballast than halibut trawlers as they only needed about 10 tons of ice for their short trips while the halibut schooners often loaded 40 tons for the long trip to the Grand Bank.

There was great and regular loss of life in the huge schooner fleet and the stoic fatality of the life was, as usual, splendidly caught by Rudyard Kipling in *Captains Courageous*. During the 1880s opinion swung against the extreme schooners, stimulated by Captain J. Collins of Gloucester, a member of the Government Fisheries Commission, and in 1884 the schooner *Roulette* was launched at Boston as an improved type. She was designed and built by Dennison Lawlor, a Nova Scotian who commenced building fishing schooners at Boston in 1850, though he was best known for fast, seaworthy Pilot Schooners. *Roulette* was over 2 feet deeper than the clipper type, sharper in the bottom, but was fast and went to windward better than her contemporaries. This improved pilot boat type was built elsewhere; in 1885 Arthur D. Story, of Essex launched one named for himself and, as a banker, she sometimes sailed to Iceland, cod fishing,

eventually being lost with all hands. These schooners were longer, the *Story* being 85ft x 23ft 4in x 9ft 7in depth of hold, and initiated a modified narrow-stern with elliptical-shaped transom set at considerable rake. Plumb stems became fashionable with shroud spreaders on the long bowsprits. Several of Lawlor's designs raced in the first fishing schooner race off Boston in 1886, organised by Thomas McManus, later a noted schooner designer. The Boston pilot schooner *Hesper*, designed by Lawlor, also raced, and won, and a centreboard fishing schooner also went round the course for 'fun'.

In 1886 the Boston schooner *Carrie E. Phillips* was designed by the noted yacht architect Edward Burgess and built by Arthur Story. Her dimensions were 93ft 6in x 24ft 11in x 11ft depth of hold, giving her a sharper bottom and a very rounded keel profile. She was the first American fishing schooner with wire standing rigging, and introduced the spike bowsprit with shroud spreaders at the stem, improved ironwork and a shorter foremast than usual. Her best improvement was to set the forestay to the stemhead, converting the old single headsail into a more seaworthy and easily handled staysail and jib. Below deck the fo'c'sle was lengthened and reduced the hold. She was extremely fast to windward and once logged 13 knots on the wind in smooth water. Three years later Burgess designed *Fredonia* for Boston owners, built by Moses Adams of Essex, while an almost identical schooner, *Nellie Dixon*, was built at East Boston for Thomas McManus. They were refinements of the *Phillips* but with more cutaway forefoots, shorter bowsprits and dimensions of 99ft 7in x 23ft 7in x 10ft 4in depth of hold. The graceful appearance and seaworthiness of these two schooners influenced the design of subsequent fishermen, though they had comparatively small carrying capacity.

Many of the schooners were still designed by half model and the fastest and most attractive clipper-bowed schooners of the 1880s and 1890s were so designed by Captain George (Mel) McLain from Friendship, Maine; a noted master in the Gloucester fleet who modelled his first schooner in 1880 and sought improvement in stability of the beamy and shallow schooners then fashionable in the fisheries. In 1886 he designed *J. B. Merrit, Jnr.* to those ideals and she was so successful that over 100 subsequent schooners were built to the design.

Gloucester, Massachusetts, fishing schooner circa 1890.
Note large wood stocked anchor

In 1890 McLain modelled the slightly larger, clipper-bowed *Lottie Haskins*, the most perfectly balanced of the fleet. She once beat up Gloucester Harbour and berthed alongside under fisherman's staysail alone; a remarkable feat. Many craft were built from her lines and the larger *Marguerite Haskins*, designed by McLain in 1893, proved another favourite prototype. Although untrained in naval architecture, McLain's reasoning and ability greatly improved the seaworthiness of the New England schooners and fewer were lost at sea after his principles of deeper, well ballasted craft took hold. He also designed the noted fast schooners *Puritan* (1887), *Eliza B. Campbell* (1890), *Marguerite Haskins* (1893), *M. Madelene* (1894), *Gertrude* (1902), *Avalon* (1903) and the last of the clipper-bowed Gloucestermen built, the fast, *Arthur James* (1905). George Willard was another noted captain, fond of model-making, who whittled out a design for the speedy 60 ton schooner *Georgia Willard*, built for him at

Bath, Maine and crack of the mackerel fleet in the 1880s.

In 1890 Gloucester had more than 500 sail in various fisheries: mackerel seining, cod, haddock and halibut dory trawling for salted and fresh fish; hand lining; dory hand-lining; herring net fishermen; gill netters and sword fishermen. By then the clipper stem, without trailboards, was fashionable, and long bowsprits, with their complicated rigging, went out as forefoots became more cut away, permitting a shorter and stronger bowsprit. The keel had much more rake with greater draft aft and the stern changed to a more graceful counter with smaller elliptical transom and tumbled-in quarters. The development of schooners in the 1890s coincided with a decline in demand for salt cod at Gloucester and Boston due to improved storage facilities. Large schooners continued to be built in great numbers for fishing on George's Bank or off the Maritime Provinces of Canada, but they did not need the capacity for salt banker fishing, so were of lighter displacement and faster for their length, which continued to increase.

Gloucester held a fishermen's regatta in 1892 to celebrate its 250th anniversary. The schooners raced 50 miles in two classes, 65ft to 80ft and 80ft and above, in a fresh NE gale. Many lost spars and sails, and some found they were less able than they thought; especially *Nannie C. Bohlin*, whose boastful, Norwegian-American skipper and his fine schooner were truly humbled. Even the swift *Elizabeth B. Jacobs* burst her mainsail gybing round a mark. *Lottie Haskins* won the small class and the fine *Harry L. Belden* won the large, and carried a broom at her masthead, in memory, for years afterwards.

The summer fresh mackerel fishery produced the fastest fishing schooners which worked a purse seine to surround schools of mackerel, spotted by lookouts at the mastheads. This method was introduced at Southport, Maine about 1862 and spread to Boston two years later. By 1870 250 schooners were seining, cruising up to 50 miles offshore between the St Lawrence Gulf and Cape Hatteras, in summer, often alone, sometimes in company and occasionally in fleets. In 1881 300 mackerel-seining schooners were in sight at once from Monhegan Island, and Boothbay Harbour was so full their crews walked

across it over their decks. Home ports of the fleet were Portland, Southport, Boothbay, Deer Isle, Swan Island and Vinal Haven, in Maine; Provincetown, Wellfleet, Dennis, Harwich, Barnstable, Gloucester and Boston in Massachusetts, Gloucester owning many. Mackerel seiners carried 20 to 28 men and, when cruising, the lookouts' cry of 'School-oh!' brought a rush of hands to man the 38 to 40 foot double-ended rowing seine boat which was towed astern in fine weather but on passage or in bad weather was carried on deck, to windward. These schooners were commonly 125 foot long and a principal requirement was for the standing rigging to be far enough apart to swing the seine boat in on deck between them by masthead tackles, sometimes in a hurry. Freeboard had to be low to facilitate this and for handling the large seine net which was flaked over the rail on a roller into the seine boat's stern, with great care. A typical mackerel purse seine was 225 fathoms long by 22 fathoms deep. Having found a school the seine boat pulled swiftly for it, the skipper steering with a long oar, perched on the net, nine hands rowing as hard as possible, followed by two men in a dory. An ellipse was made around the school, paying out the net which surrounded it like a curtain, until the seine boat

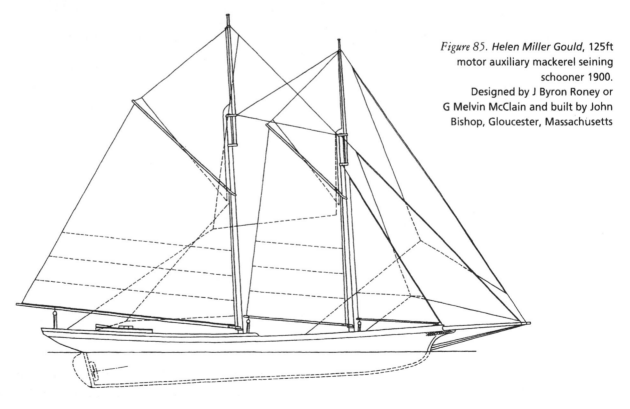

Figure 85. Helen Miller Gould, 125ft motor auxiliary mackerel seining schooner 1900. Designed by J Byron Roney or G Melvin McClain and built by John Bishop, Gloucester, Massachusetts

completed the ellipse at the dory holding the other end and the pursing lines which ran through eyes in the net's bottom and, when pulled tight, closed the net like a bag, trapping what was often tons of fish. The seine boat crew quickly reduced the size of the bag by heaving in some net, as one held an oar aloft to signal the schooner, which was left in charge of the sailor-cook in fine weather, to be sailed up and laid alongside the seine, hove-to on the starboard tack. All hands 'dried-in' the net and the milling mass of mackerel were bailed out to the schooner's deck by scoop, to be split and salted, barrelled and stowed below, or if the trip was coming to an end and a good market port near, put on ice and, setting all sail, the schooner headed fast for market, sometimes with barrelled mackerel and tons of fresh fish cluttering the deck.

Naturally, noted sail carriers were bred in this fishery and the most famous Gloucester mackerel seiner was Captain Solomon Jacobs, an adventurous and progressive fisherman who owned and commanded many fishing vessels, of which the *Ethel B. Jacobs* was favourite and fastest, queen of the mackerel fleet and capable of sailing at 15 knots for six hours, in a breeze. In 1899 mackerel were scarce off New England and he sailed *Ethel B. Jacobs* to seine mackerel off the west Scottish and Irish coasts. She had little luck and found the lee side of the Atlantic harsher than home waters, and was wrecked on the Irish coast. Even in winter the waters sailed by these schooners are not as rough as the fishing grounds of Britain and Europe, which are in comparatively shoal waters and exposed to extremely strong winds. Undaunted, Captain Jacobs returned home to have the 125 foot *Helen Miller Gould* designed by J. Byron Roney of Lynn, Massachusetts. She was typical of the fastest mackerel seiners, and the first to have an auxiliary motor; a 150bhp petrol engine, useful in keeping up with the mackerel shoals in light summer winds typical of the NE American coast. Her rig was typical of the mackerel schooners (figure 85), and appears proportionately lower than the older schooners' rigs because of the increased length of hull and base of sail plan, without increased mast height, for stability considerations, particularly with all inside ballast and the extremely fine hull form of the seiner. This fine and successful schooner was destroyed by fire at Sydney, Nova Scotia when only 18

months old. Captain Jacobs afterwards had the steam mackerel seiner *Alice B. Jacobs* built, but power made little impression on the schooner fleets until 1912, although *Thomas S. Gorton* of 1905 was the last Gloucester schooner built without auxiliary power.

Fishing schooner construction was traditional, with very heavy scantlings and sawn frames. They were cheaply and roughly built with a comparatively short expectation of life in a hard trade where after 20 years a schooner was reckoned old. These vessels were the pride of many communities but their fishing methods were often backward, and wasted many lives. On board, the crew were generally well fed and snug in the long fo'c'sle with its tiers of bunks and scrubbed table tapering away into the forward gloom. The snoring sleepers, swinging yellow oilskins, and cheerful competent cook, who got little sleep but produced wonderful meals on a tiny roaring stove and could sail the schooner singlehanded in light weather; and the all-pervading smell of fish was background to the enterprise which had the indefinable atmosphere of a craft earning a living from the sea under sail.

Around 1900 numbers of schooners were built to original designs by Thomas McManus, a Boston fish buyer, yachtsman and son of a Boston sail-maker. As a young man he met Dennison Lawlor and Edward Burgess, and this background led to his designing fishing schooners, at first as a hobby and eventually as full-time business. His first designs were *James S. Steele* and *Richard C. Steele*, built in 1892 with very cutaway forefoots similar to those then coming into vogue in yachts. Other designs followed and he blossomed out in 1898, introducing the curved stem as a characteristic of his work in a series of small schooners named after Indian chiefs, causing the profile to be known as an indian head. His early schooners also had a curved keel profile making them difficult to slip but quick in stays, and their form and rig had almost reached perfection for their work. However, mainbooms soared outboard, needing footropes for reefing, and for stowing headsails on the spike bowsprits, locally called widow makers, from the number of hands swept off them each year when changing headsails. McManus designed a schooner with an elongated bow replacing the bowsprit but retaining the staysail, jib and jib topsail, head rig. He exhibited the

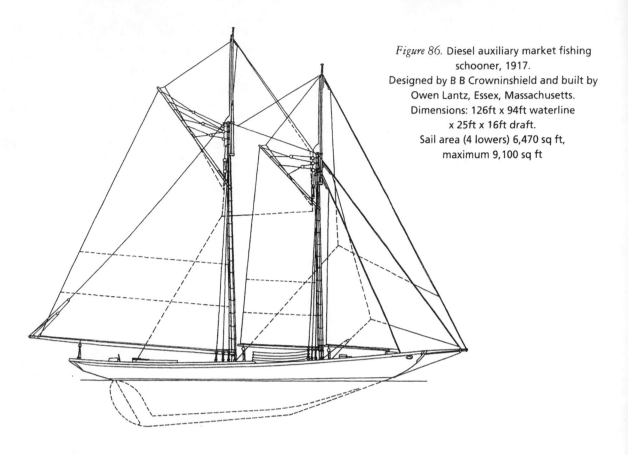

Figure 86. Diesel auxiliary market fishing schooner, 1917.
Designed by B B Crowninshield and built by Owen Lantz, Essex, Massachusetts.
Dimensions: 126ft x 94ft waterline x 25ft x 16ft draft.
Sail area (4 lowers) 6,470 sq ft, maximum 9,100 sq ft

model for a year without exciting interest as schooner pricing was based on length of hull and if this was increased, a more costly vessel resulted. Late in 1901 Captain Thomas and a partner tried the design and Oxner and Story of Essex built *Helen B. Thomas*, the first knockabout bowed fisherman, which proved handy, fast and practical. Three years later Mr Oxner modelled the larger knockabout *Shepherd King*; fast in heavy weather and very seaworthy. After 1906 many more knockabouts were built, usually with a short bow-overhang and a tall, narrow rig to achieve correct sail balance. In 1907 McManus designed the handsome, fast knockabout *Arethusa*, built at Essex by Tarr and James. Her principal dimensions were 114ft x 25ft 7in x 12ft 6in depth of hold. Mainmast 79ft deck to cap, foremast 74ft, mainboom 71ft. She was skippered by Captain Clayton Morrissey and frequently fished off the Canadian coast where American schooners often poached inside fishery limits. In 1908 she was caught, with two others, but sailed away from the steam patrol vessel *Fiona* and escaped her fine. After years of successful fishing *Arethusa* was sold for rum running in 1922; a sordid fate which

claimed many fast fishing schooners. After 1912 motors of considerable power were increasingly fitted in the schooners, but fully rigged Boston and Gloucester vessels continued to be built into the 1920s.

After 1900 many fine fishing schooners were designed by B. B. Crowninshield of Boston, the 119 foot *Tartar* of 1905 being his best known. He also designed a series of 60 foot waterline schooners built at Essex, Massachusetts, for the Gulf Fisheries Company of Galveston, Texas, which fished the Gulf of Mexico for red snapper, and was designing fast, auxiliary, fishing schooners in 1917 (figure 86). Other of his small schooners were built in Canada whose Atlantic fisheries expanded rapidly after 1900 with Lunenberg and Shelburne, Nova Scotia being the principal building centres. At first, the Canadian schooners were generally smaller than their American counterparts and were principally salt banker fishing. Some larger schooners were designed by F. A. Warner of Halifax for the Grand Bank fishery which was their mainstay, also sailing salt fish to the Caribbean, returning with salt, rum, and sometimes sugar.

In 1910, Mr Brittain of the Maritime Fish Corporation donated a cup for the winner of an annual race for fishing schooners of 90 tons and above at Digby, Nova Scotia regatta. For three years it was won by the *Albert J. Lutz*, a 90 foot waterline schooner built in 1908 at Shelburne, and sailed by Captain John D. Apt of Port Wade, Nova Scotia, on the Bay of Fundy.

Gloucester and Boston owed much to the sturdy Nova Scotia and Newfoundland fishermen who formed the backbone of skippers and crews for their schooner fleets for many years. Strangely, the captains at the wheels of both Canadian and American schooners in the controversial cup races between 1920-38 were often Nova Scotians or Newfoundlanders, as were many of the 'American' crews, though this did not lessen their rivalry.

Halifax owned many fishing schooners, rivalling Lunenberg, a smaller town along the coast, which had cooperative fisheries and launched 135 fishing schooners between 1903 and 1928. In 1920 Halifax held a fishing schooner race and *Gilbert B. Walters*, sailed by Captain Angus Walters, was leading but lost a topmast, allowing *Delawanna* to win. Fishing interests and a local newspaper proprietor originated the Dennis Challenge Cup for competition between Canadian and American fishing schooners. By then the Gloucester fleet was almost all auxiliary schooners, many reduced in rig. Gloucester chose the 107 foot McManus-designed *Esperanto*, built in 1906, which, refitted hastily from salt cod fishing, sailed to Halifax under Captain Martin Welch. She beat *Delawanna* sailed by Captain Himmelman, in two races which commenced an intense revival of schooner racing. The construction, rig and handling of these fishing schooners did not approach that of contemporary large racing yachts, which would have beaten them out of sight in a few hours. They were fishermen first and the races were a holiday, with sport of a robust order, roused by native pride and spirit. Captain Walters once invited a Nova Scotian prime minister to join *Bluenose's* crew if he doubted sailing her was a working sport! These races after 1920 engendered as much dispute as the America's Cup but as they prolonged the building and sailing of these fine schooners into comparatively recent times, are worthy of study.

There was no time allowance for difference in size but schooners were restricted to 140 foot maximum length, with all inside ballast and having fished for a trip; all intended to exclude yacht-like freaks. In 1921 the Canadians built their most noted fisherman: the schooner *Bluenose*, the first fisherman by their young yacht designer W. J. Roue and built at Lunenberg by Smith and Rhuland. Intended for other work than the Gloucesterman, she was built primarily to regain the fisherman cup, and also to be competitive as a salt banker making two trips a year, besides a winter voyage to the West Indies with salt cod cargo, returning with salt from Turk's Island. Her dimensions of 142ft overall, 111ft 9in waterline x 27ft beam x 15ft 8in draft and sail area of 10,970 square feet were typical of the ultimate development of these fishing schooners between 1920 and 1930, when they were, paradoxically, anachronisms. *Bluenose* was very powerful and fast. She displaced 285 tons and could load 210 tons of cargo. Her mainmast and main boom were both 81 feet long. Throughout her long career she was skilfully sailed by Captain Angus Walters with a crew of 28. To meet her, Gloucester fishermen and yachtsmen built the 143 foot schooner *Mayflower*, designed by W. Starling Burgess, a yacht designer following his father's fame. She was built at Essex by J. F. James, cost $51,000 and had so many yacht-like features that the Canadians refused to allow her to race, claiming she was not a true fishing vessel, and Gloucester fell back on *Elsie*, a fresh-market fisherman designed for short speedy trips by Thomas McManus in 1910, and again sailed by Captain Welch. The larger *Bluenose* naturally outsailed the 119 foot *Elsie* setting 8,500 square feet. During one race the disqualified *Mayflower* sailed in company for part of the course, bound out to the banks in fishing trim. She had earlier raced the fast Canadian schooner *L. A. Dunton* from Gloucester to Shelburne, Nova Scotia, beating her by seven hours.

Racing fever gripped Gloucester and in 1922 the ill-fated challenger *Puritan* was built to regain the Cup. Designed by Burgess and Paine, this tall-sparred schooner achieved $14\frac{1}{2}$ knots for five hours, but was lost on Sable Island six weeks after launch.

Undaunted, Gloucester matched the new schooner *Henry Ford*, (139ft, 110ft 11in x 25ft 2in x 15ft 2in, 9,640 square feet) against the indomitable *Bluenose*, which touched 13 knots in one race and won the series which was marred by the *Ford's* unwarranted protests,

142ft Nova Scotia fishing schooner *Bluenose* 1921. Photo taken at Spithead naval review, 1935. *Beken*

leading to a tense bitterness in subsequent racing. *Henry Ford* was the 427th fishing schooner designed by Thomas McManus and in 1923 she won the Gloucester centennial fishermen's race beating the new schooner *Shamrock*, sailed by Captain Welch, and *Elizabeth Howard* sailed by Captain Ben Pine, whose initiative maintained interest in schooner races until 1938.

In 1923 Captain Pine sailed Gloucester's new hope the 141 foot schooner *Columbia* against *Bluenose* off Halifax. *Bluenose* won the first race but lost the second on protest and sailed off to her fishing. This ended the series but *Columbia* beat *Henry Ford* racing off Gloucester in 1926; and her profile is sculpted on her captain's grave.

In 1929 Captain Pine had the old fishing schooner *Mary* rebuilt and renamed *Arthur D. Story*. He organised a race with the schooners *Elsie*, *Progress* and *Thomas S. Gorton* off Gloucester, won by *Progress*, with smart windward sailing of this Grand Banks halibut trawler. Wealthy Louis Thebaud was guest aboard *Story* during this race and so enjoyed it that he gave Captain Pine a cheque to help finance a new schooner which was designed by Frank C. Paine, built at Essex by Arthur Story and launched in 1930; the famous *Gertrude L. Thebaud*, the last of the North American sailing fishermen, capable of averaging $13\frac{1}{2}$ knots on passage. Later that year she challenged *Bluenose* to a private match off Gloucester, which *Thebaud* won, largely because *Bluenose* came ill-prepared from fishing, but in 1931 in the last race for the Dennis Cup, *Bluenose* defended her title successfully by beating *Thebaud* in all three races. The rivals met for the last time in 1938 when racing was again marred by disputes. Each won two races and the final one went to *Bluenose* who retained her championship title and became a Canadian legend.

She had also beaten the fast *Haligonian*, pride of Halifax, and since launch she had worked in the salt banking and West Indies cargo trades. She was almost lost anchored off the north-west bar of notorious Sable Island, in an onshore gale. The cable parted and she drifted towards the lee shore where hundreds of her kind had been wrecked. Under reefed foresail and 'jumbo', Captain Walters tried to fetch clear but had to wear round, losing ground in the snow squalls, and as she came to the wind again the seas broke yellow with sand. She clawed clear, her decks swept. *Bluenose* represented Canada at the 1935 King's Jubilee Review at Spithead and cruised in Solent waters for several weeks, attracting much attention, before sailing home from Falmouth in September. The mainsail was stowed off the Lizard, the riding sail set, and she sailed slowly westwards. A three-day gale commenced and *Bluenose* wore ship and stood SSW, labouring heavily in the unaccustomed conditions on the lee side of the Atlantic. She ranged about under shortened canvas, pounding her counter badly. A squall smashed the fore gaff, and her crew's temper was unimproved by sighting a French tunnyman close-hauled and bounding along under well-reefed mainsail and spitfire jib, and later a topsail schooner running for the Channel under considerable sail. The wind increased to a hurricane and *Bluenose* rode with but 12 feet of riding sail luff above the main boom. Swept by heavy seas and leaking badly, Captain Walters decided to bear away and run back, narrowly missing a French tunnyman hove-to, close-reefed. *Bluenose* made Plymouth, was repaired and made a comparatively easy passage home. Afterwards, she was sold to the West Indies for trading, and she was lost on the coast of Haiti in 1946.

In 1963, a replica, *Bluenose II*, was sponsored by Olands, a Halifax brewery, and built to the same design in the same yard and had her trial sail with her original designer and captain on board. She remains, a tribute to her predecessors and the courageous men who sailed them.

THE FISHING BOATS OF NORTH WEST ENGLAND

THE SAILING VESSELS which fished the coasts and estuaries of Lancashire, Westmorland, Cumberland, the Solway Firth, Cheshire, North Wales, St George's Channel and Irish Sea evolved slowly until about 1895, when they adopted features from current yacht practice which developed and radically altered the type, making it among the most interesting in Britain.

These boats worked on an exposed coast, often in shallow water with strong tides, where a trawling breeze raised a steep, breaking sea. The Morecambe Bay shrimp fishery was well established by the 1850s and the boats also trawled for flat fish, and raked up cockles in season. Brown shrimps, and the pink shrimps, locally called shanks or prawns, were trawled mainly in the channels between the miles of desolate sands which came a-dry at low water. Early boats had plumb stems, almost square-toed forefoots, transom sterns, and long straight keels and sloop rigs, with draft varying according to their home port. By the 1870s a typical Fleetwood shrimper was of about 5 to 6 tons and had the features of the earlier boats but with a short, square counter stern, slightly raked sternpost, and had a fine entrance and was built to take the ground when necessary. The boats were decked except for a well amidships and were known locally as half decked boats. All ballast was inside and they were rigged as cutters or sloops, with a boom mainsail, large jib and rectangular gaff topsail. They were sailed by one man, sometimes with a boy, and worked a conventional beam trawl, called a shank trawl at Fleetwood, with a short ground rope and $1/2$-inch mesh for shrimps. The cod end was emptied on deck and the catch culled between hauls, 25 to 30 quarts a day being good work. They were boiled in a copper in the hold and bagged up ready for market, usually in the industrial areas or London. As boats increased in size, the usual crew was two men who divided the catch into five shares; each taking two, leaving one for the boat which 'found' two nets, and the men one net each.

About 1896 the hull shape changed to improve speed and weatherliness. The forefoot was cut well away and the bow sections were fined, sternpost rake increased and the keel profile rounded as outside ballast began to be fitted. The hull beam was maintained but the bilge hardened and the bottom sections became very hollow with much reverse turn. These features were copied from contemporary racing yachts and were fully developed by about 1903 when shrimpers ranged between 35ft to 48ft length overall with from 3ft 8in to 6ft 6in draft, depending on their port and grounds. Counters were refined from the old square type to a longer, flat sectioned, narrow, elliptical shape which increased speed but slammed in a sea and, being so fine at the sheer, suffered so much damage from trawl heads that a watertight bulkhead or transom, locally termed a tuck, was fitted to stop leaks from the counter filling the bilge. The best boats were very stiff. Although the asymmetry of their shape had little effect in smooth water, in a breeze and seaway their broad quarters caused griping and erratic motion, and the low aft freeboard brought water on deck but did not seem to worry the crews, though the boats were very wet in a breeze. The bow was rounded in profile and had hollow waterlines to avoid pounding when going to windward in a sea too short for the boat to ride properly, but had sufficient flare to lift in larger waves and was deliberately kept free of weight. Early boats had all ballast inside, concentrated amidships but later a 35-footer would have $1\frac{1}{2}$ tons on the keel and another $3\frac{1}{2}$ inside. The shrimpers progressed rapidly in design between 1895 and 1914, when their building virtually ceased and it is interesting to speculate what further evolution would have developed.

The rig was a pole-masted gaff cutter with the mast stepped about $^1/_3$ length aft of the stem and supported by two shrouds each side and a forestay, but no backstays were fitted. The staysail was hoisted by a single whip purchase and the jib halliard rove through cheek blocks at the masthead sides, the fall leading to the port pinrail and the standing part made fast to the starboard pinrail, to form a purchase. Double topping lifts were often fitted and a fish tackle from the masthead handled the trawl. A typical 35ft x 10ft 6in x 5ft draft boat built in 1907 by Crossfield at Arnside, Westmorland, set 900 square feet working area, with a 500 square feet mainsail having a square-headed gaff almost as long as the boom and typical of these boats' rigs; a feature contributing to their hard steering in a breeze when, after a long run, some crews would come ashore with their backs black and blue from the tiller. The rig was heavy but many were fast on or off the wind and, whenever possible, carried a big jib on the long bowsprit to aid balance though tiller tackles were sometimes needed. No bobstay or shrouds were fitted to the long bowsprit. The owners often made their own sails and frequently measured a similar boat's spars to make a suit for a new hull and little thought appears to have been given to correct relation of hull and sail plan for balance; surprisingly, in view of the bold experiments in hull shape. The very long gaffs sagged on the wind and wrenched the gaff jaws, often springing them. Patent jaws were tried, without success. In contrast, many Mersey boats set well-peaked mainsails and smart yard topsails. Few of the shrimpers ever seem to have carried topmasts, and yard topsails were set from the pole mastheads, mainly for passage making because of frequent rucking of the long gaff when working the trawl. Three jibs were carried; a large, working and small spitfire.

Below the flush deck the smaller shrimpers were typically arranged with a gloomy, two-berth cuddy forward, with a coal stove, and the sole stepped to suit the round of the stem. Two pipe cots were fitted and a sliding door through the bulkhead stepped up to the cockpit platform which was raised at its after end for the helmsman. The shrimp copper stood to port and a large

thwart was fitted just aft of amidships with a powerful wooden box pump attached to it. The anchor, bucket and ropes were stowed under a trap in the platform. The long, narrow cockpit had 6 inch high coamings well rounded at each end, often having a canvas dodger rigged from the fore coaming. Quarter posts and bowsprit bitts handled mooring and trawl ropes but no windlass was carried for the 20 fathoms of $^3/_8$-inch chain cable. The hulls were usually painted pale blue or grey and a 9 foot rowboat served as tender and was carried on deck in a breeze.

Shrimping under sail thrived immediately before 1914 but became very competitive. Boats often carried four hands; one at the helm and the others principally to haul, boil and prepare the catch before landing while the black-sailed cutters thrust through the short, sandy seas, racing to land their three or four dozen boxes of prawns for the early morning London train. During this period some owners had more than one boat, but the trade contracted between 1914-1918 when many boats laid up, and after the War many were sold for conversion to yachts. Until 1914 the fastest half-decked and channel type fishing boats were keenly raced at Heswall, Southport, Morecambe, Ramsey and Fleetwood. *Two Brothers* of Rock Ferry, and Southport's *Majestic* and *Camilla* were among the champions.

In 1904, 40 shrimp trawlers worked from Fleetwood, and 70 from Marshside and Southport during part of the year, and the average age of the fleet was six years. A typical Fleetwood boat would be 39ft overall, 32ft waterline x 10ft 6in beam x 5ft 6in draft and setting 750 square feet, including topsail. During summer a good many Fleetwood boats supplemented the fishing by chartering for a weekend to a fortnight, the skipper going with her, to business men from Manchester and the cotton towns who were not yachtsmen, and who never expected to own a boat, but were happy pottering around in their braces helping to sail the boat for a dozen weekends each year and probably a longer cruise for a week to the Isle of Man or the Solway with a party of friends. The Fleetwood pleasure tripping boats of the 1890s, working from the beach near the ferry slip, had similarly shaped hulls with a long open well for passengers, but were rigged with a single large standing lugsail; a most unusual combination of hull and rig.

Lancashire trawler *Margaret* FD 208, built 1903.
Photo courtesy Peter Horsley

Larger shrimpers from Fleetwood, Southport and further down the coast varied between 40ft to 48ft length with 6ft to 6ft 6in draft and these boats retained a plumb stem above water but the forefoot was well cut away to a rockered keel to the heel of the sternpost. An iron keel of from 1 to 3 tons was fitted to some but the majority had very little outside ballast but plenty inside, and their bottoms were not so hollow as the smaller boats. However, the older type of smack with a deep forefoot and square counter stern was still built in 1906 as many fishermen preferred them for channel fishing, and form also varied with locality, the Lytham-St Annes Boats having very round bottoms to suit taking hard ground. About 1908 a 35 foot shrimper cost between £150 and £200 and many were built at Fleetwood by Gibsons, and

by Armour Brothers; some to advanced designs by W. Stoba, of whom too little is known. The Lancashire owners and builders were fond of experiment. In 1895, Gibsons of Fleetwood launched the 15 ton *Eureka* which was fitted with a centreboard. This does not seem to have been repeated in later boats and she was sold for a yacht in 1900.

The Crossfields of Arnside, Westmorland are among the best remembered builders of the small shrimpers. The family commenced building there in 1853 and the business, known variously as Crossfield Brothers, William Crossfield, and William Crossfield & Son, was set in sylvan surroundings on the Kent estuary. They designed by half models consisting of a backboard, to which was screwed a half-deck plan and four transverse sections of the hull shape. This shadow model appears to have been common practice on that coast. Building methods were primitive; the centreline structure being set up and the four mould frames shaped from the model were set up and ribband battens fitted around them to the stem and counter, the shape of the other frames being lifted from these with a chain like a giant wooden bicycle-chain, soaked in water before use to make it stiff-jointed to retain the curves of the hull for transfer to the oak sawn frame timber, a most wasteful method. Planking was larch or pine but generally these shrimpers were not strongly built compared with other fishing vessels, and many hogged their sheer in way of the chainplates after a few years. By 1914 Crossfield Brothers had moved to Hoylake, Cheshire, and continued building fishing vessels and yachts.

By 1909 the Lancashire fishing boat type was working all over the Irish Sea. On the Welsh coast they were owned at Deganwlly, Pwllheli, Barmouth, Aberystwyth, Conway and Caernarvon, where they proved capable of beating out over the bar with a south-west gale starting, and the certainty of a bad night at sea meant little to them. The Pwllheli boats were reputed to be capable of going through Jack Sound in any weather, and everywhere their crews had great confidence in their ability. John Crossfield, who moved from Arnside to Conway, developed fine boats of the type in North Wales, and these generally had fuller bows than the Lancashire cutters, as local seas were longer. However, most were built at Fleetwood by Armour Brothers or Gibsons, and a few at Welsh ports. Besides fish trawling these Welsh cutters drift-netted for herring in the autumn, following the fish along the Welsh coast.

In 1864 a Fleetwood boat dredged oysters in the Solway Firth so successfully that, within two months, 30 large cutters from the Essex Colne, and the Channel Islands were basing themselves on the Isle of Whithorn, gradually working out the grounds. Morecambe fishermen are reputed to have settled on the Solway a century ago taking their boats and methods with them and by 1900 20 boats of advanced design fished from Annan where their speed assisted working the strong local tides. Sounding poles were commonly used to feel along the steep-to channels of the Solway's desolate waste of sands, relieved only by gaunt stakes of the salmon kettle nets on the Scottish shore. They sailed on the ebb, fished down on the tide and returned on the flood, boiling the catch aboard. In summer they were often singlehanded, but had at least two men in winter when fish trawling was the principal work. Most of them were built by Wilson at Annan, often being temporarily ballasted and sailed to Barrow to be fitted with iron keels. The well-cut sails were usually made by the owners from coarse calico, dressed with a mixture of linseed oil and black lead which made them very heavy and shiny, but waterproof. Seventy years ago the half decked boats and the larger channel boats from Annan raced in the annual regatta, but bowsprits and topsails were discarded 75 years ago with the first auxiliary engines. Similar shrimpers also sailed out of Maryport, on the Cumberland shore, and in 1904 could earn £1 a day, then exceptional earnings fishing under sail.

On this coast the term smack was only used to refer to the large ketch and cutter rigged sailing trawlers owned at Liverpool, Hoylake and Fleetwood and, during the mid-19th century, at Whitehaven, Blackpool and Southport. In 1874, 35 of them were owned at Liverpool and 53 at Fleetwood, with small numbers at other ports. They trawled for bottom fish in the Irish Channel around the Isle of Man, and towards Campbeltown and the Clyde; while others fished the Irish coast. The Liverpool and Hoylake smacks frequently worked in Caenarvon and Cardigan Bays but the Liverpool boats' activities fluctuated considerably. Several ketch smacks were designed and built for the port at Wivenhoe, Essex, and the general size was between 30 and 40 tons, working trawl beams up to 45 foot long. Some very heavy cutters and ketches fished from Caenarvon, and their enormous bulwarks and massive construction caused a Lancashire skipper to comment 'they were built like strong boxes and sailed like them too'.

Fleetwood was laid out as a town by Lord Fleetwood in the late 1840s, on barren land at the mouth of the River Wyre, the streets being marked out by plough. In 1842 four fishing cutters from the Southport suburb of North Meols were fishing from there, attracted by offers of cheap fish-carriage by the Lancashire and Yorkshire Railway. Four years later the railway company bought four old east coast smacks to encourage use of the port which became thriving by the 1850s, but most of its smacks were old craft bought from away, and when the Hull owners took up steam trawlers in the 1880s many of the lumbering old ketches were sold to Fleetwood. Gradually local builders launched new trawlers and those evolved rapidly during the 1890s to about 1906, when some very fine ketches were built for the port and were among this country's ablest and fastest sailing fishing vessels. Development of hull form followed that of the shrimpers, but they were wet in a seaway. The owners of one issued a challenge for a considerable sum to race any sailing trawler in Britain which, unfortunately, was never taken up. They were built to stand anything, including lying aground, and one of them was reputed never to have been caulked for 30 years.

The 70ft x 18ft x 8ft 9in draft ketch *Reliance* was typical of the many Fleetwood trawlers built by Liver and Wildy at Wyre Dock, Fleetwood. Launched in 1903, she had a captain's cabin aft with crew's quarters forward of this, with a store room, fish-hold amidships and a crew's fo'c'sle. Design advanced in the 68ft x 18ft *Louis Rigby* whose forefoot was well cut away and the hull arranged with a crew's cabin aft, galley amidships, and the large hold extending from amidships to the chain locker.

By 1904 50 trawlers registered at Fleetwood, Hoylake and Liverpool, worked from the port and berthed alongside the pier, but despite this brave flowering, after about 1905 the Fleetwood fishermen preferred to work in steam trawlers rather than struggle to own smacks, and sail died early at the port.

CHAPTER NINETEEN

THE LITTLE SHIPS OF DENMARK

AMONGST THE BETTER DEFINED types of sailing craft which have emerged as vessels offering charter holidays or instructional cruises during the past 30 years, the wooden-hulled ketches and schooners colloquially known as 'Baltic traders' are to me most appealing and practical in appearance.

These chubby little cargo carriers were frequent visitors to British ports in earlier times, bringing various cargoes, often timber, and usually leaving with holds filled with coal. These ships have a long history of evolution reaching back at least to the beginning of the 18th century and probably earlier.

I recall seeing a few of these small sailing ships coming up our river in the late 1930s. We children were always puzzled by the ensign at the mizzen mast head, the white cross on a red ground, for this was the same flag flown at the sailing club at the next village, on which was superimposed for good measure the black raven of the Vikings, to further emphasise our old links with Scandinavia. However, the Danish sailing ships were relatively few amongst the scores of red ensigns and Dutch flags in that heyday of the motor coaster, and it was not until the 1960s that the round bowed and transom sterned wooden hulls started arriving in numbers as the ships were sold off from their old Baltic trades which were taken by new steel motor coasters built with encouragement from the Danish Government's progressive shipping policies.

The Danish gaff rigged cargo carrying craft were of several principal types. Smallest and perhaps the oldest was the sloop, and either evolved from it or closely related to it, the jacht. Both were rigged with a single mast on which was set a gaff mainsail having a loose foot, a fore staysail and a jib set to a sturdy bowsprit. The sloop was a well established type when naval architect Frederik Chapman, Swedish son of English parents, drew the plates for his splendid *Architectura Navalis Mercatoria* in 1768. Several plates show sloops of this type and Plate XXIX, No. 10 is typical.

A typical mid-19th century Danish sloop was about 50ft overall length, 16ft beam with a hull depth of 7ft 6in and a draft of about 6ft to 6ft 6in. The hull had a rounded bow profile, a straight keel and a heart shaped transom with pleasing rake. The hull sections throughout were heart shaped and she would sail better than would be expected, despite the full bow, as the buttocks and waterlines show a sweepingly fair form. As with most true sloops, the bowsprit was fixed and could not be run inboard. Otherwise the rig was that of a gaff cutter, having the usual mainsail, foresail and jib set to the bowsprit end and not usually on a traveller. In some, a flying jib could also be set on a jib boom rigged out through an eye at the end of the bowsprit. Many sloops carried a squaresail and had the curious arrangement of the bowsprit being housed on one side of the stem and the heel of the jib boom on the other.

Most of the hull was occupied by a single hold, served by a main cargo hatch amidships and two smaller ones forward and aft. A tiny fo'c'sle had two bunks for the usual crew of a man and a boy. There was also a small two berth cabin aft for the skipper, who was often accompanied by his wife. Steering was with a long wooden tiller working over the raised quarter deck above the aft cabin. There was a handspike windlass forward and a hand cranked cargo winch amidships to hoist cargo in and out.

A typical Danish sloop under sail, early 19th century. The unusual arrangement of a raffee headed squaresail is complemented by the triangular studding sail. (From a painting by C W Eckersberg 1836, in the possession of the Hirschsprung Collection, Copenhagen.)

Construction was sturdy with sawn frames and substantial scantlings for long life and hard work. The ship's boat was hung in fixed wooden davits, across the stern, and there were relatively deep bulwarks around the deck. The hull was usually painted white or black, with varicoloured wale strakes just below the sheer. Altogether these were handy and useful little vessels. They were well kept as one would expect in a country like Denmark, where a national tradition of neatness and order was followed by the sailing ship crews. Often one family would man the ship and either own her or have many shares in the vessel.

The origins of the jacht as a type are linked to the sloop but it may have developed as a faster sailing version with finer underwater lines. Jachts had the same ample freeboard and bulky hull above water, ending in a transom stern. At first this was probably a form of the 'lute' stern much used in English small craft into the 19th century such as smacks and brigs. In these Baltic craft it was often arranged as a flat lower transom which extended well above water. Then an upper transom was set sufficiently aft of this to allow the sternpost and the mainpiece of the rudder to be shipped up inside it. The narrow transverse aperture between the top of the lower transom and the bottom of the upper was planked-in to make a watertight stern, leaving a trunk for the rudder mainpiece.

However, in all the drawings, paintings and photographs of jachts that I have seen, the transom was in one plane with the rudder fitted on its after side – a simpler and less expensive construction. The stern of a jacht in figure 87 is of this type. The hull of the Danish jacht is distinctive and was in use at least by the early 18th century. The keel was straight and often had little declivity (i.e. was almost level when the vessel was not loaded and probably changed its attitude little when she was). The amidships sections set the hull form and were of heart shape, fuller or finer depending whether the hull was intended to carry much cargo or was to be a fast

Figure 87. The typical heart-shaped stern of the Danish Jacht hull, complete with wooden stern davits for the ship's boat

sailer. The bow shape was bold. The stem curved down to meet the forward end of the keel in a toe, which slightly simplified building and provided maximum lateral plane for sailing but was subject to damage if she struck bottom. It also slowed turning and probably created turbulence under the forefoot. The bow sections were full but were suited to the Baltic seas, providing ample reserve buoyancy in bad weather and adequate deck space for working the windlass and anchors, besides ensuring reasonable space in the fo'c'sle.

The after body of the hull drew out in a finer version of the amidships sections with a fair and easy run ending at the transom. The larger and longer jachts had remarkably sweeping buttock lines, and those and the fairing diagonals of the lines, which are usually a truer guide to a hull form's potential sailing performance than the sections and waterlines, promised speed when reaching or sailing just off the wind. Figure 88 illustrates this form and is worth studying as a practical hull shape for a small sailing cargo vessel.

These hulls appear to have derived from sailing ships rather than from 'boat' origins. When considering the sailing performance of jacht-shaped hulls under various rigs it is tempting to imagine that the almost total lack of tidal stream in the Baltic may have affected their hull shape, particularly the full bow. However, the seas are quickly generated there and that probably led to the generous freeboard and height of bulwarks in even quite small craft and also perhaps to the small hatch openings. Until the 1870s jachts were of fairly fine hull form, which afterwards tended to become fuller in section and with less rake to the stem.

The potential speed of the jacht type is illustrated by some passages of craft of this form under various rigs, discussed later in this chapter.

The vessels were carvel planked on sawn frames and were in essence small ships, many continuing to carry their stocked anchors on catheads into the 20th century in an unaffected and practical way. The hulls were (and still are) usually painted black with a white stripe around the bulwark, and continued to curve around the edge of the transom, with pleasing effect. The atmosphere in which the design, building and operation of these Danish craft was carried out typified the logical Scandinavian approach to maritime matters and to life itself; modest but achieving with limited resources what others often did with excess.

Jachts were also built along the north coast of Germany at ports such as Lubeck. Many were from the district of Schleswig-Holstein, which had been taken from Denmark by the Germans in the war of 1864.

As with most types of commercial sailing ships of the 19th century, the Danes gradually built larger vessels with this hull form. Ketch rigged jachts became more common and were known as a Galeas, evidently from the rig as a ketch. Some were also built as schooners and later also as brigantines and barquentines. Many of these larger

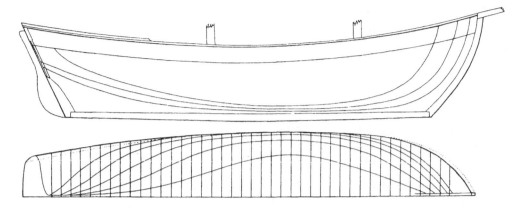

Figure 88. Lines of jacht hulled cargo schooner

craft made long sea and ocean passages. Some traded across the north Atlantic to Greenland, Denmark's largest colony, and others regularly sailed salted fish from Newfoundland and Nova Scotia to Spain, Portugal and Italy. Some took cargoes to and from central American countries. There were also occasional voyages to south America and China by these little Danish ships.

Many of the larger vessels, schooners, brigs, brigantines and barquentines also frequently carried timber cargoes from Sweden and Finland to ports in Denmark, Britain and France during the ice-free season in the Baltic, which is usually from April until November. When this timber season ended many of the ships returned to their home ports to lay up for the winter. (See photograph, page 210).

After the 1880s there were many three-masted schooners built in Denmark. Some had counter sterns and clipper bows, the others were of the jacht type. Numbers of these ships were built and owned at the small towns of Svendborg and Faaborg, in South Funen, others at the town of Marstal, on the island of Aerø. Craft from Marstal usually had the round bow and transom stern jacht hull while many Svendborg-built vessels had a counter stern and a finer bow, though these were general distinctions.

Until the 1930s Marstal owners had a fleet of schooners and some craft of other rigs. It was known throughout Denmark as *Sejlskibenes* or 'sailing ship town'. The place has a long history of seafaring since it was founded 500 years ago. Its first sailors were farmers who also built their own vessels — a long tradition in Baltic countries. The port and its ships grew over the years and

reached a zenith when the three-masted schooner rig was adopted. By 1900 Marstal owners had about 300 vessels, many of which were sailing in the deep sea trades. The shipyard of Christensen was the principal shipbuilder in the town.

Svendborg, on the south shore of the island of Funen, was another well known sailing ship port. From the 13th century Svendborg ships and sailors have ventured far away and ships have been launched from its yards. In 1369 29 small vessels sailed from there, and by 1798 there were 80 totalling 720 tons. By 1883 279 sailing ships were owned at Svendborg, the majority being three-masted topsail schooners having counter sterns and clipper bows. Most of these were built in Svendborg where there were then eight shipyards. The largest of these was that of J. Ring Andersen and this has remained in business into the present day.

Many wooden sailing and auxiliary ships were built at Svendborg during the 1914-18 war but after it numbers of fine schooners were sold to Swedish owners when the economic climate changed. Although Denmark was neutral, the 1914-18 war resulted in the loss of many Marstal ships, sunk by German submarines or by other warships. There were many risks for everyone afloat under any flag during those years but for neutrals the risks paid well and much money was made in shipping by Denmark, Norway and Sweden. At the war's end freight rates remained high for a time and numbers of ships were ordered. In Denmark many of these were large, wooden three-masted schooners, several having an auxiliary motor. (See photograph, page 211).

These were the finest working sailing vessels in the

A newly completed, two-masted Danish schooner with jacht hull form, ready for sea with sails bent and the squaresail yard cockbilled on the foremast

Baltic and were launched rigged as topsail schooners and barquentines. They were well found and their crews took pride in their upkeep and appearance. Despite the auxiliary engines the deadeyes and lanyards setting up the standing rigging and the wooden stocked anchors which were often still carried at the cat heads of some of the older ships gave this fleet an almost 18th-century atmosphere. In their usual trades these ships were loaded in the holds and above the deck and hatches with a stack of timber several feet high which extended to the bulwarks. This was loaded and discharged by hand and a vessel might take ten to fourteen days to unload her cargo, then perhaps have to wait two weeks while a freight was fixed for her next voyage. This method of working lasted into the 1950s. Some idea of the value of this trade to the Danish schooners is given by the thirty or so Scandinavian timber carriers berthing at the Carron Dock, Grangemouth, Scotland in the 1920s. The crews

manning these little Danish ships had a hard life. Most were young, many only boys, and they lived in fairly primitive style. In contrast the captains and mates were usually elderly seafarers who almost always had shares in the ship. By 1921 shipping was being affected by a slump which quickly grew worse. Many of the new Danish vessels were sold at low prices to owners in Sweden who somehow continued to find cargoes for them.

The three-masted topsail schooner *Elisabeth* of Marstal was one of the Danish vessels ordered at the end of the First World War. She was built at Marstal in 1919, with a tonnage of 270 and principal dimensions of 120ft 7in length x 28ft 7in beam x 12ft 4in depth. She found cargoes in the timber trade for some time but in 1925 was carrying stockfish from Newfoundland across the Atlantic. By 1931 the *Elisabeth* had been re-rigged as a barquentine, an unusual change for that date, and sailed

A two-masted, jacht hulled Danish schooner under sail during the 1914-18 war. The Danish colours are painted on the bulwarks to proclaim her neutrality

from Sundsvall to Jersey, Channel Islands, in fourteen days.

But there was often disaster and death in such vessels. The *Elisabeth* sailed from Hermorand with a timber cargo for Jersey but got ashore on the Danish coast. After refloating and repair she resumed the passage but met strong gales in the North Sea and two men were lost overboard before she reached the Channel Islands. However, the traditional trades and hopes were sustained through the 1920s. The little schooner *Activ* of Marstal was launched in 1925 and sailed for some time in the stockfish carrying trade. This was the carriage of dried fish across the north Atlantic, between Newfoundland, Spain, Italy and Portugal in the case of the Danish ships of that time. The *Activ* once made the passage across the

Atlantic to Oporto in 21 days. She remained in this trade into the 1930s, during 1931 crossing from Ramea in Newfoundland to Oporto in 32 days. However, this trade faltered soon after and in 1932 she was laid up for a year at Marstal while the world slump deepened. In 1933 the *Activ* was fitted out for trading in the Baltic, still without an auxiliary engine. Eventually she was sold to Swedish owners and an engine was installed. She continued to trade into the 1950s.

The three-masted, 130-ton schooner *Alf* was another of the venturesome Danish vessels of the 1920s. Launched at Marstal in 1920 she was 123ft 3in long x 28ft 3in beam x 12ft 8in depth, and was mainly employed in carrying timber from Canada to British ports. Her passage making was average and sometimes good. In 1933 she sailed from Cadiz, in Spain, to Gaultois, Newfoundland, in 46 days, then on to Miramichi in 5 days to load timber for Newry in Ireland. She made the transatlantic passage in 37 days. Soon after she sailed from Methil with a coal cargo for Tonsberg in Norway, making a fast passage.

The passages of the *Alf* under sail for the years 1934 and 1935 were typical of the work done by these little ships.

Frederikstad to Ipswich, 13 days
Ipswich to Marstal, 8 days
Windau to Loctudy, France, 14 days
Loctudy to Cadiz, 8 days
Cadiz to Miramichi, 42 days
Miramichi to Newry, 22 days
Newry to Hull, 7 days — an exceptional passage
Hull to Tonsberg, 6 days
Tonsberg to Frederikshavn, 4 days

Figure 89. Sail plan and general arrangement of the schooner *Lilla Dan*

The *Alf* was a good sailer and delivered her cargoes without damage. During the usual winter laying up period in 1936 an auxiliary engine was installed and she was sold to owners at Asajerø in Denmark, to be renamed *Inge*. However, she was to be managed by A. E. Sorensen, whose company owned sailing vessels at Svendborg. Her cargoes were now timber from Sweden and Finland to small ports in Britain and France. Although her topmasts were shortened when the engine was installed she retained a good spread of canvas. The *Inge* survived the 1939-45 war and continued to trade until January 1950 when she sprang a leak on passage in the Baltic and foundered very quickly, ending a successful career as one of the last merchant sailing ships.

After 1930 most of the remaining schooners from Svendborg were fitted with auxiliary diesel engines and the last purely sailing vessel owned there was the three-masted topsail schooner *Vera*, which was wrecked in 1935. At that time the Sorensen company of Svendborg operated a fleet of two four-masted and ten three-masted auxiliary schooners in the timber trades to Britain and France. After 1945 these ships were sold, mostly to owners in the Faroe Islands, who were then frequent buyers of old and small sailing ships and fishing smacks. The Sorensen company wisely invested in a fleet of modern motor vessels but retained a few of the auxiliaries.

At Marstal E. B. Kroman was a principal shipowner during the 1930s, owning many auxiliary schooners. When World War II started in 1939 Kroman arranged to buy four barquentines from owners at St Malo, France, hoping to repeat the commercial success of a neutral Denmark. Unfortunately German occupation ended those hopes.

The jacht type hull, rigged as a galeas or a schooner, continued to be built until about 1942 and was given impetus by the prospect of wartime fuel oil shortage, which became acute after German forces occupied Denmark in May 1940. Sail then suddenly became important again for many cargo and fishing vessels of Denmark and Sweden. A number of motor auxiliaries were built in both countries during the war years. Craft such as the 89ft ketch *Havet*, which had then just been completed, the 89ft counter-sterned

The two-masted topsail schooner *Lilla Dan* was built in 1951 by J Ring-Anderson as a training ship
for Copenhagen shipowners J Lauritzen

ketch *Talata* (later re-rigged as a three-masted schooner), the 94ft three-masted schooner *Esther Lohse* and the fine, counter-sterned auxiliary *Lehnskov* were launched from Ring Andersen's yard as late as 1944. Shortages of materials delayed their building but these were amongst the last well finished sailing cargo ships to be built and were worthy examples of the type.

In 1950 J. Lauritzen, shipowners of Copenhagen, ordered the two-masted topsail schooner *Lilla Dan* to be built as a training ship for their fleet of modern motorships which burgeoned in the heady days of post World War II shipping. This schooner was designed and built at the yard of J. Ring Andersen of Svendborg. She

was to be 'Jacht built', with the rounded stem profile and distinctive transom stern typical of these craft. Her rig and deck arrangements are shown in figure 89. An auxiliary diesel engine was installed and accommodation was provided for two officers and fourteen boys. (See photograph, this page).

The *Lilla Dan* operated as a seagoing training vessel when such ships were very uncommon for merchant seamen and only a few were owned by navies. She served in conjunction with Lauritzen's private training school for apprentices, near Svendborg. Soon after her completion in 1951 the *Lilla Dan* was challenged to a sailing race by A. E. Sorensen, owner of the three-masted fore and aft schooner *Peder Most* of Svendborg. She also had been built by J. Ring Andersen, in 1944, during the severe shortage of diesel fuel in Denmark under German occupation. By 1950 she was engaged in carrying cargo to the Danish colony of Greenland but a year later had returned to the Baltic and North Sea trades. Although a commercial vessel, the *Peder Most* remained fully rigged with a topmast rigged on each mast and a jib-headed topsail carried on each. There was little difference in size between the contestants, the *Peder Most* being about 160 tons deadweight and the *Lilla Dan* about 140.

The race comprised a run from the start to the distance mark and a beat back, a total distance of 30 miles. Each schooner had a crew of apprentices and during the race the wind fell light. The *Lilla Dan* finished first in 57 hours 43 minutes over the course, the *Peder Most* crossing 12 minutes later which, considering she was a commercial vessel, was creditable. This was perhaps the last race between truly commercial schooners ever sailed.

Danish vessels lasted for many years and the large number of galeas and several schooners which were sold out of trade during the 1960s and 1970s were usually sound craft, legislated out of cargo carrying in prejudice against their wooden hulls. The changing pattern of trade also affected their future, as did container traffic, the building of new road bridges connecting islands and the extension of ferry services in conjunction with growing road transport linking many parts of Denmark, all contributed to their end in trade. Many were brought to Britain and became charter vessels for pleasure sailing. Some were sold to America and to other countries, sometimes being given unsuitable rigs to impress or to provide pulley-hauley work for the charter crews. It was a strange end for originally handsome and practical little ships.

CHAPTER TWENTY

ON THE WEST COAST OF FRANCE

FEW AREAS of the world produced greater diversity of fishing and cargo carrying craft than the west coast of France. In the north it faces the English Channel and further south the deeper waters open to the full sweep of the Atlantic Ocean. This long coast produced skilful and hardy fishermen and sailors, who used the gaff sail and the lug in various combinations to do their work. Several types of their craft have become well known, such as the Breton tunny fishing cutters and dandies, the crabbing cutters from the same area and the splendid pilot boats serving Le Havre (described in chapter 13). However, there were many other types which are worth recalling, only a few of which can be reviewed in this chapter.

The upper part of the English Channel meets the lower part of the North Sea at the Straits of Dover, the closest France comes to England. It is a channel of swift tides and turbulent waters, needing able sailing vessels and good seamen for its regular navigation. The packet cutters which made fast, regular passages between the Channel ports of England and France were amongst the fastest of their type until steamships replaced them in the early 19th century. The long wars with France at the end of the 18th century and beginning of the 19th interrupted the trade of the packets, though fast smuggling luggers and cutters continued an illegal connection. When Napoleon was defeated in 1815 and peace resumed, new cutters were ordered in both countries and the sail plan and lines of one built at Calais in 1816 are shown in figures 90 and 91.

This cutter had a pleasing hull form with principal dimensions of 57ft 9in length overall, 51ft 8in waterline length x 18ft 4in beam x 8ft 6in draft. The stern was of the 'lute' form, a basic style of counter much used in contemporary fishing smacks and other craft. The rig was typical of cutters of the time but the mast was further forward than in contemporary English cutters and consequently the foresail was small. The bowsprit was also shown steeving upwards, in contrast to English practice where it is bowsed down parallel to the waterline to obtain a taut jib luff. The square topsail would often be carried but the lower squaresail or 'course' less so.

Such craft were handled by a crew of about six under a captain who usually considered that being commander of a Channel packet was second only to being in command of a man of war, and it was usually a more rewarding occupation. Seaworthiness and speed were the aims of the packet designers and this French cutter appears likely to have had both.

The old port of Gravelines lies on the French coast between Calais and Dunkirk and was home port for hardy fishermen. The sail plan of a Gravelines ketch shown in figure 92 may seem vaguely familiar and with reason, as these French craft fished on the Dogger Bank and elsewhere in the North Sea alongside the hundreds of British cutter and ketch rigged trawling smacks from Lowestoft, Great Yarmouth, Grimsby, Hull and Scarborough. The Gravelines men seem to have adopted many of the features of the British smacks to produce seaworthy ketches (figure 93).

They had greater beam than their British contemporaries of the same length, and the Gravelines ketches had a pole mast with a yard topsail above instead of a topmast usual in British practice. The mizzen was also large in area and the mainsail smaller, while the bowsprit appeared to be very long. Although a topsail is shown above the mizzen the use of this sail was limited to running and reaching. Like other French fishing craft these Gravelines ketches superseded an earlier and

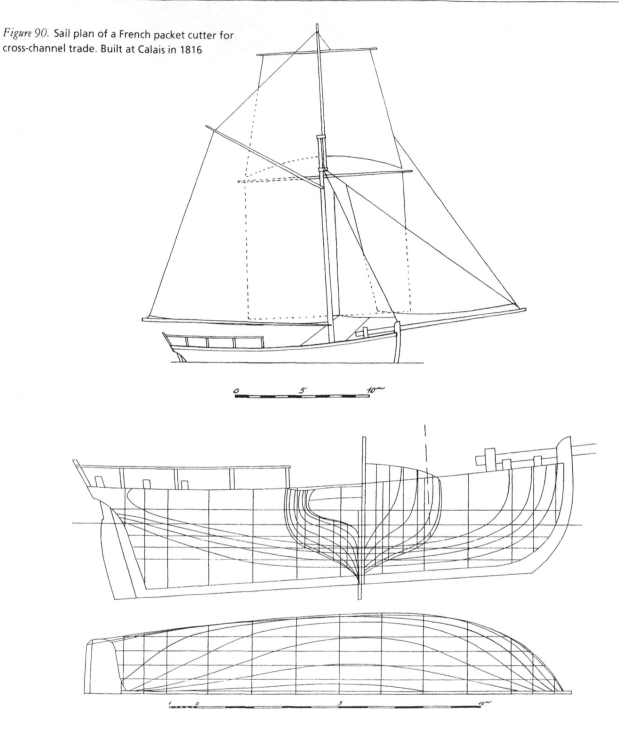

Figure 90. Sail plan of a French packet cutter for cross-channel trade. Built at Calais in 1816

Figure 91. Lines of a French packet cutter, 1816. These were fast and able craft

smaller type which had a lugsail rig.

There were also much larger ketches built and owned at Gravelines for cod fishing off Iceland. One of the last was built in 1912, without auxiliary power. Her dimensions were 100ft 9in length overall, 88ft waterline length x 25ft 3in beam x 12ft draft. She was amongst the largest sailing fishing ketches built. Figure 94 shows her lines.

At Dunkirk many of the fishing fleet were engaged in drift netting for herring for much of the year. At first the boats were rigged with lugsails but gradually the rig changed to a gaff yawl or 'dandy'. The example shown in

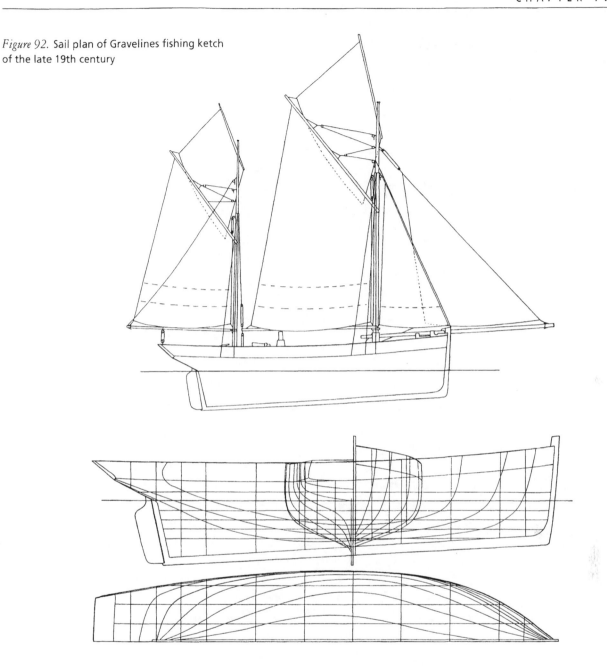

Figure 92. Sail plan of Gravelines fishing ketch of the late 19th century

Figure 93. Lines of Gravelines fishing ketch, late 19th century

figures 95 and 96 was one of many sailing and motor fishing craft designed by G. Soé a talented French naval architect who contributed considerably to the understanding of the design of working craft from many European countries at the beginning of the 20th century.

Before steam trawlers ousted sail at Boulogne a fleet of sailing smacks worked from the port and most were built there. Originally these were rigged as cutters but, as with most trawling under sail, the smacks grew in size

and in about 1892 the ketch rig was adopted for them at Boulogne. The plans of the trawler in figures 97 and 98 show a craft 75ft 6in long overall, 65ft 8in waterline length x 21ft 4in beam and drawing 12ft 6in. She had the long straight keel and deep forefoot of most sailing trawlers and the broad counter desirable for handling the after trawl head and associated gear. These were powerful boats and compared well with contemporary English sailing trawlers. The ketch rig was typical in arrangement

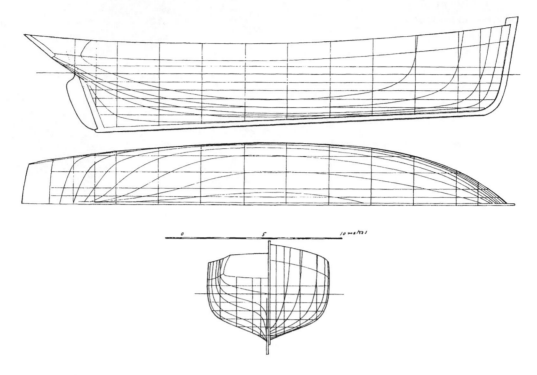

Figure 94. Lines of a 100ft 9in long Gravelines fishing ketch built in 1912 for cod fishing off Iceland

but the Boulogne ketches abandoned a fidded topmast in the 1890s for the arrangement shown here with a strangely old fashioned yard topsails set above the pole mastheads. These were fine smacks, fit to face the severe weather often encountered in their fishing voyages.

The last sailing trawler was built for owners at Boulogne in 1904. By the beginning of 1911 the last sailing craft were sold and all fishing from the port was done by steam trawlers.

Cutter rigged fishing craft were built and owned at the small port of Honfleur, on the south bank of the entrance of the River Seine, opposite Le Havre and close to Trouville. Some were still sailing in 1939. In earlier times these were sometimes sizeable craft and, in the same way as smacks from other places, sometimes carried cargoes as well as being used for fishing. A Honfleur cutter of the 1860s had hold and deck arranged for either trade. Carriage of cargo resulted in her full body, which besides giving good capacity also enabled her to take the ground at a reasonable angle for discharge. Hull dimensions were 67ft 6in length overall, 19ft beam and drafts varied from 7ft 6in to 9ft when loaded. Only modest space for crew accommodation was allowed as this cutter was primarily a cargo carrier.

The rig was that of a typical cutter of the mid-19th century, with the usual standing and running rigging common in British practice. The squaresail yard and its running sail was carried by many French craft after it had been discarded in Britain in favour of the triangular spinnaker, which could also double as a reaching jib. A storm trysail and its short gaff were carried for bad weather use. The total sail area of the mainsail, foresail and working jib was typically 2,300 square feet. A large foresail of about 750 square feet could be set besides a topsail and a jib topsail. Such craft made frequent coastal voyages with varied cargoes, not only on the French coast but to Britain and European countries.

Sail lingered in French fishing and trading vessels longer than in many countries and a sailing cargo vessel was still occasionally built in the early years of the 20th century. The plans in figures 99 and 100 are of a 99ft 6in long coasting ketch which could carry 180 tons of cargo. She was built as late as 1911. The counter sterned and full bodied hull was well formed and she sailed reasonably fast for such a craft. The entry and run were as fine as possible on the designed displacement, length and trim.

The rig was arranged with a standing bowsprit on

Figure 95. Sail plan of a Dunkirk fishing dandy, 1902

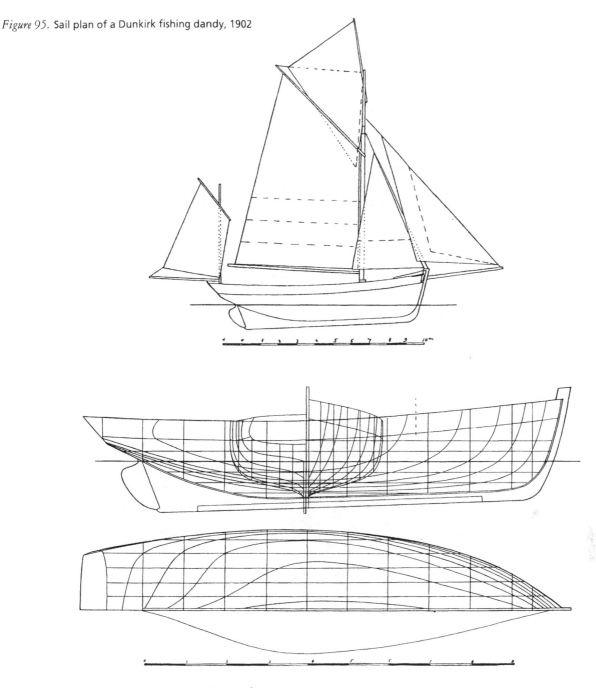

Figure 96. Lines of a Dunkirk fishing dandy, 1902

which an inner and an outer jib could be set, also a jib topsail, or 'flying jib' to square rig seafarers, set to the bowsprit end. The main topsail was jib-headed but that on the mizzen was spread by an almost vertical topsail yard and was probably rarely set. A squaresail yard was carried on which a half squaresail could be set on the weather side for running with the mainsail set. Alternatively a full squaresail could be carried on it

before the wind. A mizzen staysail was provided but was probably little used. This useful little trader had a total displacement of 255 tons when loaded.

Trouville would perhaps not now he regarded as a likely home port for pilot craft or sailing smacks but a century ago swift cutters were built and owned there, rigged and manned to face the rough and tumble of the English Channel. The cutter shown in figures 101 and

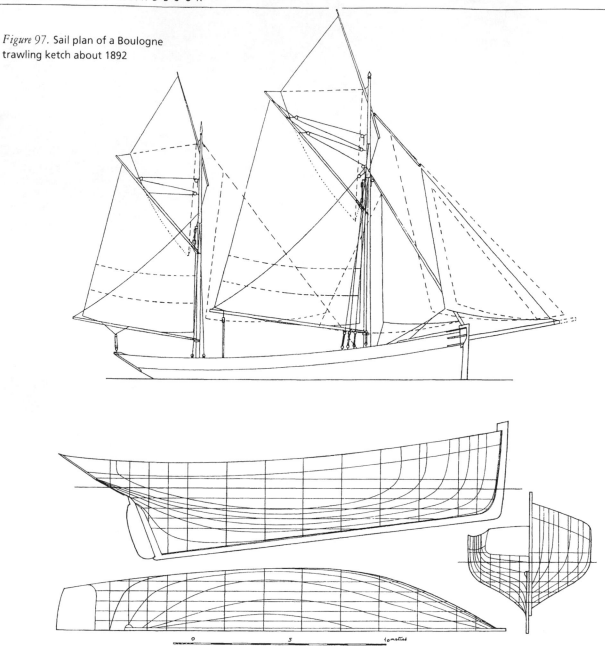

Figure 97. Sail plan of a Boulogne trawling ketch about 1892

Figure 98. Lines of a Boulogne trawling ketch about 1892

102 was typical of the Trouville pilot boats serving shipping to and from the River Seine and particularly traffic to Rouen and to the great port of Le Havre, across the estuary of Seine from Trouville. The deep hull had a plumb stem, raked keel and sternpost and a short, broad counter stern, all set off by a sweeping sheer. The principal dimensions were 44ft 3in length overall, 33ft 3in waterline length x 13ft 6in beam x 7ft 6in draft. The hull sections show considerable rise of floor and were almost vee shaped, similar to those of British racing yachts of the period 1860-80. She would be a fast craft to windward in a seaway, with a fresh breeze, despite the relatively low rig, which was set on a pole mast. The yard topsail shown on the sail plan would be almost ineffective to windward but probably added to her speed on a reach or run. Why Trouville cutters did not carry a topmast and thus set a larger topsail is unknown but probably the time spent passagemaking was much less than that spent

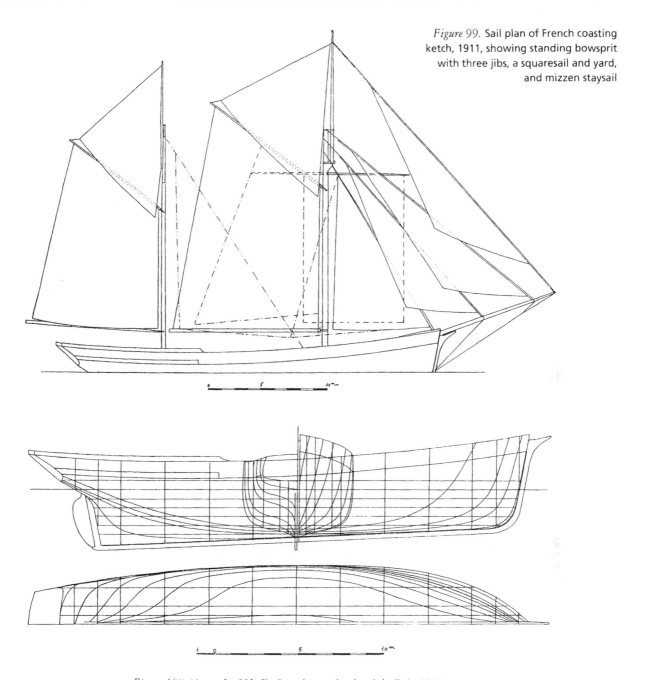

Figure 99. Sail plan of French coasting ketch, 1911, showing standing bowsprit with three jibs, a squaresail and yard, and mizzen staysail

Figure 100. Lines of a 99ft 6in French coasting ketch built in 1911

hove-to on the lookout for inward bound ships and the weight and windage of a topmast aloft was considered undesirable. The ballast was carried in the bilge, probably a mixture of iron pigs, old chain cables, stones and shingle.

Fishing smacks were also built and owned at Trouville and a typical trawling cutter was 37ft 6in overall length, 12ft 6in beam and drew 6ft 6in. The characteristics of hull form and rig were similar to the

later pilot cutters from the port, which had a reputation for building fast craft.

The hull form of the pilot cutters serving shipping in the approaches to the River Loire early in the 20th century contrasted with others in France by being developed to a yacht-like, almost rakish style with a profile very cut away to increase speed and handiness. These boats sailed seaward to meet shipping bound for the Loire ports of St Nazaire and the greater town of

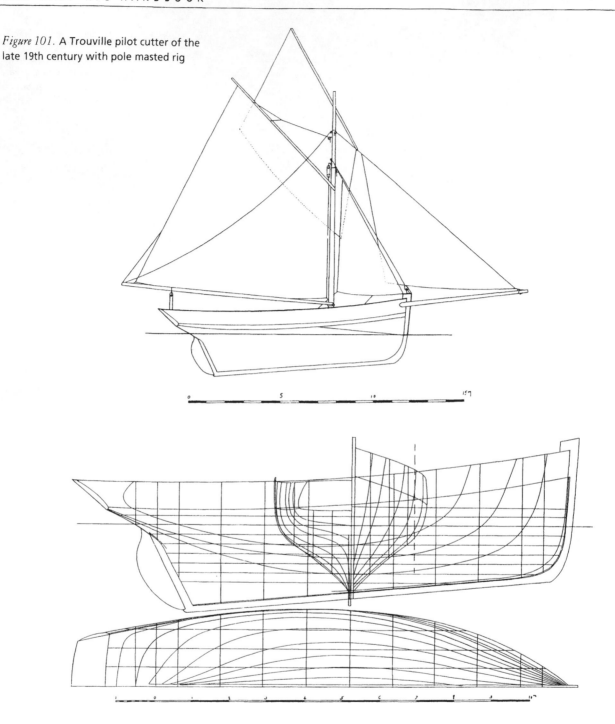

Figure 101. A Trouville pilot cutter of the late 19th century with pole masted rig

Figure 102. Lines of a Trouville pilot cutter of the late 19th century. Note the considerable rise of floor and similarity to many mid 19th century cutter yachts

Nantes. Many cutters were built at St Nazaire and the lines of one launched as late as 1909 are shown in figure 103. Influence of the contemporary racing yacht is evident in the bow, the considerably raked keel, which had a ballast keel incorporated, and in the rake of the sternpost. The counter and run were characteristically

French and similar to those of many Breton craft. These Loire boats were fast and handy cutters but like all their kind were soon to be replaced by motor propelled craft or by steam cutters which were operated by syndicates of pilots or harbour authorities.

The pilot boats of Royan, a small port at the estuary

222

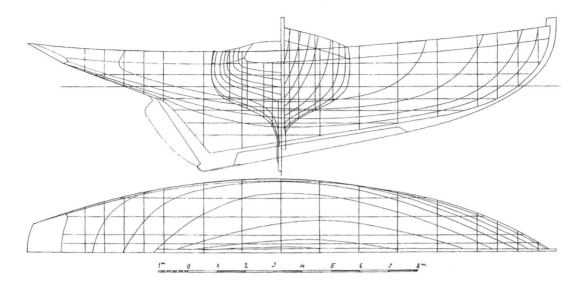

Figure 103. Lines of a Loire pilot cutter built at St. Nazaire in 1909. An unusually yacht-like hull

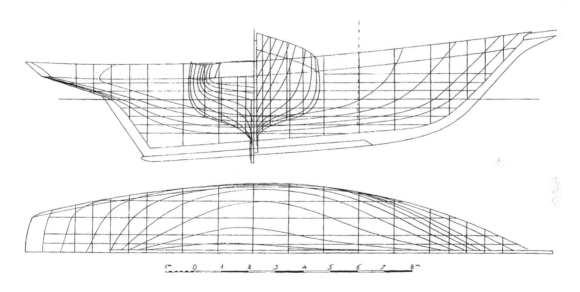

Figure 104. Lines of a pilot 'yawl' from Royan, serving the port of Bordeaux. Early 20th century

of the River Gironde, which leads to the port of Bordeaux, famous for its exported wines and other produce from south west France, had to be capable of facing the rough seas of the Bay of Biscay and strong onshore winds. So, it is surprising that these pilot boats developed to a more yacht-like form, having a clipper bow and long, flat sectioned counter. Figure 104 is of the lines of a typical Royan pilot 'yawl', which was similar to the fishing craft of that port. The ends of the hull are unusually fine for rough weather service in all conditions

and the flat sections of the counter would pound in a seaway or swell. However, the pilots seem to have been satisfied with their craft, which sailed well. Almost all the ballast was in the bilge except for a small ballast keel fitted into the wood keel.

In contrast to the many fast cutters used by pilots in many areas there were other stations where the requirements were not only seaworthiness but ability to remain on station in reasonable comfort, with windward ability secondary.

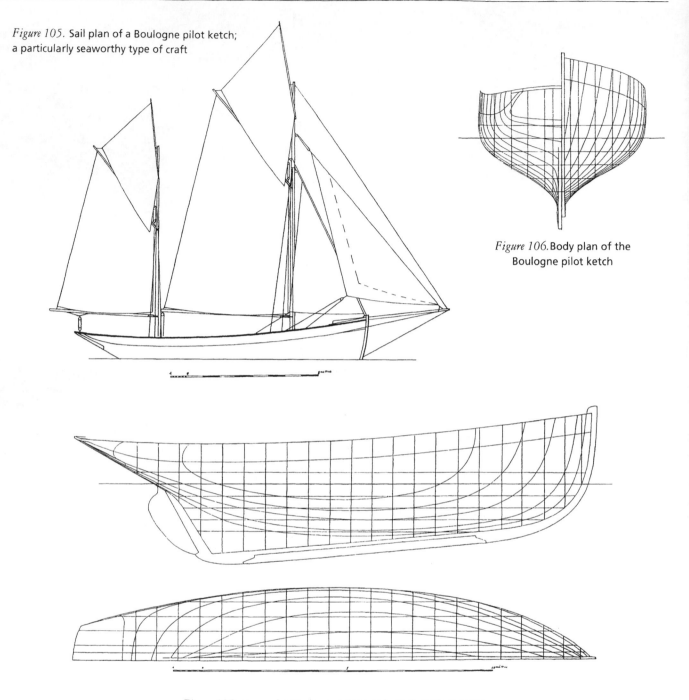

Figure 105. Sail plan of a Boulogne pilot ketch;
a particularly seaworthy type of craft

Figure 106. Body plan of the
Boulogne pilot ketch

Figure 107. Lines of a Boulogne pilot ketch. Early 20th century

The Boulogne-sur-Mer pilot ketches were of this type; able sea boats with comfortable motion and a rig which could be more easily handled with a small crew. The lines in figure 107 show a hull 59ft length overall, 49ft waterline length x 16ft 5in beam and drawing 9ft. The displacement was 51·3 tons, including an iron ballast keel of 4 tons, the remainder of the ballast being in the bilge. The forefoot was well rounded and the counter typically French. The rig totalled 1,954 square feet and she would sail fast in strong winds. The hull lines repay study by those seeking a seaworthy cruiser of traditional style.

224

EPILOGUE

THE GAFF RIG REVIVAL has inspired a segment of the sailing word which has now grown so much that is is rare to find a yacht anchorage anywhere without some craft setting it. They may be older boats, individual and restored, or new ones built to today's designs which appeal to many people. Whatever the type, the enthusiasm of their owners and crews seems set to carry the rig far into the future. The advantage of new craft is that they may, if desired, be not just replicas but can combine the appearance of tradition with advances in hull form, construction, sailing ability and have reduced needs for maintenance. They may also, perhaps, have lighter masts and spars, rigging and sails which reduce the numbers required to handle them. Many new gaff rigged craft have been and are being built and this brief review shows the diversity of size and type and may inspire others.

Gaff rig is rarely applied to boats shorter than about 19ft, below which one of the various lugsail, spritsail or bermudan rig combinations is usually preferred. But when Dr Fred Callahan, who sails from Osterville on Cape Cod, one of the original homes of the catboat type, wanted a new day-sailer he commissioned yacht designers Sparkman and Stephens to produce a boat based on a traditional shoal draught sloop with characteristics suited to his sailing area. But she was to be built from and rigged with contemporary materials and techniques. The designers considered several types as inspiration and chose the New York Bay Sloop or catboat which was once much used for oystering and sometimes for pleasure sailing. A fast and handy boat was needed and the result is known as the Stealth Dinghy. She is 15ft long overall and has 6ft 4in beam. A dinghy style centreplate is fitted and displacement is 545 pounds. She carries a gaff mainsail of 133 square feet and a jib of 41 square feet set to a steeved down bowsprit reminiscent of the oyster sloops. The boat was intended to be sailed single handed if desired and fits in well with her surroundings, is exciting to sail and will plane in suitable conditions. With a large asymmetrical spinnaker set from the bowsprit end she seems to fly, but then needs a crew of three. The Stealth Dinghy is certainly one of the liveliest gaff rigged boats built recently.

The gaff rig revival has seen the restoration of many attractive craft including *Storm*, the 24ft bawley yacht once owned by author and designer Maurice Griffiths. *Photo: Peter Chesworth*

In contrast one of the largest gaff rigged vessels to be built in recent times is the two masted topsail schooner *Pride of Baltimore II*, a wooden hulled near-replica of the swift naval, privateer, slave carrying or blockade running schooners which were built and manned from the Chesapeake Bay area until the mid 19th century. An earlier schooner of the same name and type was lost at sea off Puerto Rico in 1986. She too had been built in recognition of the past maritime greatness of the port of Baltimore, a tradition continued with her replacement. The new schooner was designed by Profession Thomas Gillmer, a naval architect who has specialised in warship design. He carried out a thorough analysis of the Baltimore Clipper type schooner from historical sources before designing this second *Pride*. She is 99ft 8in length

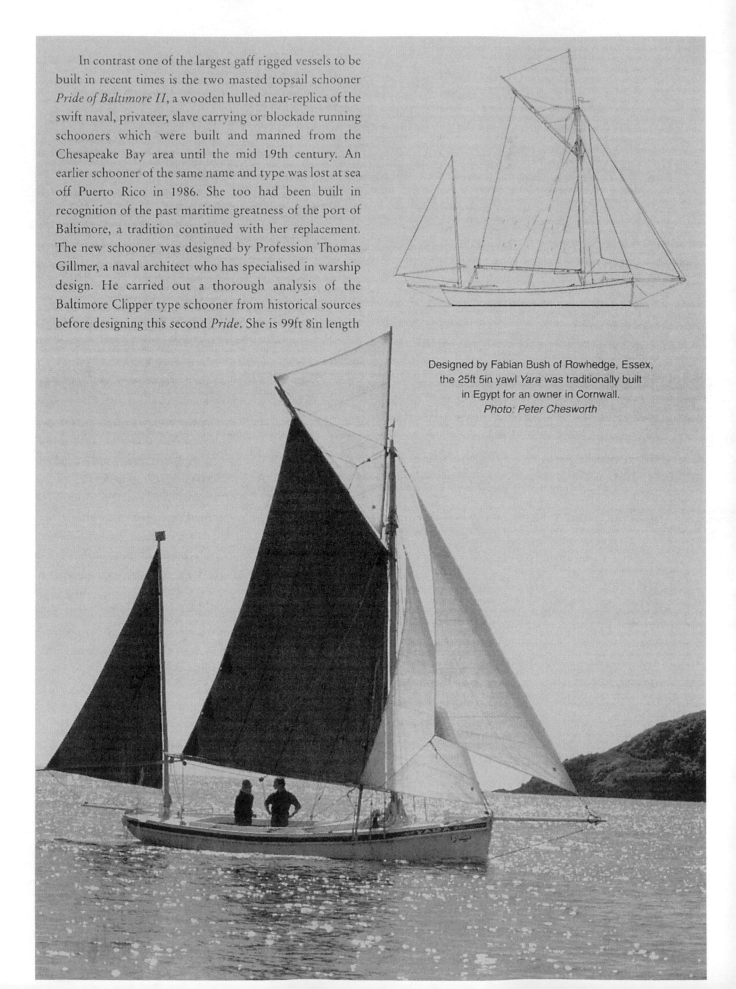

Designed by Fabian Bush of Rowhedge, Essex, the 25ft 5in yawl *Yara* was traditionally built in Egypt for an owner in Cornwall.
Photo: Peter Chesworth

The 29'2" LOD *Skoit Trekka* was inspired by American schooners, designed by Lunstroo Custom design of Amsterdam and is built of steel in Hungary. *Photo: Peter Chesworth*

overall, 91ft waterline length, 26ft 5in beam and draws 12ft 5in. Displacement is 190 tons. Sails area is 10,442 square feet and she has square topsails forward, studding sails when off the wind in lighter breezes and can set a ringtail down the leach of the mainsail when off the wind. With well raked masts, standing bowsprit and jib boom, apparently low freeboard and period colours, she has the rakish looks of her ancestors and is one of the most spectacular gaff rigged craft now sailing. Her sawn frames and pine planking provided a great contrast against the background when she was built near the Baltimore business district to remind the city of its past.

At the other extreme of design is the 21ft canoe yawl *Sea Otter*, built by David Moss at his yard on Skippool Creek, on that fascinating part of the Lancashire coast near Blackpool. The hull form of the pointed stern craft was created by the empirical means of a series of carved half models. She was originally to be a dayboat but was changed during building into a small cruising yacht. The *Sea Otter's* dimensions are 21ft length overall, 18ft waterline length, 6ft 9½in beam and 2ft hull draught. A centreplate extends this to 4ft and the stated displacement is 3,920 pounds. The rig of gaff mainsail, foresail set to a short bowsprit and a standing lug mizzen is typical of many canoe yawls of a century ago. Below a snug cabin is laid out for two with warm colours of varnished hardwoods complementing soft furnishings. Altogether a delightful little cruiser which will be cherished by owners into the future.

The centuries-old traditions of designing, building and sailing yachts and commercial vessels at the maritime village of Rowhedge, on the River Colne in north east Essex, is now kept alive by one small boatbuilder. Fabian Bush designs and builds small craft in wood, from dinghies to cruising yachts and racers. An owner in Cornwall commissioned the design and construction of a day sailing keelboat which was to have characteristics of the Falmouth watermen's boats, but not their freeboard, slab sided appearance or transom stern. Fabian designed the *Yara* with a yawl rig after convincing the owner of its virtues for short handed sailing. The mainmast is stepped well forward and the mainsail has a short boom. The

mizzen has its clew spread by a boom set up to the mast with a rope snotter and sheeted to a bumkin. Hull dimensions are 25ft 5in overall length, 22ft 8in waterline length, 7ft 9in beam and 4ft 1in draught. The displacement is 7,350 lbs and under a sail area of 495 square feet. The lines show a hull with reasonable potential for speed, which has been proved in service.

The hull of the *Yara* was, unusually, built in Egypt during her owner's sojourn there on business. She was shipped to Rowhedge for completion. The sail plan illustrates this interesting yawl which has proved sufficiently seaworthy to voyage around the coast from Rowhedge to the Fal, where she fulfils her owner's expectations as a comfortable day sailer.

Small schooners are always interesting but are usually difficult to get going well to windward. The 29ft 2in Skoit is a good practical example designed by Lunstroo Custom Design of Amsterdam in the

The Broads racing yacht *Luna*, designed at the turn of the century by the American designer Charles D Mower, provided the inspiration for the 37ft yawl *Fawn* recently designed by Andrew Wolstenholme

Class racer produced by Guy Ribadeau Dumas of Paris. The combination of a cutter rig and a hull which has carried several different rig arrangements in its eight decades of existence results in a very fast and nimble gaff rigged boat which is credited in one handicap race of beating an International Rating Class 6 Metre without time allowance.

More sedate yachtsmen who like the atmosphere of the Essex smacks but do not relish the amount of work and the cost of restoring an old vessel or building a new one in wood can find a practical solution in the 27ft cutter rigged Smack Yacht built by International Marine Designs at Abergroes, Wales. The hull mould was lifted from the wooden smack yacht *Secret*, build in the twenties by Shuttlewood at Paglesham, an Essex village

Tyrrell and Young of Faversham, Kent, build this GRP smack yacht from a mould taken off the 27ft *Secret* built by Shuttlewood at Paglesham in the 1920s.
Photo: Peter Chesworth

Netherlands and built of steel in Hungary. Her rig has much of the character of eighteenth century schooners, for example the small (60ft) pilot boats of Virginia with foresails and mainsail of almost equal area and shape and a large jib set to a well steeved bowsprit. Waterline length is 26ft 3in and beam 10ft 10in. Hull draught is 3ft and extends to 5ft 11in with centreplate lowered. Displacment is about 7-5 tons. As a cruising yacht the Skoit is attractive looking and practical and has berths for 5-6 in the cabins and a generous cockpit centred around the mainmast. Constructed in steel and panelled and outfitted in wood she makes best use of both materials and has the further advantage of attractive individuality under sail or at anchor - a quality lacking in many other new cruising yachts.

Many gaff rigged yachts have been designed in recent years by Andrew Wolstenholm of Cotishall, Norfolk and one of the most spectacular is the 37ft yawl *Fawn* which was commissioned by a Norfolk yachtsman to participate in the Broads cruiser/racer class.

Those of a competitive nature who pine of the days of fast, sleek 'Raters' of a century ago with fin and bulb keel and separated rudder can now have a gaff rigged Star

Designed in the 1940s in America by John Atkin,
the 20ft gaff cutter *Robelia* was built in the 1990s
in Britain by boatbuilder Euan Seel of Ipswich.
Photo: Peter Chesworth

Renard of St Malo in France is a modern reproduction of a topsail cutter of 1808. *Photo: Association Corsaire*

with extensive and centuries-old involvement in the oyster fisheries. She is, of course, smaller than a working sailing smack with waterline length of 22ft 10in, beam of 8ft 6in and draught of 3ft 10in but she is of handy size for cruising. Displacement is 5.9 tons and sail area is 662 square feet. She can be arranged with a bowsprit which may be lifted on a pivot bolt when not in use, reducing length over spars in these days of high charges for berthing. A comfortable four berth cabin arrangement provides snug accommodation and with a coal stove in the saloon and a bridge deck dividing it from the cockpit, these little cutters continue the old time cruising ways, besides sailing as well as full bodied craft of this length usually do.

When American yachtsmen became acquainted with the English cutter yacht during the 1870s and 1880s they divided into some who admired them and other who did not but after a few years of rivalry, some characteristics of the cutter were mingled with the shallower draught type which was desirable for many American waters and a compromise shape evolved, though the Americans retained their term of sloop for the rig of the two headsail cutter. Many beautiful, fast and useful yachts descended from that union of sailing cultures and one proof of this is in many of the yachts created by John Atkin of Noroton,

Connecticut, and his late father, William. One of the most attractive of John's little sloops is the *Maid of Endor,* which he designed in 1953 as a weekend and coastal cruiser. A fixed but moderate draught was desired and the 20ft 3in overall length hull draws 3ft 4in. Beam is 7ft 8in and a lead ballast keel of 1,100 pounds is supplemented by 400 pounds of internal ballast. The hull is shapely and with fine ends below water and a delicately formed transom. A two berth layout below deck is complemented by a roomy cockpit. The "sloop" rig has a boomed staysail with its tack set to the bowsprit forward of the stem head and the bowsprit sensibly follows the line of the sheer to keep the jib clear of water in a seaway. The mainsail has a long luff and for my taste, not enough peak to the gaff. It is a handy and sensible rig set on a small cruiser which is a delight to see, whether under sail or at anchor.

Early cutters were fascinating vessels and a replica of one used as a privateer, *Le Renard* of 1808, has been built at Saint Malo in Brittany and is as close in detail as possible under present conditions of use. She is the only example afloat of these small ships which were so much used until a century and a half ago, both as naval,

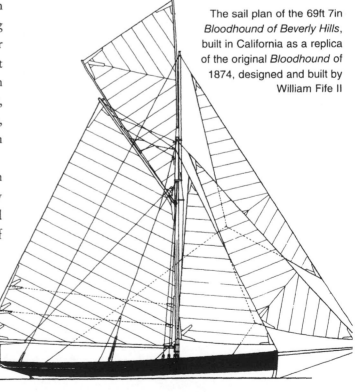

The sail plan of the 69ft 7in *Bloodhound of Beverly Hills,* built in California as a replica of the original *Bloodhound* of 1874, designed and built by William Fife II

coastguard, revenue and customs cutters, also for carriage of passengers and light goods, for smuggling and for the questionable privateering against enemy merchant shipping and fishing vessels which often bordered on licensed piracy. The plans for *Le Renard* were prepared from study of several designs of French cutters of that time. She was built by Chantiers L'Abbé to scantlings typical of her type and is now rigged and fitted out for charter to provide unique opportunity to experience the handling of a small historic ship under sail and gain some understanding of the type of vessels which inspired many of the characteristics of the early cruising and racing yachts two centuries ago.

American Robert Gilbert of Marina Del Rey, in California, has taken the cutter concept back more than a century in a replica of the famous 40 Ton racing cutter *Bloodhound* which was designed and built in 1874 by William Fife (I) for the Marquis of Ailsa. Seeking a worthy gaff rigged yacht to reproduce, Robert Gilbert decided on the famous vessel, whose racing days spanned the years 1874 to 1914 in both the 40 Ton Class and handicap racing. After research, Naval Architect Michael Richardson prepared the design largely as she was rebuilt in 1908-9, after which she was wonderfully successful in the handicap classes. The replica, built in wood at the Ventura, California yard of Fine Yachts Inc. by Harvey Swindall, is one of the finest examples of a recreated Victorian yacht. Dimensions are 69ft 7in overall length, 59ft 3in datum waterline length, 10ft 6in beam (a striking feature of the contemporary rating rule which penalised beam but not depth or draught) and 10ft 3in draught. The lead keel weighs 23 tons. Sail area is 3,576sq ft.

The new *Bloodhound of Beverly Hills* has achieved speeds of up to 12 knots under mainsail, jib and reaching staysail. She is well balanced on the helm, tacks quickly and provides her owner with the unique experience of handling a Victorian racing cutter and while offering visual pleasure to many on the coast of California.

One of the best known builders of modern gaff rigged yachts and boats is Cornish Crabbers Ltd of Rock, North Cornwall. For the past years they have moulded and completed several hundred craft with the rig, all designed by Roger Dongray, starting with the 24ft Cornish Crabber gaff cutter. The 19ft sloop-rigged Shrimper has been built in considerable numbers and

Probably the most commercially successful modern gaffer, Cornish Crabbers Ltd have built over 870 of their 19ft Shrimper since its appearance in 1979. *Photo: Peter Keeling*

they have introduced many to the gaff rig. The 24ft Cornish Yawl and the 30ft Pilot Cutter are also good examples of using current materials to create efficient and characterful cruising yachts with reduced demands on the owner for maintenance, which can be such a deterrent to many with limited time or who live a distance from their mooring.

Similar advantages can be claimed for the 30ft Norfolk Smuggler cutter which is built by Charlie Ward Traditional Boats at Morston, Norfolk, to the design of Andrew Wolstenholme. With 10ft 4in beam and draughts of 3ft 9in and 5ft with the centreboard lowered and a sail area of 607 square feel, this is a cruiser capable of sea passages

Pocket cruisers can lend themselves to a sensible use of gaff rig. The 18ft sloop rigged Winkle Brig is one, built with a glass reinforced plastic hull by the Ferry Boatyard at Fiddlers Ferry in Cheshire as a two berth boat which may also be arranged as an open dayboat.

These are just a few of the many new gaff rigged craft which builders offer as standard types or which are built to individual order. Very many other yachts, old sailing fishing boats and other craft are painstakingly maintained or are restored and sailed by their owners, often on limited income. Some are of substantial size and quality, such as the Camper and Nicholson built cruising cutter *Marigold*, a 60ft yacht launched in 1892 with a sail area of 2,691 square feet. Besides cruising she was occasionally raced in the smaller handicap classes, for at 30 tons Thames Measurement she was then regarded as fairly small. As the years passed the *Marigold* became outmoded, but survived breaking up for the value of her lead ballast keel; the sad fate of most of her contemporaries Her sound hull was converted to bermudian cutter rig in 1933 with a sail area half the original; an interesting comparison. She retained this rig

in most conditions yet able to lie in shallow water anchorages in places far from the crowd; a capability now sought by many.

Other gaff cutter yachts and dayboats with glass reinforced plastic hulls are built by Martin Heard of Gaffers and Luggers at Mylor in Cornwall. The yard's 23ft cutter was designed by the late Percy Dalton and has 8ft beam and 4ft draught, total sail area is 421 square feet including a topsail set on a yard. Many have been built and have accommodation for four people and space for an inboard auxilliary engine if desired. They also built a 28ft gaff cutter cruising yacht based on a Falmouth Working Boat hull and 18 and 20ft Toshers with gaff rig, usually as open dayboats.

The elegant cutter yacht *Marigold* was designed and built in 1892, for cruising, by Camper and Nicholson Ltd. After being re-rigged with a Bermudian mainsail for several years she was restored to gaff rig in 1993. *Photo: S. Philp*

into the 1950s. Rebuilt and re-launched in 1993, the *Marigold* has reappeared very closely resembling her original appearance.

Several interesting replicas have been built, notable among them the 27ft 4in clipper bowed cutter *Seabird* of 18ft 5in waterline length and built from the plans of the *Seagull*, designed by William Fife III in 1889. She is a miniature of many larger racing yachts of that time and was built for Hubert Stagnol at Benodet, in Brittany. The hull is strip planked in cedar and is sheathed with glass cloth and resin. Sails and running rigging are of synthetic fibres and standing rigging of stainless steel; an old yacht reproduced in modern materials thus reducing maintenance. Her sail area is 485 square feet. The *Seabird*

is believed to be the first replica of a Fife yacht to be built.

I hope the owners and crews of these many and varied craft will enjoy the same sense of freedom and adventure which I experienced as a twenty year old, hoisting the mainsail of my newly completed conversion of a ship's lifeboat to the 24ft gaff cutter *Venture* for the first time, to set course seaward in my own small cruiser.

Books can provide useful information but we must always remember Hilaire Belloc's sound injunction to "Read less and sail more".

INDEX

Printed and bound by CPI Group (UK) Ltd, Croydon, CR0 4YY

11/10/2024

01043551-0001